The Rat Brain
IN STEREOTAXIC COORDINATES

COMPACT SEVENTH EDITION

The Rat Brain
IN STEREOTAXIC COORDINATES
COMPACT SEVENTH EDITION

George Paxinos

Neuroscience Research Australia,
The University of New South Wales
Sydney, Australia

g.paxinos@neura.edu.au

Charles Watson

Neuroscience Research Australia,
The University of New South Wales
Sydney, Australia

c.watson@curtin.edu.au

ACADEMIC PRESS
An imprint of Elsevier

Academic Press is an imprint of Elsevier
125 London Wall, London EC2Y 5AS, United Kingdom
525 B Street, Suite 1800, San Diego, CA 92101-4495, United States
50 Hampshire Street, 5th Floor, Cambridge, MA 02139, United States
The Boulevard, Langford Lane, Kidlington, Oxford OX5 1GB, United Kingdom

First edition 1982
Second edition 1986
Third edition 1997
Fourth edition 1998
Fifth edition 2005
Sixth edition 2007
Compact Sixth edition 2009
Seventh edition 2014
Compact seventh edition 2018

Copyright © 2018 George Paxinos and Charles Watson.
Published by Elsevier Inc. All rights reserved

The right of George Paxinos and Charles Watson to be identified as the authors of
this work has been asserted in accordance with the Copyright, Designs and Patents
Act 1988

Cover design by Lewis Tsalis
Book design by Lewis Tsalis

No part of this publication may be reproduced or transmitted in any form or by any
means, electronic or mechanical, including photocopying, recording, or any information
storage and retrieval system, without permission in writing from the publisher.
Details on how to seek permission, further information about the Publisher's permissions
policies and our arrangements with organizations such as the Copyright Clearance
Center and the Copyright Licensing Agency, can be found at our website: www.elsevier.
com/permissions.

This book and the individual contributions contained in it are protected under copyright
by the Publisher (other than as may be noted herein).

Notice

Knowledge and best practice in this field are constantly changing. As new research and
experience broaden our understanding, changes in research methods, professional
practices, or medical treatment may become necessary.

Practitioners and researchers must always rely on their own experience and knowledge in
evaluating and using any information, methods, compounds, or experiments
described herein. In using such information or methods they should be mindful of their
own safety and the safety of others, including parties for whom they have a
professional responsibility.

To the fullest extent of the law, neither the Publisher nor the authors, contributors, or
editors, assume any liability for any injury and/or damage to persons or property as a
matter of products liability, negligence or otherwise, or from any use or operation of any
methods, products, instructions, or ideas contained in the material herein.

Library of Congress Cataloging-in-Publication Data
A catalog record for this book is available from the Library of Congress

British Library Cataloguing-in-Publication Data
A catalogue record for this book is available from the British Library

ISBN: 978-0-12-814549-4

For information on all Academic Press publications
visit our website at https://www.elsevier.com/books-and-journals

Printed and bound in China

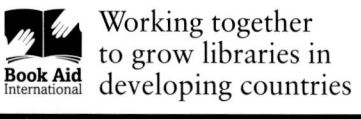

Typeset by SPi Global, India

Dedicated to Cem Hizli

Contents

Key features of the Compact Seventh Edition IX

Reproduction of figures by users of the atlas IX

Preface to the seventh edition IX

Acknowledgements X

Introduction X

Methods XI

Stereotaxic Reference System XIII

Nomenclature and the construction of abbreviations XV

The basis of delineation of structures XVI

Mini-atlas of the rat brain XXVI

References XLI

List of Structures XLVII

Index of Abbreviations LV

Figures 1

Key features of the Compact Seventh Edition

- 161 thoroughly revised coronal diagrams and accompanying photographic plates spaced at 120 μm intervals
- Photographic plates printed in color from high resolution digital images
- The most accurate and most widely used stereotaxic coordinate system
- An introductory 'mini-atlas' to assist beginning students of neuroanatomy
- 1000 structures identified
- Electronic diagrams available to purchasers of this book via a password-protected web site

Reproduction of figures by users of the atlas

As authors, we give permission for the reproduction of any figure from the atlas in other publications, provided that the atlas is cited. Permission from the publisher may be sought directly on-line via the Elsevier homepage (http://elsevier.com/locate/permissions) or from Elsevier Global Rights Department in Oxford, UK: phone: (+44) 1865843830, fax: (+44) 1865 853333, email: permissions@elsevier.com. The same procedure should be followed for the reproduction of figures from *The Mouse Brain in Stereotaxic Coordinates* (Paxinos and Franklin, 2013), *Atlas of the Developing Rat Nervous System* (Ashwell and Paxinos, 2008), *Atlas of the Developing Mouse Brain* (Paxinos et al., 2007a), *Atlas of the Human Brainstem* (Paxinos and Huang, 1995), *Atlas of the Human Brain* (Mai, Majtanic and Paxinos, 2016), *The Chick Brain in Stereotaxic Coordinates* (Puelles et al., 2007), *The Rhesus Monkey Brain in Stereotaxic Coordinates* (2009), *The Marmoset Brain in Stereotaxic Coordinates* (Paxinos et al., 2012), *MRI/DTI Atlas of the Rat Brain* (Paxinos et al, 2015). We recommend that you use the nomenclature and abbreviation scheme that we developed for this book. This scheme has been developed on a systematic basis (see Introduction), is widely recognized and is identical to that used in brain atlases of many other species (see above) including humans.

How to cite this book

Paxinos G and Watson C (2018). Paxinos and Watson's *The Rat Brain in Stereotaxic Coordinates Compact,* 7th Edition. Elsevier Academic Press, San Diego.

Preface to the seventh edition

It is thirty-two years since we published the First Edition of *The Rat Brain in Stereotaxic Coordinates*. During that time there has been an explosion in neuroanatomical information, and we have tried to incorporate advances in knowledge in each new edition of the atlas. Over the past decade anatomical discoveries related to developmental gene expression have challenged many traditional beliefs about the organization of the brain. With the help of Professor Luis Puelles of the University of Murcia the Seventh Edition incorporates new concepts, particularly those relating to neuromeres. We have checked and revised every diagram, and have added neuromeric boundaries to eight of the sagittal diagrams.

As ever, we welcome suggestions, criticisms, and corrections that might improve the atlas in the future. Over the years, we have received advice from many scientists, and we are very grateful for their willingness to share their expertise. We welcome any further advice that might improve the accuracy of our diagrams in the future. Please email us on g.paxinos@unsw.edu.au or c.watson@curtin.edu.au.

The Compact Seventh Edition has the coronal diagrams and plates of the full Seventh edition (Paxinos and Watson, 2014) with only minor modifications in the introduction.

George Paxinos and Charles Watson

Acknowledgements

This work was supported by the Australian Research Council Centre of Excellence for Integrative Brain Function (ARC Centre Grant CE140100007), National Institutes of Health BRAIN Initiative award (U01MH105971) and project grants (APP1086083 and APP1086643) from the NHMRC. George Paxinos is supported by an NHMRC Senior Principal Research Fellowship (APP1043626).

Introduction

For many decades, the rat has been the most popular subject for research in mammalian neuroscience. There are many reasons for this: rats are a suitable size for accurate stereotaxic localization of discrete brain areas; they are hardy and easy to maintain in colonies; and inbred strains are available, which increases stereotaxic accuracy.

The Rat Brain in Stereotaxic Coordinates was the first rat brain atlas to be based on the flat skull position. Moreover, it was the first to offer accurate stereotaxic coordinates, allowing a choice of bregma, lambda, or the midpoint of the interaural line as the reference zero point. Although the coordinates were developed from study of adult male Wistar rats with weights ranging from 270 to 310 g, the atlas can be successfully used with male or female rats, with weights ranging from 250 to 350 g (Paxinos *et al.,* 1985).

The segmentation of the First Edition was based on both Nissl and acetylcholinesterase sections. The importance of acetylcholinesterase as a guide to segmentation cannot be overestimated, and it has since formed the basis of atlases of the mouse (Paxinos and Franklin, 2013), the human brain stem (Paxinos and Huang, 1999), the rhesus monkey (Paxinos *et al.,* 1999), and the chick (Puelles *et al.,* 2007). Since the First Edition, we have progressively made use of more marker stains to assist with segmentation. We have made extensive use of a parallel series of rat brain sections stained with a series of eight different markers (Paxinos *et al.,* 2009). Discoveries in gene expression patterns during development have seriously altered our appreciation of the subdivision of the brain into different regions. It is now clear that the rhombomeric subdivisions in the hindbrain and the prosomeric subdivisions in the forebrain are present in mammals as well as in birds. The result is a new developmentally based hierarchical classification (ontology) of structures in the brain. We have organized our text on delineation of structures on the basis of this developmental ontology. This ontology is based on the work of Luis Puelles of the University of Murcia. It first appeared as an avian brain ontology (Puelles *et al.,* 2007) and Puelles has modified it to form a complete mouse brain ontology that can be accessed on the Allen Brain Institute web site (www.brain-map.org/; www.developingmouse.brain-map.org). A summary of the developmental ontology can be found in an article by Puelles, Harrison, Paxinos, and Watson (2013), and in Watson et al. (2017).

Methods

A fresh brain from a male Wistar (290 g) rat was frozen, and sections were cut at 40 μm thickens at right angle to the horizontal plane joining bregma and lambda.

Stereotaxic surgery

We placed an anesthetized rat in a Kopf small-animal stereotaxic instrument and adjusted the incisor bar until the heights of lambda and bregma were equal in order to achieve the flat skull position. This position was achieved when the incisor bar was lowered 3.3 ± 0.4 mm below horizontal zero (Table 1). Because the point of intersection of the lambdoid and sagittal sutures is variable, we have chosen to define lambda as the midpoint of the curve of best fit along the lambdoid suture (see skull diagram). This redefined reference point is considerably more reliable than the true lambda (the true point of intersection of the sagittal and lambdoid sutures), and it is located 0.3 ± 0.3 mm anterior to the interaural line. We defined bregma as the point of intersection of the sagittal suture with the curve of best fit along the coronal suture. When the two sides of the coronal suture meet the sagittal suture at different points, bregma usually falls midway between the two junctions. The anteroposterior position of bregma was 9.1 ± 0.3 mm anterior to the coronal plane passing through the interaural line, but for the brain represented in this atlas bregma is deemed to lie at 9.0 mm. The top of the skull at bregma and lambda was 10.0 ± 0.2 mm dorsal to the interaural zero plane.

To confirm the stereotaxic orientation of sections, reference needle tracks were made perpendicular to the horizontal and coronal planes. One horizontal needle insertion perpendicular to the coronal plane was made from the posterior of the brain at 4.0 mm above the interaural line and was 2.0 mm lateral to the midline. The reference track from the horizontal needle appears as a small hole in coronal sections. Two vertical electrodes were inserted to confirm the brain was in fact cut in the coronal plane in which it was placed in the stereotaxic instrument. For the brain sectioned in the sagittal plane, vertical needles were inserted in both hemispheres at 3.0 mm posterior to the interaural line and at 1.0 and 2.0 mm lateral to the midline. A second pair was inserted 11 mm anterior to the interaural line at 1.0 and 2.0 mm lateral to the midline. Horizontal needle tracks perpendicular to the coronal plane were made at 5.0 mm and 6.0 mm dorsal to the interaural line and 2.0 mm lateral to the midline. Horizontal needle tracks perpendicular to the sagittal plane were made at 5.0 mm dorsal to the interaural line and 2.0 mm and 8 mm anterior to it. Following surgery, the rats were decapitated and the whole head frozen on dry ice. The frozen skull was then prised off the frozen brain, and the brain was carefully mounted on the stage of microtome so that the sections would be cut in the appropriate stereotaxic plane.

For the coronal set, every third section was used for preparation of the atlas diagrams, so that the interval between atlas diagrams is 0.12 mm. Exceptions to this rule are found in the region rostral to the rostrum of the corpus callosum (Interaural AP 11.28) and in the region of the medulla caudal to the inferior olive (Interaural AP -5.76 mm). In these two regions, sections were selected for presentation in the atlas at 0.24 mm intervals. Finally, the olfactory bulb is depicted at only three representative levels. For the original horizontal and sagittal section sets published in the First Edition in 1982, sections were taken at 0.5 mm intervals. Since then, we have added some intervening sections that were stained but not mapped in the First Edition, in order to make the series more comprehensive. In the Sixth Edition we added eight more sagittal sections from the original series, in addition to those added in the Second and Fourth Editions.

Histological methods

For the coronal series, the sections chosen for presentation were stained with either cresyl violet or for the demonstration of AChE on an alternate basis, so that the cresyl violet sections are 0.24 mm apart and AChE sections are also 0.24 mm apart. The two sections that intervene between sections presented in the atlas were stained with cresyl violet or for the presence of AChE or NADPH diaphorase, according to the following sequence:

1. Cresyl violet section presented in this atlas
2. AChE intervening section
3. NADPH intervening section
4. AChE section presented in this atlas
5. Cresyl violet intervening section
6. NADPH diaphorase intervening section

This sequence was repeated throughout the series of coronal section. This arrangement ensures that every section presented in the atlas is accompanied by two adjacent sections, each of a different stain. For example, the AChE section described as number four above is preceded by an NADPH diaphorase section and followed by a cresyl violet section. This arrangement gave us maximum information for mapping the section chosen for presentation. Staining was carried out on the same day as section cutting. In the sagittal and horizontal sets, we originally aimed to present alternate cresyl violet and AChE sections at 0.5 mm intervals. The intervening sections we have added were chiefly selected on the basis of stereotaxic position, and they break the sequence of alternating stains.

Quality of Sections

In some cases, the sections were stretched or compressed in the process of cutting and mounting on slides. We have compensated for this by constructing diagrams that represent, as best we can judge from the study of adjacent sections and sections from other brains, the original shape of the brain section. In a few cases, the 'atlas' section was so badly damaged that we have taken our drawing from an adjacent section.

Cresyl Violet Staining

Slides were immersed for 5 min in each of the following: xylene, xylene, 100% alcohol, 100% alcohol, 95% alcohol, and 70% alcohol. They were dipped in distilled water and stained in 0.5% cresyl violet for 15-30 min. They were differentiated in water for 3-5 min and then dehydrated through 70% alcohol, 95% alcohol, 100% alcohol, and 100% alcohol. They were then put in xylene and coverslipped. To make 500 mL of 0.5% cresyl violet of about pH 3.9, 2.5 g of cresylecht violet (Chroma Gesellschaft, Postfach 11 10, D-73257, Kongen, Germany) was mixed with 300 mL of water, 30 mL of 1.0 M sodium acetate (13.6 g of granular sodium acetate in 92 mL of water), and 170 mL of 1.0 M acetic acid (29 mL of glacial acetic acid added to 471 mL of water). This solution was mixed for at least 7 days on a magnetic stirrer and then filtered.

AChE Histochemistry

The method for the demonstration of AChE followed the procedures of Koelle and Friedenwald (1949) and Lewis (1961). Slides were incubated for 15 h in 100 mL of stock solution (see below) to which had been added 116 mg of S acetylthiocholine iodide and 3.0 mg ethopropazine (May & Baker). The slides were rinsed with tap water and the color was developed for 10 min in 1% sodium sulphide (1.0 g in 100 mL of water)

at pH 7.5. The slides were then rinsed with water and immersed in 4% paraformaldehyde in phosphate buffer for 8 h, and then allowed to dry. Subsequently, they were dehydrated for 5 min in 100% alcohol, then immersed in xylene and coverslipped with Permount. The stock solution was a 50 mM sodium acetate buffer at pH 5.0 which was made 4.0 mM with respect to copper sulphate and 16 mM with respect to glycine. This was done by adding 6.8 g of sodium acetate, 1.0 g of copper sulphate crystals, and 1.2 g of glycine to 1.0 L of water and lowering the pH to 5.0 with HC1. We found that fresh, unfixed tissue from the frozen brains showed a substantially stronger reaction for both stains than tissue fixed with formalin, paraformaldehyde, glutaraldehyde, or alcohol. A detailed protocol of the staining procedures may be obtained from our website:

http://www.neura.edu.au/research/people/profiles/scientia-prof-george-paxinos-ao.

Photography

For our earlier editions, the photographs of stained brain sections were taken with a Nikon Multiphot macrophotographic apparatus using 4"x5" Kodax Plus X film. A high contrast filter was used to print the photographs of cresyl violet sections, whereas a lower contrast filter was used to print the photographs of AChE sections. For the last edition, the biological sections were imaged for us by the Allen Institute of Brain Science, which has enabled us to reproduce the sections in color.

Drawings

Pencil drawings, which later formed the basis of the figures, were made by tracing the photographs of sections. Fiber tracts in the drawings are outlined by solid lines, and nuclei and neuron groups are outlined by broken lines. In general, each abbreviation is placed in the center of the structure to which it relates; where this was not possible, the abbreviation is placed alongside the structure and a leader line is used. In the early editions of this atlas, abbreviations for fiber tracts and fissures were positioned on the left side of the drawings of the coronal sections, and the abbreviations for nuclei and other cell groups were positioned on the right side. However, in our presentation of the coronal series (from the Fifth Edition onwards), we have broken from this tradition. We have placed most labels on both sides of the diagram for convenience. The outlines of the ventricles and aqueduct are filled in with solid color.

The Rat Brain in Stereotaxic Coordinates   Compact 7th Edition   Paxinos & Watson

Stereotaxic Reference System

The stereotaxic reference system is based on the flat skull position, in which bregma and lambda lie in the same horizontal plane. Two coronal and two horizontal zero planes are used as reference planes in these drawings. One reference coronal plane cuts through bregma and the other cuts through the interaural line. Similarly, one horizontal plane is at the level of bregma on the top of the skull and the other is at the level of the interaural line. Lambda is usually located 0.3 mm anterior to the interaural line, and it can be used as an alternative reference point in conjunction with the dorsoventral coordinate of bregma. The position of the stereotaxic reference points and planes are indicated on the skull diagram. The stereotaxic reference grid shows 0.2 mm intervals.

Drawings of coronal brain sections

In each of the coronal drawings, the large number at the bottom left shows the anteroposterior distance of the section from the vertical coronal plane passing through the interaural line. The large number at bottom right shows the anteroposterior distance of the plate from a vertical coronal plane passing through bregma. Note that these two coronal planes are 9.0 mm apart, so the two numbers on any one plate add up to 9.0 mm. The small numbers on the left margin show the dorsoventral distance from the horizontal plane passing through the interaural line. The numbers on the right margin show the dorsoventral distance from the horizontal plane passing through bregma and lambda on the surface of the skull. The numbers on the top and bottom margins show the distance of structures from the midline sagittal plane.

Table 1 Craniometrie and stereotaxic data (means + S.D.) for rats of different sex, strain and weight

Subject	Mean weight (g)*	AP I − B (mm)	AP I − L (mm)	DV I − B (mm)	AP I − Acb (mm)**	AP B − ac (mm)**	AP I − 7n (mm)**	DV I − incisor bar (mm)
Atlas' Wistar	290	9.1 ± 0.3	0.3 ± 0.3	10.0 ± 0.2	11.7	0.0	−1.3	−3.3 ± 0.4
Coronal plates	300	9.2	0.2	10.1				
Sagittal plates	270	8.9	0.0	10.0				
Horizontal plates	290	9.1	0.2	10.1				
Female Wistar	282	9.3 ± 0.2	0.5 ± 0.3	10.0 ± 0.1	11.6	0.1	−1.2	−3.2 ± 0.5
Hooded	290	9.4 ± 0.4	0.3 ± 0.6	9.8 ± 0.2	11.9	0.0	−1.2	−3.9 ± 0.6
Sprague	299	9.0 ± 0.2	0.7 ± 0.2	10.1 ± 0.1	11.7	0.1	−1.2	−3.9 ± 0.5
Juvenile Wistar	180	7.7 ± 0.4	−0.4 ± 0.3	9.9 ± 0.2	10.2	−0.1	−1.6	−2.0 ± 0.4
Mature Wistar	436	9.7 ± 0.3	0.6 ± 0.3	10.7 ± 0.4	12.4	−0.1	−0.8	−2.7 ± 0.3

*S.D.s < 20 g.
**S.D.s < 0.4 mm.

ac, anterior commissure; Acb, accumbens nucleus; AP, anterior-posterior; B, bregma; DV, dorsal-ventral; 7n, facial nerve; I, interaural line; L, lambda.
Reprinted with permission from *J. Nenroscience Methods.* 13 (1985) 139–143.

290 g Male Wistar

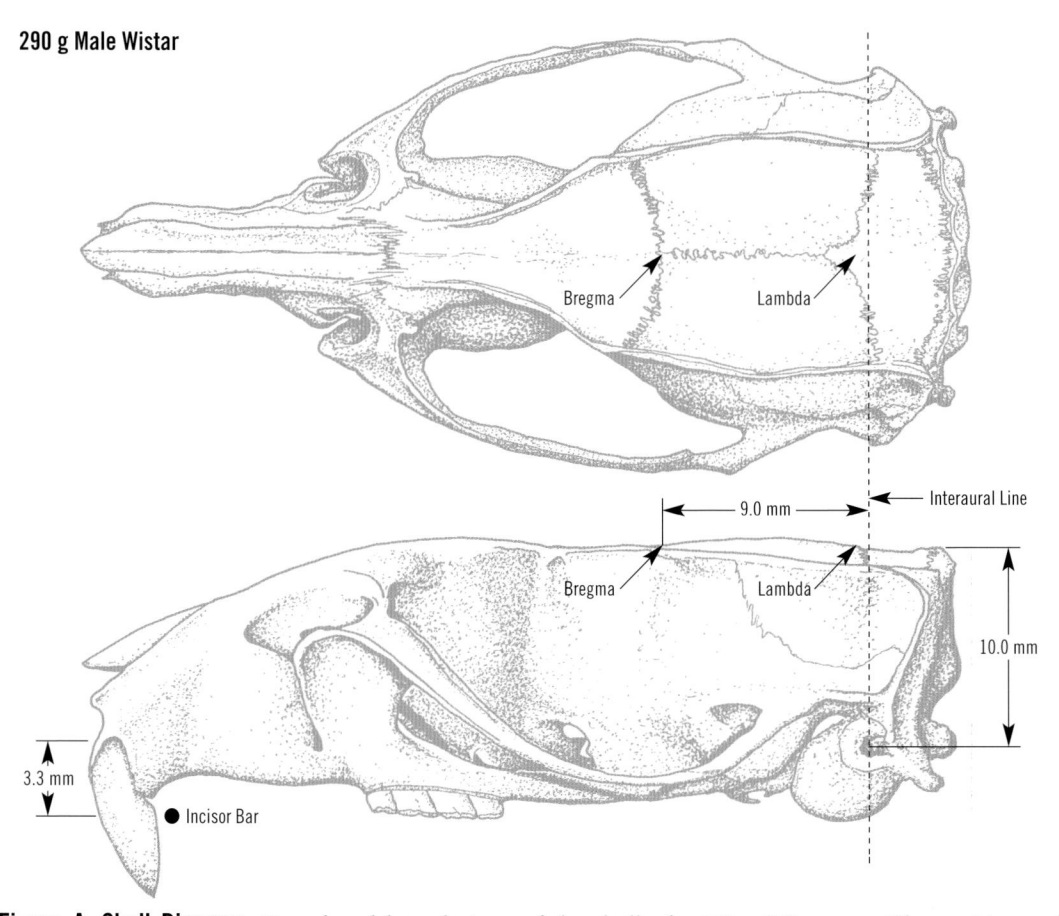

Figure A: Skull Diagram Dorsal and lateral views of the skull of a 290 g Wistar rat. The positions of bregma, lambda and the plane of the interaural line are shown above the lateral view. The distance between the horizontal plane passing through the interaural line is shown on the right of the lateral view. The distance between the incisor bar and the horizontal plane passing through the interaural line is shown on the left of the lateral view. Lambda (midpoint of the curve of best fit along the lambdoid suture) is 0.3 mm anterior to the coronal plane passing through the interaural line.

Accuracy of the stereotaxic coordinates

In almost all cases, the potential error in defining the position of any point in the brain is less than 0.5 mm. Although we used medium-sized (average 290 g) male Wistar rats in the construction of this atlas, we recognize that researchers often use animals of different sex, strain, and weight. Because of this, we have estimated the error that may occur if this atlas is used with female Wistar rats, male hooded (Long Evans) rats, male Sprague Dawley rats of 300 g weight, juvenile (180 g) Wistar rats, and mature (436 g) Wistar rats. The results of these estimations are shown in Table 1.

It is evident from these studies that no substantial stereotaxic error will occur when rats of different sex and strain are chosen, provided that the rats are of similar weight to those on which the atlas is based (290 g). For example, for rats of different sex and strain but of similar weight, the anteroposterior distance between the interaural line and bregma is between 9.0 and 9.4 mm. Similarly, the dorsoventral distance between the interaural line and the surface of the skull at bregma and lambda is very stable (9.8-10.1 mm). By contrast, craniometric data for juvenile (180 g) and mature (436 g) Wistar rats differ substantially from those of other groups. The anteroposterior distance between the interaural line and bregma is 7.7 mm in the juvenile and 9.7 mm in the mature rats (9.0 mm in 290 g male rats). Lambda is 0.4 mm posterior to the interaural line in the juvenile rats and 0.6 mm anterior to this line in the mature rats (0.3 mm anterior in 290 g rats). Unexpectedly, the dorsoventral distance between the interaural line and bregma for juvenile rats (9.9 mm) was almost the same as that of 290 g rats (10.0 mm). In the mature rats, the interaural line to bregma vertical distance was 10.7 mm. In female rats, as well as in hooded, juvenile (180 g), mature (436 g) and 290 g Wistar rats, bregma was found to be above the most forward crossing fibers of the anterior commissure. This is the point at which the posterior limbs of the anterior commissure appear. These data confirm the observation of Whishaw *et al.*, (1977) that bregma is more stable than the interaural line for positioning of electrodes in brain structures close to, or anterior to, bregma. However, data from insertion of needles aimed at the level where the facial nerve leaves the facial genu show that the interaural reference point is more stable than bregma for localization of such posterior structures. Therefore, if juvenile or mature rats are used, greater accuracy can be achieved if bregma is used as the reference point for work with rostral structures and the interaural line for work with caudal structures.

A further improvement in accuracy can be obtained by taking into account the actual location of the accumbens nucleus and the genu of the facial nerve in animals of a selected strain and weight. In agreement with Slotnick and Brown (1980), we noticed that coordinates of structures were closer to target if the coordinates given by the interaural and bregma reference systems were averaged. However, no atlas or stereotaxic instrument will compensate for inaccurate identification of bregma and lambdoid points. The reference skull mark for bregma is the midpoint of the curve of best fit along the coronal suture, and the reference skull mark for lambda is the midpoint of the curve of best fit along the lambdoid suture. These two reference marks are not necessarily the points of intersection of these sutures with the midline suture.

Nomenclature and the construction of abbreviations

There is an obvious need for a stable neuroanatomical nomenclature to accurately and efficiently convey information between neuroscientists. Despite this, many terms and abbreviations are still used in the literature to describe a single structure. In some cases, the same term or abbreviation is used for completely different structures. We urge researchers to consider the merits of our system of nomenclature because it is systematic and derived after extensive consultations with neuroanatomy experts.

Neuroscience communities concerned with different systems have developed identical abbreviations for completely different structures; for example, SO may stand for both supraoptic nucleus and superior olive, SC for suprachiasmatic nucleus and superior colliculus and IC for inferior colliculus and internal capsule. Further, homologous structures are nonetheless named or abbreviated differently in different species.

The nomenclature and abbreviations used in this atlas are those that have been employed in *The Mouse Brain in Stereotaxic Coordinates*, 4th ed. (Paxinos and Franklin, 2013), *Atlas of the Developing Rat Nervous System* (Paxinos *et al.*, 2008), *Atlas of the Developing Mouse Brain* (Paxinos *et al.*, 2007a), *Chemoarchitectonic Atlas of the Rat Brain* (Paxinos *et al.*, 2012), *Atlas of the Human Brainstem* (Paxinos & Huang, 1995), *The Marmoset Brain in Stereotaxic Coordinates* (Paxinos *et al.*, 2012) and *Atlas of the Human Brain* (Mai *et al.*, 2008). Over the last 30 years we have made a sustained effort to use identical abbreviations for homologous structures in various species so that readers are not burdened with the meaningless task of learning different abbreviations for the same structures. Bowden *et al.*, (2012) in their atlas of the monkey brain, as well as in *NeuroNames*, used the same principles for construction of abbreviations as we have used here. Foster (1998) and Morin and Wood (2001) also use this set of abbreviations.

To illustrate the need for a common system, we point out that there are more than 20 ways that have been used to abbreviate "accumbens nucleus": Acb, ACB, acb, NAS, nas, A, a, Ac, ac, NA, na, AN, an, NAC, nac, ACN, acn, ACU, acu, ACC and Acc. We used Acb in all our atlases.

In considering the merit of a particular name over synonyms, we have chosen terms that have been ratified by modern usage, unless they are illogical. We have used Anglicized versions of terms rather than older Latinized versions wherever possible and we have, in all but a handful of cases, avoided the use of eponyms.

The principles used in the construction of abbreviations are the same as those used to derive the abbreviations for the elements of the Periodic Table and for the word acetylcholinesterase (AChE).

1. The abbreviations represent the order of words as spoken in English (e.g., DLG = dorsal lateral geniculate nucleus).

2. Capital letters represent nuclei and lower case letters represent fiber tracts. Thus, the letter 'N' has not been used to denote nuclei and the letter 't' has not been used to denote fiber tracts.

3. The general principle used in the abbreviations of the names of elements in the periodic table was followed: the capital letter representing the first letter of a word in a nucleus is followed by the lower case letter most characteristic of that word (not necessarily the second letter; e.g., Mg = magnesium; Rt = reticular thalamic nucleus).

4. Compound names of nuclei have a capital letter for each part (e.g., LPGi = lateral paragigantocellular nucleus).

5. If a word occurs in the names of a number of structures, it is usually given the same abbreviation (e.g., Rt = reticular thalamus nucleus; RtTg = reticulotegmental nucleus of the pons). Exceptions to this rule are made for well-established abbreviations, such as VTA.

6. Abbreviations of brain regions are omitted where the identity of the region in question is clear from its position (CM = central medial thalamic nucleus, not CMTh).

7. Arabic numerals are used instead of Roman numerals in identifying (a) cranial nerves and nuclei (as in the Berman, 1968, atlas), (b) Rexed's laminae of the spinal cord and (c) cerebellar folia. While the spoken meaning is the same, the detection threshold is lower, ambiguity is reduced and they are easier to position in small spaces available on diagrams. In the past, naked numbers have been used to refer to cortical areas, cortical layers, spinal cord layers, cranial nerve nuclei and cerebellar folia. However, informatic systems cannot deal with the use of the same symbol for totally different things. We have, therefore, allocated the capital letter A (for area) to as a prefix to cortical areas, as detailed below and have added the suffix Cb to numbered cerebellar folia and the suffix Sp to Rexed's laminae of the spinal cord.

Given we allocated the capital letter A to cortical areas, we were forced to modify the alphanumeric nomenclature for the catecholamine cell groups. For A1 we used NA1 (noradrenaline), for CI we used Adl (adrenaline), for A8 we used DA8 (dopamine group), for All we used DAI 1. Whereas this change in nomenclature was necessary to avoid duplication, it is not without benefit because the old A1 stood for noradrenalin, something counterintuitive given the A makes one think of adrenalin. Adrenalin on the other hand was designated by CI. The Swedes named these according to the order they discovered them without knowing what catecholamine it was (A for first). DAI 1 is more explicit that it is dopamine and NA7 is more explicit that it is noradrenaline. See section on cortex for our justification for using the primate terms for the cingulate areas.

The basis of delineation of structures

For the Seventh Edition, we have reviewed our delineations of structures in all areas of the brain. Our primary guide was the comprehensive archive of marker stains in the rat brain (Paxinos et al., 2009). We have also made extensive use of other brain atlases (Paxinos and Huang, 1995; Paxinos et al., 1994; Watson et al., 2009; Watson and Paxinos, 2010; Swanson, 2004; Dong, 2008) and texts such as The Rat Nervous System (Paxinos, 2014) and The Mouse Nervous System (Watson et al., 2012). The following is a summary of the basis of delineation of structures. We have not repeated here all the rationale for the delineation of all structures presented in earlier editions.

Prosencephalon

The prosencephalon (forebrain) is made up of three components, the telencephalon, hypothalamus (including the eye), and the diencephalon (Puelles et al., 2012a,b; 2014). The telencephalon can be divided into pallial and subpallial elements.

Telencephalon - Pallium

Neocortex

For the development of the telencephalon see recent reviews of Marin (2014) and Puelles et al., (2014). There have been a number of comprehensive neocortical panellation schemes in the rat in recent decades. The first one to consider is that of Zilles (1985), which was based on the original biological sections of the First Edition of our atlas. In the Second Edition of our atlas (1986) we used the cortical parcellations of Zilles (1985). A second important and comprehensive cortical delineation scheme was proposed by Swanson (2004). The Zilles (1985) delineations differ significantly from the Swanson (1992,2004) scheme. A recent revision of the Zilles scheme has been published (Palomero-Galagher and Zilles 2014) and we have adopted many features of this subdivision scheme. The atlas of chemical markers (Paxinos et al., 1999a,b; 2009) has enabled us to make decisions on the strengths of the competing schemes. We have retained many of the features of the sensory, motor, and insular areas proposed by Zilles (1985). However, we have curtailed the rostral spread of Zilles's occipital areas and we have delineated the sensory representation of the trunk region and temporal association area in line with Swanson (2004). On the basis of the work of Palomero-Gallagher and Zilles (2014) we have expanded the cerebral real estate of Fr3 in Figures 8–11 and equivalent areas in sagittal and horizontal diagrams. The expansion has been principally at the expense of Ml which is now more restricted at these levels. We have retained the perirhinal cortex at caudal levels (along with Zilles, 1985) because there is a characteristic NADPH- diaphorase reactivity associated with this area. The area which we identify as ectorhinal cortex (Ect) is considered to be a neocortical auditory area by Palomero-Gallagher and Zilles (2014) because they argue that our ectorhinal cortex is isocortical in nature. However, we have retained the term ectorhinal cortex in this edition of the atlas. Palomero-Gallagher and Zilles (2014) have substantially revised their delineations in the non-sensory pariet al regions, and we have generally followed their lead in these regions.

We have found that the pattern of a range of immunohistochemical markers is very useful in defining many of the cortical areas (Paxinos et al., 2009). Strong parvalbumin immunoreactivity is present in layer 4 of the primary somatosensory cortex, and SMI-32 immunoreactivity formed distinctive patches in layer 4 of the barrel field and forelimb and hindlimb region. The primary auditory area was identified on the basis of reduced calbindin immunoreactivity in the deep layers. All the auditory areas were marked by the presence of SMI-32 positive cells in the superficial layers (Paxinos et al., 2009). AChE marked the location of the anterior cingulate and agranular insular cortical areas. NADPH-diaphorase assisted in defining the agranular insular, perirhinal, and retrosplenial granular cortical areas. The dorsolateral orbital cortex was delineated in accordance with the work of Ray and Price (1992). We use the term frontal association cortex (FrA) for the frontal cortex that others allocated to the secondary motor cortex (Swanson, 2004; Zilles, 1985). Our designation of FrA is in agreement with microstimulation data (Neafsey et al., 1986).

We have illustrated the cortical layers in only a few sections. Those who require a complete presentation of cortical layer distribution in the rat brain should refer to Swanson (2004) for the mouse brain to Paxinos and Franklin (2013).

The nomenclature of the rodent **cingulate cortex** and related areas has been a major problem for some decades. A central problem has been the application of the term 'retrosplenial' to areas that were clearly rostral to the splenium of the corpus callosum. We have resolved this terminological impasse by following the lead of Vogt and Paxinos (2012) to use for the rat the numerical system standard for the primate brain.

Hence, we have abandoned the nomenclature for the cingulate areas of the rodent used in previous atlases for reasons detailed in *The Rat Nervous System* (Vogt, 2014). There are three critical features of the new nomenclature. First, it is based on the same cytoarchitectonic criteria used for the non-human and human primates and, therefore, will foster a closer interaction between human imaging research and efforts to develop rodent models of human neuronal diseases. A summary of comparative studies in rodents and primates can be found in *Cingulate Neurobiology and Disease* (Vogt, 2009). Rodent and primate divisions of area 32 have also been compared (Vogt et al, 2013). Second, we abandon the terms "prelimbic" and "infralimbic." Even for those who might believe in the existence of the limbic system, these terms are inadequate, because they cruelly declare these areas as non-limbic. Third, the midcingulate concept has become an important part of primate research and was introduced in the above noted studies for the mouse and rat. This represents a major shift in our understanding of cingulate organization. Horizontal sections in the rat display the midcingulate cortex (A24al) in a way that it can be directly compared on the same section with the anterior cingulate area (A24a). Overall, AChE density is higher in area 24a' than 24a in layers 1, 3 and 5–6 (Fig 207). Only layer 2 of area 24a has a higher density of AChE than area 24a'. The Nissl-stained section shows differences between these two areas: layers 2–6 all have larger and more densely packed neurons in area 24a' than in area 24a (Fig 206). Since the connections and functions of rat area 24' are not known, these differences and localization of area 24' in the atlas will provide a substrate for future research.

From a regional perspective, anterior cingulate cortex includes A25, A32, A33 and A24, the midcingulate cortex is represented by A24a'/A24a' and the retrosplenial cortex is comprised of A29a, A29b, A29c, and A30. The laminar organization of each of these areas has been reported by Vogt and Paxinos (2014). Notice that rodents do not have a posterior cingulate cortex. The posterior cingulated cortex is present in primates and is comprised of areas 23 and 31.

In terms of equivalencies with the previous editions of our atlas, A32 is similar to the former prelimbic cortex but has dorsal and ventral divisions. A25 is similar to the previous infralimbic cortex, rostral Cg1 is similar to A24b, and Cg2 is similar to A24a. Caudal Cg1 is similar to A24b' and caudal Cg2 is similar to A24a'. Notice that the most rostral part of RSGc is actually part of A24'. Finally, area RSD is similar to A30, RSGa to A29b, and RSGc to A29c.

The sagittal sections close to the midline present a unique problem in that they traverse individual layers rather than presenting superficial and deep layers at the same level. This makes detailed delineations impossible. The segmentation of the cingulate cortex in two sagittal sections close to the midline was therefore based a rat flat map, and the borders shown are approximations of the boundaries of individual areas.

Since the previous rodent nomenclature has not been part of human cytoarchitectonic and functional imaging work, it is time to integrate the research in rodents and primates. While scientists working with rodents may find it difficult initially, their data will be more easily appreciated by those working on humans if they follow a scheme that is well established in the human brain and that of other primates.

Hippocampal Region

Refer to Cappaert *et al.,* (2014) for a general description of the hippocampal region. There is a dorsal and a ventral subiculum. We have labeled the **transition area of the dorsal and ventral subiculum** labeled as STr. We have drawn the borders of the presubiculum and parasubiculum have been drawn so as to reach the white matter as suggested by

Haug (1976) and Mulders *et al.* (1997). The **entorhinal cortex** panellation scheme of Insausti *et al.,* (1997) is appealing because each of the cytoarchitectonically distinct divisions has a different pattern of connections as they detail in their paper. The parcellation recognizes that the medial and lateral sectors of the entorhinal area are separated by two intermediate sectors obvious in our cresyl violet preparations. Insausti *et al.,* specifically identify six entorhinal fields: (1) an amygdalopiriform cortex which they termed amygdalo-entorhinal transition field; (2) a medial entorhinal field (MEnt) equivalent to the ventromedial entorhinal area of Krettek and Price (1977); (3) a caudomedial entorhinal field (CEnt), which is the classic medial entorhinal area; (4) a ventral intermediate entorhinal field (VIEnt) equivalent to the caudal ventrolateral entorhinal field of Krettek and Price (1977); (5) a dorsal intermediate entorhinal field (DIEnt); and (6) a dorsolateral entorhinal field (DLEnt). The last two fields together are equivalent to the dorsolateral entorhinal field of Krettek and Price (1977). The **postsubicular area** was identified on the basis of the work of Van Groen *et al.* (1992). **CA2** has been shown to possess a unique gene expression signature (Lein *et al.,* 2005).

Olfactory areas

Refer to Ennis *et al.,* (2014) and Olucha-Bordonau *et al.,* (2014) for a general description of the olfactory system.

The **olfactory nerve layer of the olfactory bulb** (ON) is not present in the photos of coronal sections in the new coronal set. It was apparently stripped off during removal of the brain from the skull. We have indicated its position based on the images of the original coronal set (Paxinos and Watson, 1982).

The **intermediate endopiriform nucleus** (IEn) is an area ventral to the **dorsal endopiriform nucleus** (DEn). Both DEn and IEn lie deep to the piriform cortex. We previously included this area in DEn (RBSC4); Swanson (2004) sometimes calls it DEn and sometimes includes it in the deep layer of the piriform cortex. The cells in this area are relatively sparse and smaller than those in DEn. Both DEn and IEn are NADPH positive (Fig 88, Paxinos *et al.,* 2009), but IEn has parvalbumin positive elements (Fig 89, Paxinos *et al.,* 2009).

We have identified a nucleus located caudal to the amygdala and caudal to the endopiroform nuclei and we named it the **retroendopiriform nucleus** (REn). Paxinos and Franklin (2013) have identified a homologous nucleus in their mouse brain atlas.

The Rat Brain in Stereotaxic Coordinates   Compact 7th Edition   Paxinos & Watson

Telencephalon - Subpaluum

The components of the telencephalic subpallium are the striatum, pallidum, septum, diagonal domain, bed nucleus of stria terminalis, preoptic area, and amygdala (Bardet et al., (2010)).

Striatum and pallidum

Refer to Heimer et al. (1995) and Gerfen and Dudman (2014) for a general description of the striatum and pallidum and to Olucha-Bordonau et al., (2014) for a discussion of the substantia innominata and 'extended amygdala,' which we now call the **extension of the amygdala** for reasons of logic. Immunoreactivity for parvalbumin and the neurofilament protein SMI-32 identifies the ventral pallidum (Paxinos et al, 2009). The concept of the **ventral pallidum,** first proposed by Heimer and his associates, has revolutionized our appreciation of the basal forebrain (Barragan and Ferreyra-Moyano, 1995; Heimer et al, 1997). An area previously called **fundus striati** resembles the striatum proper in some respects and the accumbens shell in others. Given that the use of the term fundus striati creates problems with primate homologues, we followed the advice of George Alheid and called it the **lateral accumbens shell.** The remaining accumbens is delineated in accordance with Zaborszky et al., (1985) and Heimer et al., (1991).

There has been confusion over the naming of parts of the **globus pallidus,** because of a previous opinion that the **entopeduncular nucleus** should be considered as a homolog of the internal segment of the primate globus pallidus. It is now thought that the entopeduncular nucleus is not part of the globus pallidus, but instead belongs to the hypothalamus (Puelles et al, 2012). However, it is clear from calbindin sections (Paxinos et al, 2009) that the globus pallidus proper in the rat can be divided into internal and external segments. We have labeled them EGP and IGP in this atlas.

The lateral stripe of the striatum (LSS) is a dense band of cells in cresyl violet stained sections. The area is negative in calbindin sections and lighter stained than the LAcbSh and striatum in TH. The distinction is very clear in Fig. 87 of Paxinos et al., (2009). In the Sixth Edition, we identified distinct dorsal and ventral parts of the claustrum (DCl and VCl) in AChE stained sections. The dorsal part is positive for AChE and the ventral part is negative.

Diagonal domain

Puelles has introduced this term to refer to structures in the basal forebrain with common gene expression, including the nuclei of the diagonal band, the navicular nucleus, and the basal nucleus (Bardet et al, 2010). Researchers at the University of Virginia and Universidad Nacional de Cordoba have carved out of the traditional substantia innominata another large structure, the sublenticular extension of the amygdala (Alheid et al, 1995). Paxinos and Franklin (2013) have used the new scheme for their mouse brain atlas. We have applied the name substantia innominata to only a small area labeled SIB. We followed Alheid et al., (1995) in the identification of the interstitial nucleus of the posterior limb of the anterior commissure (IPAC). In the Sixth Edition, we gave the name **navicular nucleus** (Nv) to an area in the basal forebrain that we previously called the semilunar nucleus. The reason is that the name semilunar nucleus has longstanding use in the avian literature and it refers to a completely different structure. The extent of the area previously identified as the semilunar nucleus was established on the basis of NADPH-diaphorase histochemistry (Paxinos et al, 2009). We acknowledge assistance of R. Harlan and P-Y. Wang in the identification of this structure (Ahima and Harlan, 1990; Wang and Zhang, 1995).

In the marmoset brain atlas, Paxinos et al., (2012) renamed the magnocellular preoptic nucleus as the lateral nucleus of the diagonal band because this nucleus is cholinergic like the other diagonal band nuclei and has nothing to do with the preoptic area.

In the mouse (as in the marmoset, Paxinos et al., 2012) there is a large zone lateral to the diagonal band nuclei and medial to the caudal part of the accumbens shell that presented us with a problem in many species (eg, Fig 24 in this atlas). This zone is not septal, but neither does it belong to the preoptic area. We first called it the paradiagonal zone (PDZ) in *The Marmoset Brain in Stereotaxic Coordinates* (Paxinos et al., 2012).

Preoptic area

In the preoptic area we have generally followed Simerly et al., (1984), except for the identification of the ventromedial and ventrolateral preoptic nuclei, for which we followed Elmquist et al., (1996) and Sherin et al., (1996). The **compact part of the medial preoptic nucleus** is negative for substance P (Harding et al., 2004). We have recognized the **anterior commissural nucleus** of Kumi Kuroda (Tsuneoka et al., 2013). The parastria area (PS) has been removed from Figure 36 and parastria and striohypothalamic areas from Figure 37, and replaced them with the anterior commissural nucleus. The dopamine group **DA14** has been inserted medial to the anterior commissural nucleus. The area previously labeled APF (anterior periformical area) has been deleted. Jutting ventrolaterally from the anterodorsal preoptic nucleus is a strip that is negative for parvalbumin which we have called the **alar nucleus.** The alar nucleus displays substance P positive cell bodies but little reactivity in its neuropil (Larsen, 1992).

Amygdala and bed nucleus of stria terminalis
Refer to Olucha-Bordonau *et al.,* (2014) for a general description of the amygdala and the bed nucleus of the stria terminalis. The anterodorsal part of the medial nucleus of the amygdala and the basomedial nucleus are defined by the presence of intense NADPH-diaphorase reactivity (Paxinos *et al.,* 1999a; 2009). The lateral part of the central nucleus of the amygdala is marked by the presence of tyrosine hydroxylase fibers and AChE negativity (Paxinos *et al.,* 2009).

We have named the **reticulostrial nucleus** (RtSt) for its position between the stria terminalis and the reticular nucleus. In calretinin stained sections it has a densely positive neuropil, whereas the reticular nucleus of the thalamus has a pale neuropil. In parvalbumin sections, RtSt is positive in neuropil while the reticular nucleus is positive for cells and neuropil (streaky and spotty). RtSt is largest at the anterior pole of the thalamus. In calbindin sections, the stria terminalis is positive, whereas the RtSt is negative (Fig 160, Paxinos *et al.,* 2009). Its medial part is negative and lateral is positive in parvalbumin (Fig 166, Paxinos *et al.,* 2009).

The **rostral amygdalopiriform area** (RAPir) is a distinct region between PLCo and Pir and has a dense layer 2 in the lateral two thirds but much less dense in medial third. We had outlined but not labeled this structure in previous editions of our atlas. Swanson (2004) calls this area the amygdalocortical area, a term which we did not adopt because it does not fit well with the names we had already given to surrounding areas.

Hypothalamus

Refer to Simerly (2014), Armstrong (2014), Saper and Stornetta (2014) and Oldfield and McKinley (2014) for a general description of these structures. Readers should consult the recent comprehensive review of the mouse hypothalamus by Puelles *et al.,* (2012a). In the lateral hypothalamus we identified a **ventrolateral hypothalamic nucleus** on the basis of NADPH-diaphorase reactivity (Paxinos *et al.,* 1999a). This nucleus is caudal to the ventrolateral preoptic nucleus and dorsal to the supraoptic nucleus. The ventral part of dorsomedial nucleus is marked by densely stained cell bodies and terminals in NADPH- diaphorase preparations (Paxinos *et al.,* 1999a). The **gemini**

nucleus is a conspicuous nest of NADPH-diaphorase positive cell bodies (Paxinos *et al.,* 1999a). The **parasubthalamic nucleus** is present in the rat (Wang and Zhang, 1995), but it is not as impressive as the homologous structure in the mouse brain (Paxinos and Franklin, 2013). The **arcuate nucleus** was delineated according to the work of Magoul *et al.,* (1994). See Paxinos and Watson (1986) for the identification of the striohypothalamic, magnocellular lateral hypothalamic, terete, and subincertal nuclei.

We have given names to different regions that are now recognized as comprising the lateral hypothalamus. The **lateral hypothalamus** has traditionally been defined as the area lateral to the fornix. However, the features characteristic of lateral hypothalamus (particularly the population of large cells) are not limited to the area lateral to the fornix. The orexin and hypocretin containing cells are not confined lateral to the fornix, but are also found medial to it (PeFLH). Consistent with this, Swanson (2004) has correctly extended the lateral hypothalamus medial to the fornix. The components of the lateral hypothalamus in our scheme are the peduncular part (PLH), tuberal part (TuLH), perifornical part (PeFLH), and juxtaparaventricular part (JPLH).

The **posterior hypothalamus, dorsal area** (PHD) was previously identified as PHA in our atlas. The area labeled **arcuate nucleus** (ArcMP) is AChE positive and this distinguishes it from the dorsomedial nucleus. We named the **episupraoptic nucleus** (ESO) on the basis of its location. Its rostral pole begins at the caudal pole of the ventrolateral preoptic nucleus (Figs 39–45). The **paraterate nucleus** (PTe) is located within the ventrolateral hypothalamic tract (Swanson, 2004) and is rostral, dorsal, and lateral to the terete hypothalamic nucleus.

Diencephalon

The components of the diencephalon are the pretectal region (prosomere 1), the thalamus (prosomere 2), and the prethalamus (prosomere 3). Note that the habenular region is classified as part of prosomere 2, and that the subthalamic nucleus is now classified as part of the hypothalamus. The terms epithalamus and subthalamus (as regions) have been abandoned (Puelles *et al.,* 2012b).

Thalamus (prosomere 2)
Refer to Vertes (2014) for a general description of thalamic nuclei. See Paxinos and Watson (1986) for the identification of the ethmoid, retroethmoid, subgeniculate, and precommissural nuclei. We have reverted to the use of the term **ventral posterior nucleus, parvicellular part** (VPPC) (Paxinos and Watson, 1982) for the nucleus that we

---

The Rat Brain in Stereotaxic Coordinates   Compact 7th Edition   Paxinos & Watson

previously named the gustatory nucleus of the thalamus (Paxinos and Watson, 1986). We made this change on the advice of Clifford Saper who pointed out that the gustatory input is restricted to the medial part of this nucleus, and that autonomic-related inputs can be found at more lateral parts of this structure (Yasui et al., 1989). We identified a **retroreuniens nucleus** (RRe) dorsal to PH, ventral to CM, and medial to VPPC and SPF. Caudally, RRe merges with the periventricular gray The **paraxiphoid nucleus** (PaXi) lies between the xiphoid nucleus (Xi) of the thalamus medially and the zona incerta laterally The **ventral limitans thalamic nucleus** (VLi) is a thin sheet between subparafascicular, parvicellular part and the medial lemniscus. Jones (2007) considers the limitans nucleus to be continuous with the suprageniculate nucleus. The area we have identified mirrors the posterior limitans thalamic nucleus of the primate brain. Palkovits has observed CGRP positivity in this nucleus (Palkovits, personal communication, 2004).

Refer to Martin et al., (2014) for a general description of the diencephalic nuclei of visual system. We have abandoned the term 'ventral lateral geniculate nucleus' in favor of the **'pregeniculate nucleus'** in order to maintain consistency with the extensive avian literature on this nucleus and to show that it is not part of the thalamus, but the prethalamus. The **intergeniculate leaf** was delineated on the basis of the work of Morin and Blanchard (1995). The medial geniculate was delineated according to the work of LeDoux et al., (1985).

Pretectal area (prosomere 1)
We have named the **lithoid nucleus** (Lth) for the Greek word for a stone. Li is a prominent group of large cells in the dorsal part of the rostral PVG (Fig 71–73). Lth lies medial to MCPC caudally, and medial to RPF rostrally. It is ventral to the PrC and dorsal to the fasciculus retroflexus. More caudally it is dorsal to the Darkschewitsch nucleus. It can be readily identified in horizontal sections (Fig 195, 196). In the pretectal area, we have replaced the name pi periaqueductal gray (plPAG) with the generic term **periventricular gray** (PVG). The previous name was unsatisfactory because the area is related to the third ventricle and not the aqueduct (which does not appear for some distance caudal to this area).

Mesencephalon (Midbrain)

Gene expression patterns indicate that the midbrain has two rostrocaudal segments - a larger rostral area called mesomere 1, and a much smaller preisthmic segment called mesomere 2 (Puelles E. et al., 2012; Puelles et al, 2014).

At the rostral border of the superior colliculus is a substantial area that we have previously included in the pretectal region, labeling it as either posterior pretectal (PPT), the nucleus of the optic tract (OT), or part of the olivary pretectal nucleus (OPT). Is now clear that this area is the mammalian homologue of the **tectal gray** (TG) of birds (Puelles E. et al., 2012), and we have identified it in, inter alia, Figs 78–82. The terms PPT and OT have been completely subsumed into the tectal gray. In birds the tectal gray occupies an extensive region caudal to the pretectal area (Puelles et al., 2007) and this correction aligns our atlas with that of the bird.

Refer to Keay and Bandler (2014) for a general description of the **periaqueductal gray** (PAG). The boundaries of the cell columns within the periaqueductal gray were drawn according to Carrive (1993), Carrive and Paxinos (1994), and Paxinos and Huang (1995). We identify the rodent homologue of the human pleoglial periaqueductal gray in Figs 77–80. Two nuclei lateral to the central gray pars alpha were identified on the basis of SMI-32 immunoreactivity - central gray pars beta and central gray pars gamma (Paxinos et al, 2009).

We have applied the term **central mesencephalic nucleus** (CeMe) to an area in the rostral midbrain lateral to the periaqueductal gray that contains a distinct population of calbindin positive cells. This area was first identified in the mouse brain atlas of Watson and Paxinos (2010). The CeMe lies rostral to the precuneiform nucleus. Based on the neuromeric nomenclature for tegmental areas in the avian brain (Puelles et al., 2007), we have named the central midbrain tegmentum the **mesencephalic reticular formation** (mRt) to distinguish it from the isthmic and rl reticular formation (isRt and rlRt).

We used Faye-Lund and Osen (1985) as well as Malmierca (2014) for the identification of areas of the **inferior colliculus.** The commissure of the inferior colliculus contains many neurons, and on the advice of Ellen Covey we have named this the **nucleus of the commissure of the inferior colliculus** (CIC). Note that the **substantia nigra** (SN) and the **ventral tegmental area** (VTA) are plurisegmental structures, which extend from the isthmus to prosomere 3 (Puelles E et al., 2012). We have used the primate terminology for the dorsal and ventral tiers of the substantia nigra and also for the part of the

VTA which is called the **parabrachial pigmented nucleus.** The names of the primate substantia nigra subdivisions have a longer history than those subsequently used in rodent studies, and the identified primate subdivisions are consistent with the degeneration patterns seen in Parkinson's disease (Halliday, personal communication, 2004). The reticular part of the substantia nigra can be divided into a ventrolateral and a dorsomedial component on the basis of parvalbumin and calbindin distribution (Paxinos *et al.,* 2009). The remainder of the substantia nigra and the ventral tegmental area were delineated according to the work of McRitchie *et al.,* (1996). We have identified a rat homologue of the human parapeduncular nucleus (Paxinos & Huang, 1995) but have named it the **parainterfascicular nucleus.** In the Sixth Edition, we suggested that the new term be also used for the human given its more descriptive nature. With the addition of the parainterfascicular term, we noted that the entire VTA can be represented by specifically named component parts. This avoids the problem of earlier editions of this atlas where the label VTA was placed only on what we now call parainterfascicular nucleus, giving the impression that it alone was the whole of the VTA. The VTA in our view consists of the paranigral nucleus (PN), the parainterfascicular nucleus (PIF), the parabrachial pigmented nucleus (PBP), and the rostral VTA (VTAR). A useful summary of the organization of the rodent VTA and substantia nigra can be found in the study of Fu *et al.,* (2012).

Rhombencephalon (Hindbrain)

We use the term rhombencephalon to refer to the collection of structures that make up the hindbrain; these are the isthmus, eleven rhombomeres, and the cerebellum (which is an outgrowth from the isthmus and rhombomere 1). The isthmus and eleven rhombomeres can be collectively referred to as the axial hindbrain, in order to distinguish them from the cerebellum.

Subdivision into pons and medulla?

The hindbrain has traditionally been divided into a rostral half, the pons, and a caudal half, the medulla oblongata. Unfortunately, this subdivision does not reflect developmental subdivisions, and so has become the source of much confusion in comparative studies. Studies of gene expression in mammalian rhombomeres show that the pontine nuclei arise in the rhombic lip in rhombomeres 6–7 and migrate to their final position in the ventral part of r3 and r4 (Farago *et al*, 2006). However the basilar pons becomes massively enlarged in primates and the pontine swelling appears to extend from the midbrain to the level of the inferior olive in the human. In animals with a modest pontine crossing, like the rat, the pontine crossing and the pontine nuclei remain restricted in the ventral part of r3 and r4. The variation in size of the external bulge of the pons makes it inappropriate to use this feature as a topographic descriptor to subdivide the hindbrain. We recommend that the term 'pons' be reserved for naming the nuclei and crossing fibers of the basilar pontine formation.

Cerebellum

We have segmented the cerebellum according to the scheme originally developed by Bolk (1906), and subsequently refined and expanded by Larsell (1952, 1970). The Bolk/ Larsell scheme has since been adopted by all major modern studies of the rat cerebellum including those of Swanson (2004) and Ruigrok *et al.,* (2014). The rat cerebellum consists of a central vermis and two lateral hemispheres. The vermis is separated from the hemispheres by paramedian sulci. The vermis consists of a series of ten lobules (lCb to lOCb) separated by named fissures. Each of the ten lobules has a traditional name which was current before the development of the Bolk/Larsell scheme: lobule 1 is the lingula; lobules 2 and 3 are the central lobule; lobules 4 and 5 are the culmen; lobule 6 is the declive; lobule 7 is the folium and tuber; lobule 8 is the pyramis; lobule 9 is the uvula; and lobule 10 is the nodulus. Lobules 6 to 10 extend laterally to form the cerebellar hemisphere: lobule 6 extends into the simple lobule and crus 1 of the ansiform lobule; lobule 7 extends into crus 2 of the ansiform lobule and the paramedian lobule; lobule 8 extends into the copula of the pyramis; lobule 9 extends into the flocculonodular lobe; and lobule 10 extends into the flocculus. While lobules 6–8 can be seen to be directly continuous with their lateral extensions, lobules 9 and 10 are more difficult to visualize. The reason is that the flocculus and paraflocculus are displaced rostrally with an attenuated connection to their parent lobules.

Axial Hindbrain-Isthmus and Rhombomeres1-11

The **isthmus** of the mammalian brain should be recognized as a distinct segment of the brain, separating the mesencephalon from the first rhombomere (Watson, 2012). The signature components of the isthmus are the trochlear nucleus, the caudal linear nucleus, and the parabigeminal nucleus (Watson, 2012).

We have followed Puelles *et al.,* (2007) and Paxinos and Watson (2007) in renaming the pedunculopontine tegmental nucleus to pedunculotegmental nucleus (PTg) for a compelling reason: the nucleus belongs to the isthmus and rhombomere 1 and not to the 'pons.' As noted above, the pons does not exist as a subdivision of the brain in the same subordination as the rhombencephalon, the isthmus, the mesencephalon, the diencephalon and the telencephalon. In any case, the pontine crossing is situated in rhombomeres 3 and 4, well away from the pedunculotegmental nucleus.

In the previous edition of our atlas we renamed the retrorubral nucleus as retroisthmic because the term retrorubral has been often confused with the retrorubral fields (the dopamine cell group), and also because of our belief the nucleus was in fact retroisthmic. However, our recent work with fgf8 expression indicates that the nucleus may lie in the caudal part of the isthmus instead of r1, in which case we were premature in naming it retroisthmic. We have not renamed it yet again, but will wait until its neuromeric location is confirmed.

We have divided the **cuneiform nucleus** into three layers based on AChE staining. We consider that the intermediate and ventral parts lie in the isthmus, but the dorsal part lies in the midbrain. The **rhabdoid nucleus** is a striking feature of AChE sections through the rostral hindbrain. It is an elongated nucleus ('rhabdoid' means 'sticklike') that is intensely AChE positive. The ventralmost part of the nucleus is continuous with the interpeduncular nucleus and the dorsal parts are closely applied to the caudal border of the decussation of the superior cerebellar peduncle. We believe that the rhabdoid nucleus represents the most caudal part of the isthmus. The **epirubrospinal nucleus,** which we first identified in the Second Edition of this atlas (Paxinos and Watson, 1986), has been shown to project to the spinal cord in the mouse (Liang *et al.,* 2011)

The **epipeduncular nucleus** (EpP) is a small but distinctive group of large cells below the peripeduncular nucleus and above the cerebral peduncle (Fig 78–80) that has no home in the surrounding nuclei. This nucleus gives rise to a tract which reaches the spinal cord in mice (Liang *et al.,* 2011). We have identified the rat homologue of the subnucleus L of the reticular tegmental nucleus (RtTgL), originally described by Olzewski and Baxter (1954).

We identified the **raphe nuclei** on the basis of 5-hydroxytryptamine reacted sections prepared by G. Halliday and I. Tork (see also Harding et al., 2004). We identified the raphe interpositus nucleus on the basis of the work of Buttner-Ennever *et al.,* (1988). Two recent studies have analyzed the components of the serotonin raphe system according to gene expression in the mouse (Jensen *et al.,* 2008; Alonso *et al.,* 2012). We have modified our naming and delineation of a number of raphe components according to the scheme of Alonso *et al.,* (2012). The most rostral part of the dorsal raphe is the **mesencephalic dorsal raphe** (mDR). We have extended the area occupied by the RMg into a space formerly given to the raphe pallidus. Note that the PDR and DRL parts of the dorsal raphe in our diagrams are collectively named the wing of the dorsal raphe (DRW) by Alonso *et al.,* (2012). According to the results of Alonso *et al.,* (2012), the raphe magnus extends to the caudal pole of the facial nucleus (the border between r6 and r7), and the raphe pallidus is found caudal to r6. The designation has been changed to be consistent with this view. Note that the supragenual nucleus (SGe) has been identified by Alonso *et al.,* (2012) as part of the serotonin raphe. We have renamed the SGe as 'the **supragenual nucleus of the raphe.**' Refer to Aston-Jones (2004) for the delineation of the locus coeruleus. We delineated the **catecholamine cell groups** by following Hökfelt *et al.,* (1984) with assistance from our own tyrosine hydroxylase preparations (Paxinos *et al.,* 2009). There is a great deal of potential for confusion in the use of different systems for naming the brainstem catecholamine groups. We have followed the nomenclature of Paxinos *et al.,* (2012) in their atlas of the marmoset brain in naming dopamine groups with the prefix DA, noradrenalin groups with the prefix NA, and adrenaline groups with the prefix Ad. However, we have retained the name of locus coeruleus for the previously named A6 group and the name supralemniscal nucleus (SuL) for the B9 dopamine group. Similarly we have retained the names retrorubral field (RRF), substantia nigra compact (SNC), and ventral tegmental area for the groups previously defined as A8, A9, and A10.

The **intermediate reticular zone** was first identified in the rat (Paxinos and Watson, 1986), and this subdivision is seen to advantage in the human brain (Paxinos and Huang, 1995). The intermediate reticular zone at levels of the caudal pole of the facial nerve nucleus is marked by NADPH-diaphorase positive cells. The **lateral paragigantocellular nucleus** is conspicuous in NADPH-diaphorase preparations (Paxinos *et al.,* 2009). We have identified the **parapyramidal nucleus of the raphe** as the cell group dorsolateral to the pyramidal tract, which is outlined but not named in the Second Edition of this atlas (Paxinos and Watson, 1986). The identification of the **epifascicular nucleus** is based on the description of this nucleus in the human brain (Paxinos and Huang, 1995).

We have identified the **conterminal nucleus** (Ct) in the medulla close to the inferior olive. This group was originally identified by Olzewski and Baxter (1954) and is clearly present in the human brain stem (Paxinos and Huang, 1995). The nucleus is seen as two separate AChE positive cell groups, one lateral to the inferior olive (caudal pole of IOA) and a second group medial to the IOA.

Refer to Lundy and Norgren (2014) for a general description of the **hindbrain viscerosensory nuclei.** The posterior part of the **nucleus of the solitary tract** was delineated in accordance with the work of Whitehead (1990), Herbert *et al.,* (1990), McRitchie (1992), and Altschuler *et al.,* (1989). The rostral part of the nucleus of the solitary tract was difficult to delineate, but we recognize a rostrolateral subnucleus on the basis of NADPH-diaphorase positivity. The **parabrachial region** was delineated in accordance with Fulwiler and Saper (1984), Herbert *et al.,* (1990), Whitehead (1990), and Herbert and Saper (1990). The external part of the lateral parabrachial nucleus and medial parabrachial nucleus are marked by NADPH- diaphorase positive cells and fibers (Paxinos *et al.,* 2009). Refer to Beckel and Holstege (2014) for the organization of the neural pathways that innervate the lower urinary tract to control micturition.

The **medullary respiratory groups and the Botzinger complex** areas were delineated in accordance with Ellenberger *et al.,* (1990), Kanjhan *et al.,* (1995), and Cox and Halliday (1993). Refer to Travers (2014) for a description of the **oromotor nuclei.** Refer to Ruigrok *et al.,* (2014) for a general description of the **precerebellar nuclei of the hindbrain.**

We reallocated part of the magnocellular medial vestibular nucleus to the lateral vestibular nucleus (where we had it in the First Edition) on the basis of the work of Vidal *et al.,* (2014) and Liang *et al.,* (2011,2013).

Refer to Ebner and Kaas (2014) for a general description of the **somatosensory nuclei of the hindbrain** and to Westlund and Willis (2014) for a description of the pain system. The general basis of delineation of these structures is described in Paxinos and Watson (1997). However, we followed Marfurt and Raj chert (1991) for the borders of the spinal trigeminal nucleus. We identify the NADPH-diaphorase positive area medial to the principal sensory trigeminal nucleus as the **trigeminal transition zone** (5Tr) given its juxtaposition between the trigeminal and the parabrachial nuclei.

The **trigeminosolitary zone** (5Sol) commences caudal to the trigeminal transition zone, and extends as far caudal as the level of the area postrema. The rostral part of this zone was previously identified by Paxinos and Huang (1995) in the human brainstem and named the subsolitary nucleus. We note that in the medulla, as in the thalamus (VPM, VPPC - Lundy, and Norgren, 2014), there is a progression of functional areas from the trigeminal concerned with somatosensory function to the solitary concerned with gustatory function from receptors of the same peripheral structures. Paxinos and Huang (1995) identified in the human the pericuneate nucleus and peritrigeminal

matrix. We observed a similar structure in the rat and identified it as the **matrix zone of the medulla** (Mx). We note that, as in primates, the rat has a **rotund part in the cuneate nucleus** (CuR) that almost certainly represents the forepaw area.

Refer to Malmierca (2014) for a general description of the brainstem auditory system. **The nucleus of the central acoustic tract** (CAT) has been identified in the cat, and Ellen Covey has delineated this structure in our atlas. For additional details on the basis of delineation of the components of the auditory system refer to Paxinos and Watson (1986). We have adopted new abbreviations for the layers of the **dorsal cochlear nucleus:** DCDp is the **dorsal cochlear deep layer; Lokesh** DCFu is the **dorsal cochlear fusiform layer; Lokesh** DCMo is the **dorsal cochlear molecular layer.**

The periolivary region has a dorsolateral protrusion positive in parvalbumin (Paxinos *et al,* 1999b), which we termed the **periolivary horn** (POH).

We have identified major blood vessels according to Scremin (2014).

Spinal Cord

We refer the reader to the chapters on the spinal cord, including the substantia gelatinosa of Sengul (2014a,b), Sengul and Watson (2014) and Ribeiro-da-Silva (2014). In this edition we have reproduced the rat spinal cord diagram series from our own atlases of the spinal cord (Watson *et al.*, 2009; Sengul *et al.*, 2013).

We identified laminae in the spinal cord gray matter according to the schema developed by Rexed (1952), using the prefix 'Sp' to indicate each lamina so as to allow computer searching systems to distinguish between this set of digits and those used in other areas (e.g., layers of cerebral cortex). Certain nuclei have been identified within the laminae of Rexed. Some of these are well known (such as the dorsal nucleus), but some others are less well recognized. In the latter category are the nuclei of origin of the ventral spinocerebellar tract. We have identified these as the **lumbar** and **sacral precerebellar nuclei** (LPrCb and SPrCb). We have labeled the sympathetic preganglionic cells as the **intermediolateral cell group** (IML), but we chose the name **'sacral parasympathetic nucleus'** for the parasympathetic preganglionic neuron groups. We have not separately identified the lateral outlying parts of either of these nuclei, but it should be noted that some authors name these lateral outlying cells in the thoracic region as the funicular part of the IML.

We have attempted to resolve an area of nomenclatural confusion involving the lumbar and sacral intermediate gray area. The name **'dorsal commissural nucleus'** has been applied by different groups of authors to two completely different nuclei, one lumbar and one sacral. We have named the former the **lumbar dorsal commissural nucleus** (LDCom) and the latter the **sacral dorsal commissural nucleus** (SDCom).

The motor neuron clusters in lamina 9 have been named for the muscles or muscle groups they supply. We have not attempted to subdivide these motor nuclei into groups that supply individual muscles because such data are limited, and in any case there is considerable overlap between the groups of motor neurons supplying adjacent muscles. We were not able to identify the many of long tracts in these spinal cord sections. The tracts we were able to identify were the dorsolateral fasciculus, the gracile fasciculus, the cuneate fasciculus, the postsynaptic dorsal column pathway, the rubrospinal tract, and the dorsal corticospinal tract. The position of other tracts can be inferred from the position they appear to occupy in the mouse (Watson and Harrison, 2012).

Mini-atlas of the rat brain

This section of the book is a set of simplified drawings of brain sections (a mini-atlas) to assist those who are unfamiliar with brain anatomy. The mini-atlas contains twelve diagrams of coronal sections from one end of the rat brain to the other. The major regions of the brain and some important nuclei and tracts are labeled. Each drawing is accompanied by a short description of the main features. Following the twelve coronal diagrams are diagrams of two sagittal sections. These latter sections show some areas of the brain to advantage, and are particularly useful for understanding the relationships between diencephalic, midbrain, and hindbrain regions.

The brain is made up of three main parts, the forebrain (the prosencephalon), the midbrain (the mesencephalon), and the hindbrain (the rhombencephalon). The forebrain is divided into the telencephalon (which includes the cerebral cortex and the subpallium), the hypothalamus, and the diencephalon. Each of these main subdivisions is labeled in the lower image (b). The midbrain joins the hindbrain to the forebrain. The hindbrain has three major parts, the isthmus, the rhombomberes 1-11, and the cerebellum. The end of the rhombencephalon joins with the spinal cord.

(a) is a photograph of a Nissl-stained sagittal section of a rat brain. The section is close to the midline. The dark folds of the cerebellum are seen on the right, and the dark blue stripe on the far left is a part of the olfactory bulb. The dark diamond-shaped patch ventral to the middle of the brain is the pituitary gland.

(b) is a color-coded diagram of the photograph seen in (a). The major parts of the brain are labeled. Each of the structures shown here will be discussed in herein.

The bottom diagram shows the position of the coronal section diagrams which form an atlas series in this chapter (Figures 1 to 12). Each vertical line shows the position of the relevant coronal section.olfactory tubercle

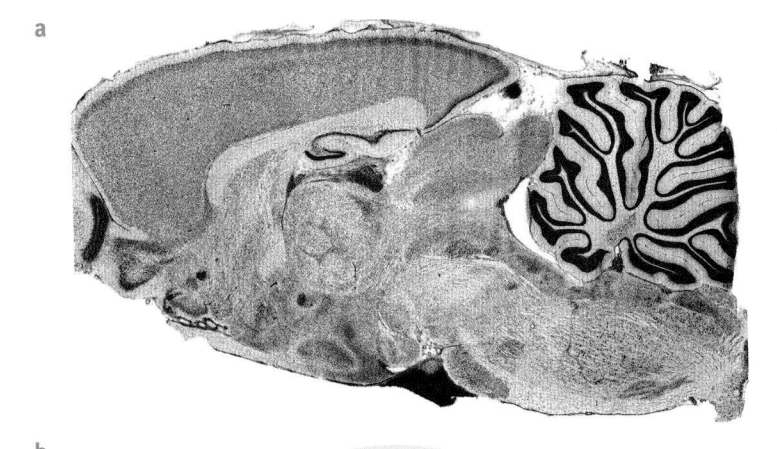

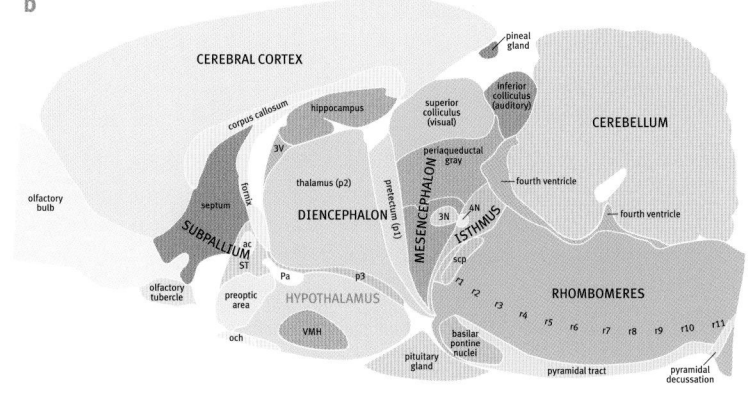

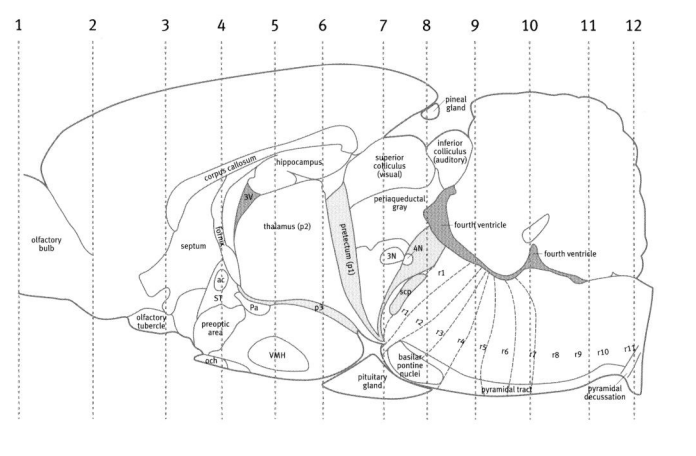

The Rat Brain in Stereotaxic Coordinates   Compact 7th Edition   Paxinos & Watson

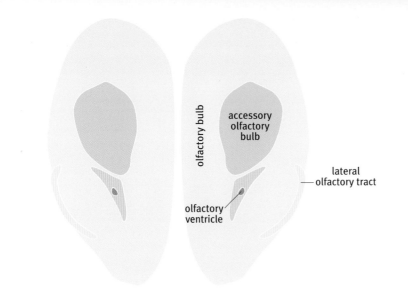

Figure 1 The olfactory bulb

The most rostral part of the forebrain is the olfactory bulb, which receives the olfactory nerve fibers from the roof of the nose. The fibers leaving the bulb form the lateral olfactory tract. Embedded in the dorsal part of the main olfactory bulb is the accessory olfactory bulb; it receives inputs from the vomeronasal organ, a pheromone receptive area in the nasal cavity. The vomeronasal organ and the accessory olfactory bulb are not found in the human brain. A small extension of the lateral ventricle, called the olfactory ventricle, is seen in the center of the olfactory bulb.

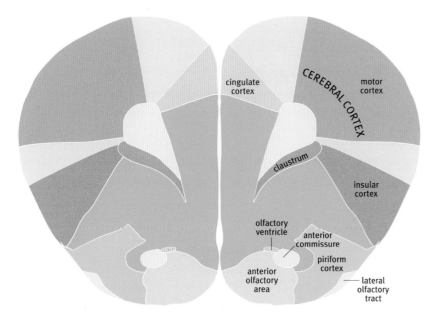

Figure 2 The forebrain at the level of the frontal pole of the cerebrum

The section cuts through the rostral end of the motor cortex. Below the motor cortex are the insular cortex and the piriform cortex. On the surface of the piriform cortex is the lateral olfactory tract, which connects the olfactory bulb to the piriform cortex. Medial to the piriform cortex is the anterior olfactory area, which also receives fibers from the olfactory bulb. Medial to the motor cortex is the cingulate cortex.

The anterior limb of the anterior commissure is embedded in the dorsal part of the anterior olfactory area. A small extension of the lateral ventricle, called the olfactory ventricle, lies adjacent to the anterior limb of the anterior commissure and extends forward into the olfactory bulb. Deep to the insular cortex is an area of white matter and below this is the claustrum.

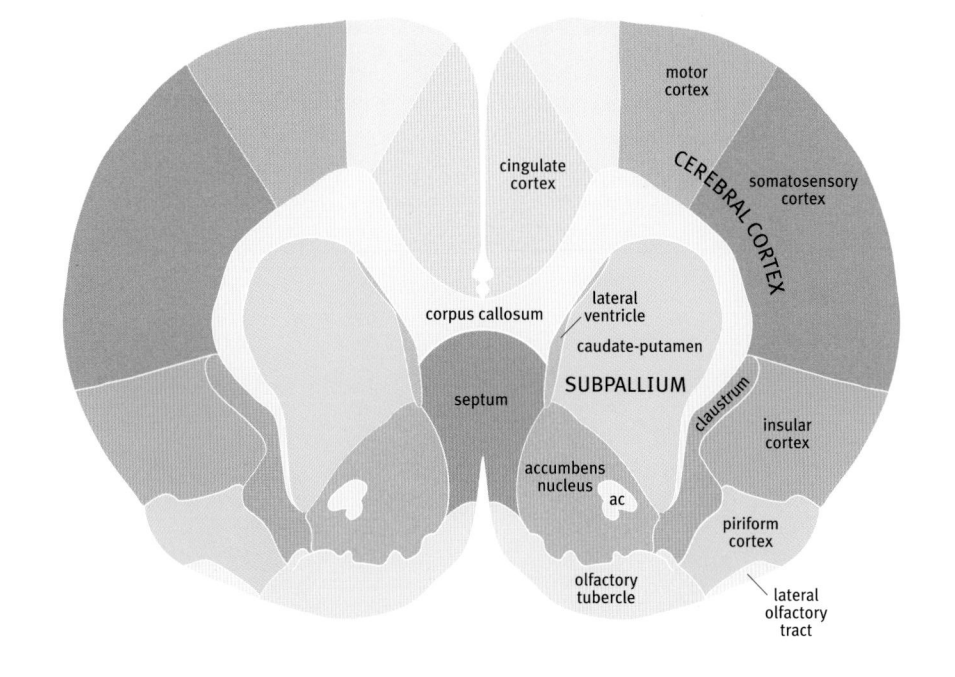

Figure 3 *The forebrain at the rostral end of the corpus callosum*

This section passes through the rostral end of the corpus callosum and cuts through the rostral end of the somatosensory cortex and the middle of the motor cortex. Medial to the motor cortex is the cingulate cortex. Below the somatosensory cortex are the insular cortex (taste and visceral sensation) and the piriform cortex. On the surface of the piriform cortex is the lateral olfactory tract which connects the olfactory bulb to the piriform cortex. Medial to the piriform cortex is the olfactory tubercle, which also receives fibers from the olfactory bulb.

The layer of white matter lateral and dorsal to the caudate-putamen is called the external capsule. Ventral to the corpus callosum is a large group of nuclei called the septum, which is the largest of the medial group of deep cerebral nuclei. Medial to the external capsule lies the caudate-putamen. The caudate-putamen is the largest of the lateral group of deep cerebral nuclei. The lateral ventricle lies between the septum and the forceps minor (fibers of the of the corpus callosum that are traveling toward the frontal pole) medially, and the caudate-putamen laterally. Lateral to the most ventral part of the external capsule is another deep nucleus of the cerebrum called the claustrum.

The anterior limb of the anterior commissure (ac) is prominent bundle of fibers dorsal to the olfactory tubercle. Surrounding the anterior limb of the anterior commissure is the accumbens nucleus.

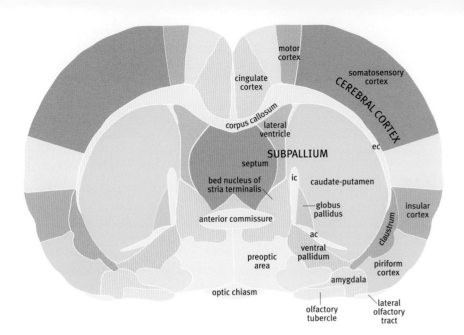

Figure 4 The forebrain at the level of the anterior commissure

This section is taken immediately in front of the rostral pole of the thalamus. In the center of the section is the anterior commissure, with the septum above, the preoptic area below, and the bed nucleus of the stria terminalis around its lateral edge. In the midline of the preoptic area is the third ventricle, which links the lateral ventricles to the cerebral aqueduct. Immediately below the third ventricle is the optic chiasm, in which optic nerve fibers cross the midline.

The areas of cerebral cortex seen in this section are the somatosensory cortex and the motor cortex. Below the somatosensory cortex is the insular cortex (taste and visceral sensation) and the piriform cortex. On the surface of the piriform cortex is the olfactory tract, which connects the olfactory bulb to the piriform cortex. Medial to the piriform cortex is the rostral part of the amygdala. The cingulate cortex is seen on the medial side of the cerebrum above the corpus callosum.

Medial to the external capsule (ec) lie representatives of the lateral group of deep cerebral nuclei, the caudate-putamen and the globus pallidus. Along the medial border of the globus pallidus is the rostral part of the internal capsule (ic). The internal capsule is a large sheet of fibers which connects the cerebrum with the thalamus, brainstem, and spinal cord.

The thin layer of white matter lateral to the caudate-putamen is called the external capsule. The lateral ventricle lies between the septum and the caudate-putamen in this section. The anterior commissure (also labeled ac) is a large bundle of crossing fibers, which connects the olfactory bulbs and parts of the cerebrum to the same areas on the opposite side.

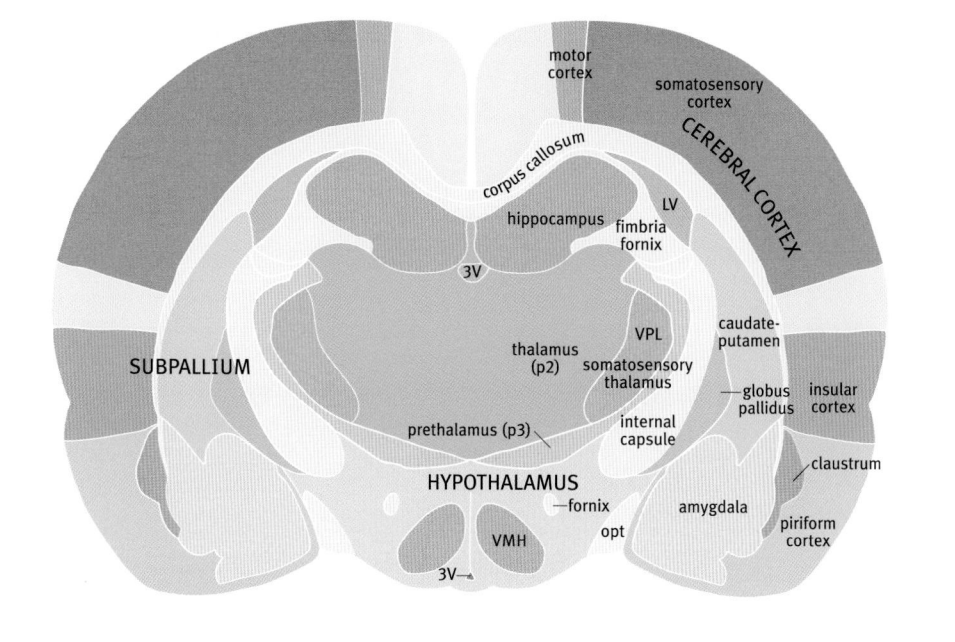

Figure 5 The forebrain at the level of the internal capsule

This section shows the thalamus and hypothalamus in the center covered above and to the side by the cerebrum. In the midline is the third ventricle, a cavity that links the lateral ventricles to the aqueduct. Separating the thalamus from the cerebrum is the internal capsule, a very large sheet of fibers that connects the cerebral cortex with the thalamus, brainstem and spinal cord. The caudal end of the internal capsule forms the cerebral peduncle.

The section cuts through the somatosensory (touch sensation) cortex and the caudal end of the motor cortex. Below the somatosensory cortex is the insular cortex (taste and visceral sensation) and the piriform cortex. Medial to the piriform cortex is a large group of nuclei called the amygdala.

The hippocampus is an easily recognizable part of cerebral cortex involved with memory registration. In this section, the hippocampus is seen between the corpus callosum and the thalamus. The thin layer of white matter lateral to the ventricle and the caudate-putamen is called the external capsule which is continuous with the corpus callosum, the major commissure of the forebrain.

Between the internal capsule and the external capsule are members of the lateral group of deep cerebral nuclei, the caudate-putamen and the globus pallidus. These nuclei are involved in motor control, particularly semi-automatic movements and locomotion. The thalamus (prosomere 2) receives input from the somatosensory (touch), auditory, and visual pathways, and sends projections to almost all the cortical areas. In this section, the somatosensory nucleus (VPL) is a half-moon shaped structure at the lateral edge of the thalamus.

The hypothalamus lies ventral to the thalamus, and between the thalamus and hypothalamus is the prethalamus (prosomere 3). The ventromedial hypothalamic nucleus (VMH) is a prominent landmark in the hypothalamus. Between the hypothalamus and the amygdala is the optic tract (opt), a bundle of fibers running from the optic chiasm to the dorsal lateral geniculate nucleus of the thalamus.

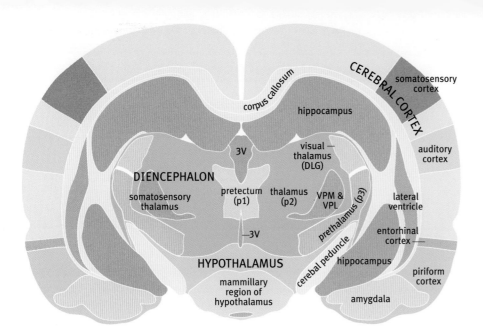

Figure 6 The forebrain at the level of the caudal end of the thalamus

This section shows the thalamus and hypothalamus in the center, covered above and to the side by the cerebrum. In the midline is the third ventricle, which links the lateral ventricles to the aqueduct.

The thalamus (prosomere 2 of the diencephalon) lies dorsal to the hypothalamus. The thalamus receives input from touch, auditory, and visual pathways, and sends fibers to the cerebral cortex. The part of the thalamus that receives sensory information from the eyes is called the dorsal lateral geniculate nucleus (DLG). The dorsal lateral geniculate connects with the visual cortex of the cerebrum. The somatosensory (touch sensation) nuclei of the thalamus (VPL and VPM) lie ventral to the DLG. On the medial edge of thalamus is part of the pretectal area (prosomere 1 of the diencephalon), and the lateral edge is the prethalamus (prosomere 3).

The hypothalamus is located ventral to the thalamus. At this level of the hypothalamus, there is a pair of bulges called the mammillary bodies. At the lateral edge of the hypothalamus is the cerebral peduncle. The cerebral peduncle contains the pyramidal (corticospinal) fibers as well as the fibers traveling from the cortex to the pons.

The part of cerebral cortex seen here is about midway between the occipital and frontal poles of the cerebrum. The most obvious cortical areas at this level are the somatosensory cortex above, the auditory cortex laterally, and the piriform (olfactory) cortex below. Above the piriform cortex is the rostral end of the entorhinal cortex. Medial to the piriform cortex is a large group of nuclei called the amygdala. The amygdala is associated with emotional responses and a number of survival behaviors Medial to the lateral ventricle is the hippocampus, an easily recognizable part of cerebral cortex involved with memory registration. The corpus callosum is a large sheet of fibers connecting the left and right sides of the cerebrum.

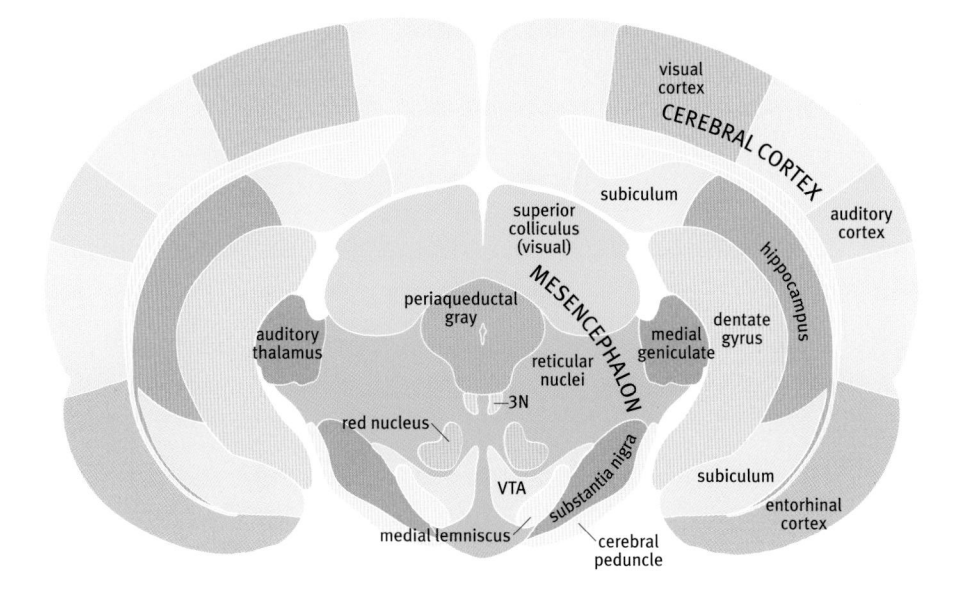

Figure 7 *The rostral end of the midbrain*

This section shows the midbrain in the center, covered above and to the side by the bulging caudal end of the cerebral cortex. The superior colliculus occupies the dorsal part of the midbrain, but the lateral part contains the most caudal part of the thalamus - the medial geniculate nucleus, which is the auditory thalamic nucleus.

At this level, the pyramidal (corticospinal) fibers and other descending fibers form a bundle called the cerebral peduncle on the ventral aspect of the midbrain. Coating the inside (medial) border of the fibers of the cerebral peduncle is the substantia nigra, a large cell group that is important for motor control. Fibers from the substantia nigra project to the deep nuclei of the cerebrum, especially the striatum. Just above the medial end of the substantia nigra is the medial lemniscus, a fiber bundle carrying touch information from the body and face to the thalamus. On the medial side of the medial lemniscus is the ventral tegmental area (VTA).

Surrounding the aqueduct is a thick layer of neurons called the periaqueductal gray. Below the periaqueductal gray is a prominent pair of motor nuclei called the oculomotor nuclei. Each oculomotor nucleus (3N) supplies four of the six muscles that move each eye. In addition, the oculomotor nerve supplies the main muscle of the upper eyelid and the muscles of the pupil and lens. Between the periaqueductal gray and the medial lemniscus is the red nucleus. This large group of cells gives rise to a fiber bundle that travels down the spinal cord, called the rubrospinal tract. Between the two cerebral peduncles at the ventral margin is the interpeduncular nucleus. The area between the red nucleus and the superior colliculus is filled with reticular nuclei. Attached to the lateral edge of the midbrain is the medial geniculate nucleus of the thalamus.

The part of the cerebrum seen here contains the visual cortex above and the auditory cortex laterally. The ventral part of the cortex is occupied by the entorhinal cortex, and the medial part is occupied by the hippocampal structures (hippocampus, dentate gyrus, and subiculum).

The Rat Brain in Stereotaxic Coordinates   Compact 7th Edition   Paxinos & Watson

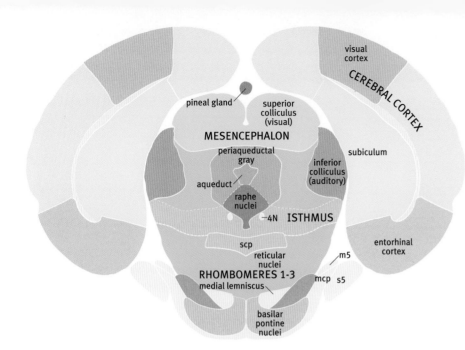

Figure 8 The brainstem at the level of the midbrain

This section shows the brainstem in the center covered above and to the side by the bulging caudal end of the cerebral cortex. Because the neuraxis is flexed in this region, the top half of the brainstem section includes the midbrain, whereas the bottom half includes the rostral parts of the hindbrain, such as the pons. The pons are made up of the basilar pontine nuclei and the fibers that connect to them. The fibers that arise from the pontine nuclei cross to reach the opposite side of the cerebellum (pons is Latin for bridge). The fibers of the pyramidal tract are embedded in the pontine nuclei. Just above the pons is the medial lemniscus, a fiber bundle carrying touch information from the body and face to the thalamus. Sitting just lateral to the pons is the sensory trigeminal nerve (s5), a large bundle of fibers that carries sensory information from the skin of the face to the trigeminal sensory nuclei.

In the center of the midbrain is the aqueduct, a small canal along which cerebrospinal fluid (CSF) flows from the ventricles of the forebrain to the fourth ventricle of the hindbrain. The aqueduct is surrounded by a thick layer of cells called the periaqueductal gray. Below the periaqueductal gray is the small motor neuron cluster called the trochlear nucleus (the nucleus of the fourth cranial nerve). Each trochlear nucleus supplies just one of the six muscles that move each eye. Above the periaqueductal gray is the pair of superior colliculi. Each superior colliculus receives visual sensory information from the opposite eye. Lateral and ventral to the superior colliculus is the inferior colliculus, which receives auditory information from the superior olive.

The part of the cerebrum seen here is the occipital pole, which is occupied mainly by the visual cortex. The entorhinal cortex occupies the ventral zone and the subiculum (part of the hippocampal complex) occupies the medial zone of cortex. The pineal gland is seen in the midbrain, but is attached to the dorsal surface of the thalamus by its stalk. It manufactures the hormone melatonin, which has a role in the sleep- wake cycle.

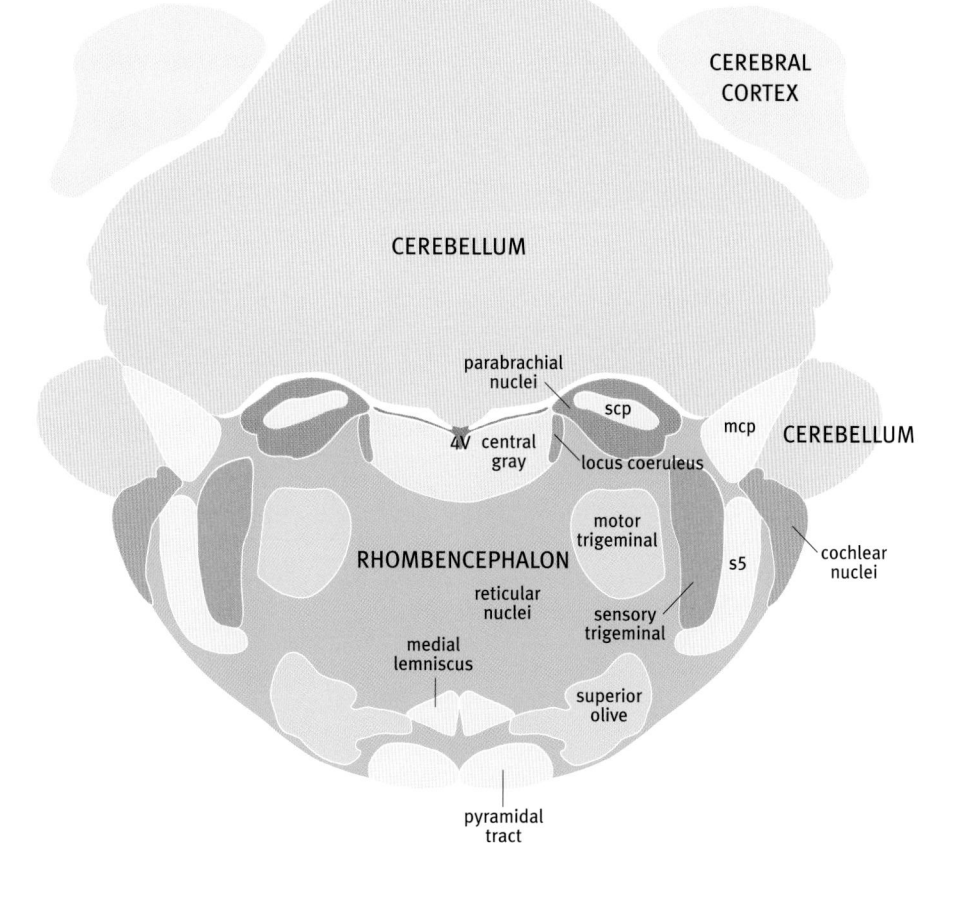

CEREBRAL CORTEX

CEREBELLUM

parabrachial nuclei

scp

mcp CEREBELLUM

4V central gray

locus coeruleus

motor trigeminal

RHOMBENCEPHALON

s5

cochlear nuclei

reticular nuclei

sensory trigeminal

medial lemniscus

superior olive

pyramidal tract

Figure 9 *The hindbrain at the level of the upper end of the fourth ventricle*

This section shows the hindbrain below and the rostral part of the cerebellum above. Dorsal and lateral to the cerebellum is the caudal end of the cerebral cortex. The rostral end of the fourth ventricle lies between the brainstem and the cerebellum. The part of the brainstem seen here is the rostral end of the hindbrain. The large trigeminal nuclei are found in the lateral part of the hindbrain. This part of the trigeminal complex is called the principal sensory trigeminal nucleus. It receives touch sensation from the face. Medial to this is the motor trigeminal nucleus, which controls the chewing (masticatory) muscles. Two large fiber bundles, the pyramids, lie next to the midline on the ventral margin of the hindbrain. Just above each pyramid is the medial lemniscus, a fiber bundle carrying touch information from the body and face to the thalamus. The thalamus then sends this information to the cerebral cortex. Lateral to the trigeminal nuclei are the cochlear nuclei. Between the pyramid and the trigeminal nucleus at this level is the superior olive, a collection of nuclei that process and analyze auditory information. Above the superior olive is an area occupied by the reticular nuclei. The cerebellum is dorsal to the core of the hindbrain. The cerebellum is joined to the hindbrain by two fiber bundles, the middle cerebellar peduncle (mcp) and the superior cerebellar peduncle (scp). Between the cerebellum and the core of the hindbrain is the fourth ventricle (4V). The fourth ventricle contains cerebrospinal fluid (CSF) that has flowed down from its origin in the lateral (cerebral) ventricles.

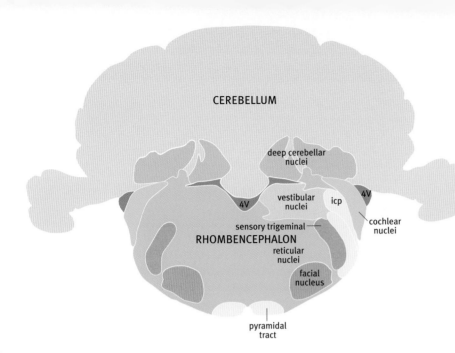

Figure 10 The hindbrain at the level of the facial nucleus

This section shows the hindbrain below and the cerebellum above with the fourth ventricle in between.

The large trigeminal nuclei are found in the lateral part of the hindbrain at this level. This part of the trigeminal sensory complex is called the spinal trigeminal nucleus because it extends into the cervical spinal cord. Two large fiber bundles, the pyramids, lie on either side of the midline on the ventral margin of the hindbrain. The pyramids contain the pyramidal (corticospinal) tracts. Between each pyramid and the trigeminal nucleus is a large group of motor neurons that supplies the muscles of facial expression on that side, called the facial nucleus. The vestibular nuclei lie at the junction of cerebellum and hindbrain at the side of the fourth ventricle. They receive information from the position sense organs of the inner ear. At the lateral edge of the hindbrain under the cerebellum are the cochlear nuclei, which receive auditory sensations from the inner ear. Most of the remainder of the section is occupied by the reticular nuclei, which extend for the whole length of the hindbrain.

The cerebellum consists of an outer layer of cerebellar cortex and a core of white matter (fibers). At this level, some cell groups can be seen in the deep cerebellar white matter dorsal to the ventricle and the vestibular nuclei. These cell groups are the cerebellar nuclei. They receive input from the cerebellar cortex and send fibers to the brainstem and thalamus.

The fourth ventricle has lateral extensions at this level. They can be seen next to the cochlear nuclei.

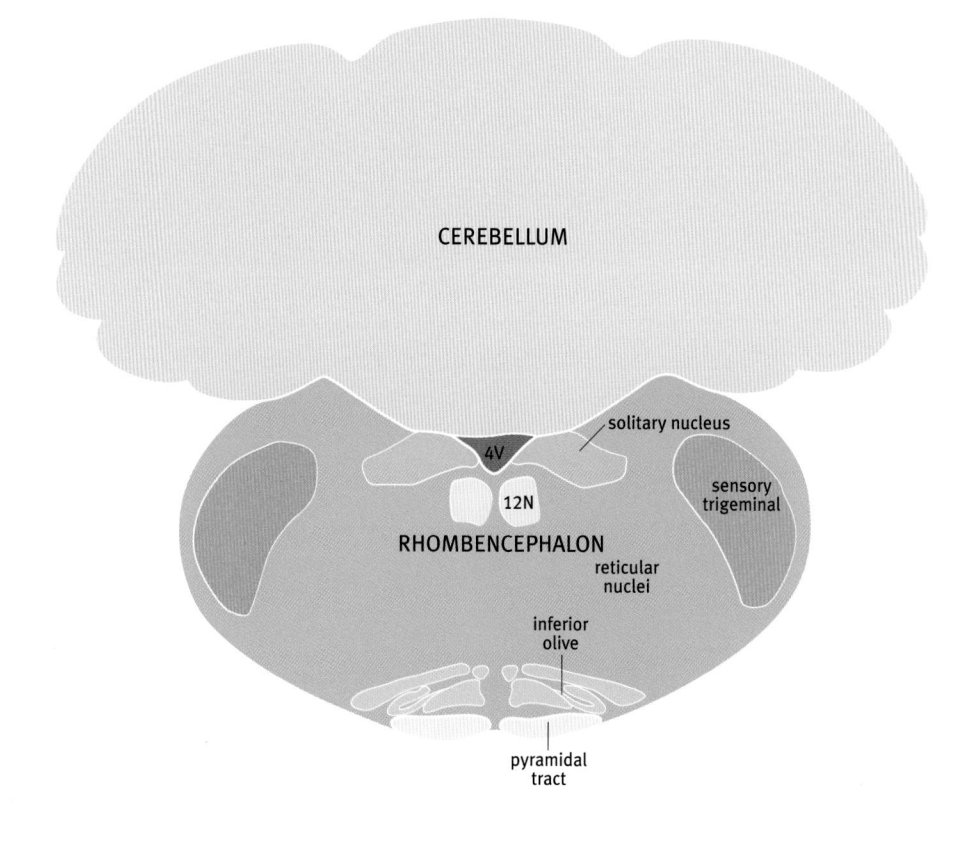

Figure 11 *The hindbrain at the level of the inferior olive*

This section shows the brainstem below and the cerebellum above with the fourth ventricle in between.

The large trigeminal nuclei are found in the lateral part of the hindbrain. This part of the trigeminal sensory complex is called the spinal trigeminal nucleus. The two hypoglossal nuclei (which control the tongue) are next to the midline under the fourth ventricle. Lateral to the hypoglossal nucleus is a cluster of cell groups called the solitary nucleus. This cluster receives taste sensation and sensory information from internal organs such as the stomach and lung. Two large fiber bundles, the pyramids, lie on either side of the midline on the ventral margin of the hindbrain. The fibers in the pyramids arise in the cerebral cortex and travel down to cross to the opposite side of the spinal cord. Next to each pyramid is the inferior olive, which is functionally connected with the cerebellum. Most of the remainder of the area of this section is occupied by the reticular nuclei, which extend for the whole length of the hindbrain. They are involved in basic sensory and motor functions.

The fourth ventricle contains cerebrospinal fluid (CSF), which has flowed down from its origin in the lateral (cerebral) ventricles through the third ventricle and aqueduct to reach the hindbrain. The CSF escapes from the roof of the fourth ventricle to fill the subarachnoid space.

The cerebellum is a large structure concerned with coordination of movement. It consists of an outer layer of cerebellar cortex and a core of white matter (fibers).

The Rat Brain in Stereotaxic Coordinates   Compact 7th Edition   Paxinos & Watson

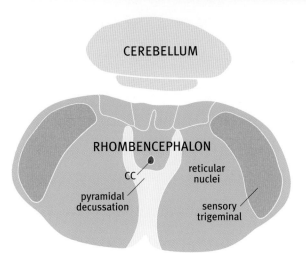

Figure 12 The junction of the hindbrain and spinal cord

This section shows the brainstem at the point where it joins the spinal cord. Sitting dorsal to the hindbrain is the caudal part of the cerebellum. In the center of the hindbrain is the prominent crossing of the pyramidal tract (pyramidal decussation). The fibers in the pyramidal tract arise in the cerebral cortex and are here seen crossing to the opposite side of the spinal cord. The large trigeminal nuclei, which receive touch, pain, and temperature sensations from the face, are found in the lateral part of the hindbrain. This part of the trigeminal sensory complex is called the spinal trigeminal nucleus because it extends into the cervical spinal cord. Most of the remainder of the area of the section is occupied by the reticular nuclei, which extend the whole length of the hindbrain. The part of the cerebellum seen here is the caudal end of the vermis, the midline part of the cerebellum. In this figure, the ventricular system of the brain is represented by the small central canal (CC), which continues down through the center of the spinal cord. The central canal extends to the lower end of the spinal cord, but it is a blind canal from which the cerebrospinal fluid cannot escape.

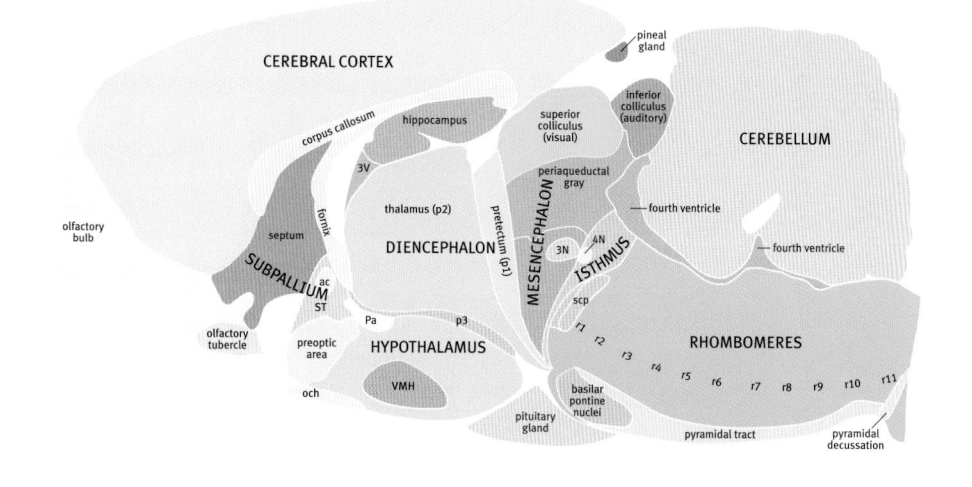

Figure 13 Sagittal section −0.40mm lateral to the midline

This section is close to the midline and shows all of the major parts of the brain. On the far left is the olfactory bulb and on the far right is the hindbrain. In between these two structures are the cerebrum, thalamus, and midbrain. The cerebral neocortex covers the centrally placed hippocampus, septum, and thalamus. Between the hippocampus and neocortex is the corpus callosum, a large bundle of fibers that connects the two cerebral hemispheres. Ventral to the thalamus and septum, from left to right, are the olfactory tubercle, the preoptic area, and the hypothalamus.

The thalamus is the largest part of the diencephalon. The two smaller parts of the diencephalon are the pretectum (between thalamus and midbrain), and the prethalamus (between thalamus and hypothalamus). The pituitary gland, which is attached to the hypothalamus, is seen ventral to the brain. A second gland, the pineal, is seen above the colliculi of the midbrain. The pineal has a long stalk (not seen here), which attaches it to prosomere 2 of the diencephalon.

Between the preoptic area and the septum is a collection of cell groups called the bed nucleus of the stria terminalis (ST). The stria terminalis is a large fiber bundle that starts in the amygdala and ends in the bed nucleus. The ST is also wrapped around another large fiber bundle, the anterior commissure. Three hypothalamic landmarks can be seen in this section. The optic chiasm (och) is stuck to the ventral surface of the hypothalamus, and two important hypothalamic nuclei - the paraventricular nucleus (Pa) and the ventromedial hypothalamic nucleus (VMH).

Squeezed between the pretectum and the hindbrain is the wedge-shaped midbrain (mesencephalon). The ventral compression of the midbrain is caused by the sharp flexion of the neuraxis in this region. The dorsal surface of the midbrain is marked by two bulges on each side, the superior and inferior colliculi. Deep to the colliculi is the periaqueductal gray Embedded in the ventral part of the midbrain periaqueductal gray is the nucleus of the oculomotor nerve (3N).

The hindbrain is formed by a series of twelve embryonic segments - the isthmus and the eleven rhombomeres. The cerebellum develops from the isthmus and the first rhombomere. The last rhombomere (r11) joins the first segment of the spinal cord.

The Rat Brain in Stereotaxic Coordinates   Compact 7th Edition   Paxinos & Watson

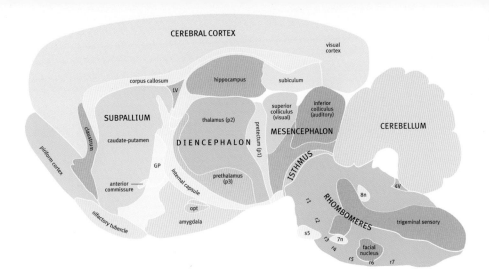

Figure 14 Sagittal section −2.62mm lateral to the midline

This section is some distance from the midline, but the major features are similar to those seen in the previous section. However, the olfactory bulb is no longer seen attached to the frontal pole of the cerebrum; the piriform cortex is there in its place. The corpus callosum forms a roof over the caudate- putamen and the hippocampus. Deep to the hippocampus is the diencephalon (pretectum, thalamus, and prethalamus). The caudal part of the cerebral neocortex overhangs the colliculi of the midbrain.

In the ventral region of the cerebrum, from left to right, we can see the piriform cortex, the olfactory tubercle, and the amygdala. The amygdala is a mass of gray matter in the temporal lobe of the cerebrum. The other masses of gray matter seen in this lateral region of the cerebrum are the caudate-putamen, the globus pallidus (GP), and the claustrum. The claustrum can be seen in this section pressed against the rostral part of the corpus callosum. The caudate-putamen fills the space between the corpus callosum and the olfactory tubercle. Embedded in the caudal border of the caudate-putamen are the fibers of the anterior commissure. Separating the amygdala from the diencephalon (prethalamus and thalamus) is a thick sheet of fibers, the internal capsule. On the right of the diencephalon, caudal to the pretectum, is the midbrain (mesencephalon), which is distinctly wedge-shaped. The large dorsal surface is occupied by the superior and inferior colliculus, but the ventral surface is a small area between the pretectum and the isthmus.

The hindbrain is made up of the isthmus and the eleven rhombomeres. The cerebellum grows out of the first rhombomere and the isthmus. In this lateral region of the hindbrain, the most obvious feature is the trigeminal sensory nucleus. It extends from the second rhombomere down to the third cervical segment of the spinal cord. The sensory root of the trigeminal nerve is seen here entering the hindbrain at the level of the second rhombomere (r2). Ventral to the sensory trigeminal nucleus in the region of the sixth rhombomere (r6) is the facial nucleus, which supplies the muscles of facial expression. Although the facial nucleus is located in the sixth rhombomere (r6), the facial nerve emerges from the hindbrain at the level of the fourth rhombomere (r4).

The cerebellum consists of a folded cortex and a central core of fibers. Within this central core are the cerebellar nuclei. Between the cerebellum and the hindbrain is the fourth ventricle, one of the reservoirs of cerebrospinal fluid in the central nervous system.

References

Ahima RS, and Harlan RE. (1990). Charting of Type II glucocorticoid receptor-like immunoreactivity in the rat central nervous system. Neuroscience 39: 579-604.

Alheid GF, De Olmos JS, and Beltramino CA. (1995). Amygadala and extended amygdala. In G. Paxinos (Ed.), The Rat Nervous System, 2nd ed. Academic Press, San Diego.

Alonso A, Merchan P, Sandoval JE, Sanchez-Arrones L, Garcia-Cazorla A, Artuch R, Ferran JL, Martmez-de-la-Torre M, Puelles L. (2013). Development of the serotonergic cells in murine raphe nuclei and their relations with rhombomeric domains Brain Struct Funct. DOI 10.1007/s00429-012-0456-8.

Altschuler SM, Bao X, Bieger D, Hopkins DA, and Miselis RR. (1989). Viscerotopic representation of the upper alimentary tract in the rat: Sensory ganglia and nuclei of the solitary and spinal trigeminal tracts. J. Comp. Neurol. 283: 248-268.

Armstrong W. (2014). Hypothalamic supraoptic and paraventricular nuclei. In G. Paxinos (Ed.), The Rat Nervous System, 4th ed. Elsevier Academic Press, San Diego.

Aston-Jones G. (2004). Locus coreuleus, A5 and A7 noradrenergic cell groups. In G. Paxinos (Ed.), The Rat Nervous System, 3rd ed. Elsevier Academic Press, San Diego.

Ashwell KWS and Paxinos G. (2008). Atlas of the Developing Rat Nervous System, 3rd ed. Elsevier Academic Press, San Diego.

Bardet, S.M. et al. (2010). Ontogenetic expression of sonic hedgehog in the chicken subpallium. Front. Neuroanat. 4, 1-16.

Barragan EI, and Ferreyra-Moyano H. (1995). Ventrostiopallidal functional interconnections with cortical and quasi-cortical regions. Brain Res. Bull. 37: 329-336.

Beckel JM, and Holstege G. (2014). The lower urinary tract. In G. Paxinos (Ed.), The Rat Nervous System, 4th ed. Elsevier Academic Press, San Diego.

Berman AL. (1968). The Brainstem of the Cat: A Cytoarchitectonic Atlas with Stereotaxic Coordinates. University of Wisconsin Press, Madison.

Bolk L. (1906). Das Cerebellum der Säugetiere. Jena G. Fischer.

Buttner-Ennever JA, Cohen G, Pause M, and Fries W. (1988). Raphe nucleus of the pons containing omnipause neurons of the oculomotor system in the monkey, and its homologue in man. J. Comp. Neurol. 267: 307-321.

Cappaert NLM, van Strien NM, and Witter MP. (2014). Hippocampus. In G. Paxinos (Ed.), The Rat Nervous System, 4th ed. Elsevier Academic Press, San Diego.

Carrive P, and Paxinos G. (1994). The supraoculomotor cap: A region revealed by NADPH diaphorase histochemistry. NeuroReport 5: 2257-2260.

Carrive P. (1993). The periaqueductal gray and defensive behavior functional representation and neuronal organization. Behav. Brain Res. 58: 27-47.

Cox M, and Halliday GM. (1993). Parvalbumin as an anatomical marker for discrete subregions of the ambiguous complex in the rat. Neurosci. Lett. 160: 101-105.

De Olmos JS, Beltramino CA, and Alheid G. (2004). Amygdala and extended amygdala of the rat: a cytoarchitectonical, fibroarchitectonical, and chemoarchitectonical survey. In G. Paxinos (Ed.), The Rat Nervous System, 3rd ed., Elsevier Academic Press, San Diego.

Dong HW. (2008). Allen Reference Atlas: A Digital Color Brain Atlas of the C57Black/6J Male Mouse. Wiley, Hoboken.

Ebner FF, and Kaas JH. (2014). Somatosensory system. In G. Paxinos (Ed.), The Rat Nervous System, 4th ed. Elsevier Academic Press, San Diego.

Ellenberger HH, Feldman JL, and Zhan W-Z. (1990). Subnuclear organization of the lateral tegmental field in the rat. II: Catecholamine neurons and ventral respiratory group. J. Comp. Neurol. 294: 212-222.

Elmquist JK, Scammell TE, Jacobson CD, and Saper CB. (1996). Distribution of Foslike immunoreactivity in the rat brain following intravenous lipopolysaccharide administration. J. Comp. Neurol. 371: 1-19.

Ennis M, Puche AC, Holy T, and Shipley MT. (2014). Olfactory System. In G. Paxinos (Ed.), The Rat Nervous System, 4th ed. Elsevier Academic Press, San Diego.

Farago AF, Awatramani RB, and Dymecki SM. (2006). Assembly of the brainstem cochlear nuclear complex is revealed by intersectional and subtractive genetic fate maps. Neuron 50: 205-218.

Faye-Lund H, and Osen KK. (1985). Anatomy of the inferior colliculus in rat. Anat. Embryol. 171: 1-20.

Fu Y, Yuan Y, Halliday G, Rusznak Z, Watson C, and Paxinos G. (2012). Cytoarchitecture and chemoarchitecture of the substantia nigra, ventral tegmental area, and retrorubral field in the mouse. Brain Struct Funct. 217: 591-692.

Fulwiler CE, and Saper CB. (1984). Subnuclear organization of the efferent connections of the parabrachial nucleus in the rat. Brain Res. Rev. 7: 229-259.

Furness J. (2014). Peripheral autonomic nervous system. In G. Paxinos (Ed.), The Rat Nervous System, 4th ed. Elsevier Academic Press, San Diego.

Gerfen C, and Dudman J. (2014). Basal ganglia. In G. Paxinos (Ed.), The Rat Nervous System, 4th ed. Elsevier Academic Press, San Diego.

Grove EA. (1988). Efferent connections of the substantia innominata in the rat. J. Comp. Neurol. 277: 347-364.

Harding A, Paxinos G, and Halliday G. (2004). The serotonin and tachykinin systems. In G. Paxinos (Ed.), The Rat Nervous System, 3rd ed. Elsevier Academic Press, San Diego.

Haug F-MS. (1976). Sulphide silver pattern and cytoarchitectonics of parahippocampal areas in the rat. Advances in Anatomy, Embryology, and Cell Biology 52: 1-73.

Heimer L, Harland RE, Alheid GF, Garcia MM, and De Olmos J. (1997). Substantia inominata: a notion which impedes clinical-anatomical correlations in neuropsychiatric disorders. Neuroscience 76: 957-1006.

Heimer L, Zahm DS, and Alheid GF. (1995). Basal ganglia. In G. Paxinos (Ed.), The Rat Nervous System, 2nd ed. Academic Press, San Diego.

Heimer L, Zahm DS, Churchill L, Kalivas P, and Wohltmann C. (1991). Specificity in the projection patterns of accumbal core and shell in the rat. Neuroscience 41: 89-125.

Herbert H, Moga M, and Saper C. (1990). Connections of the parabrachial nucleus with the nucleus of the solitary tract and the medullary reticular formation in the rat. J. Comp. Neurol. 293: 540-580.

Herbert H, and Saper CB. (1990). Cholecystokinin-, galanin-, and corticotropin-releasing factorlike immunoreactive projections from the nucleus of the solitary tract to the parabrachial nucleus in the rat. J. Comp. Neurol. 293: 581-598.

Hökfelt T, Martensson R, Bjorklund A, Kleinau S, and Goldstein M. (1984). Distributional maps of tyrosine-hydroxylase-immunoreactive neurons in the rat brain. In A. Bjorklund and T. Hökfelt (Eds.), Handbook of Chemical Neuroanatomy, Vol. 2. Part 1. Elsevier, Amsterdam.

Insausti R, Herrero MT, and Witter MP. (1997). Entorhinal cortex of the rat: cytoarchitectonic subdivisions and the origin and distribution of cortical efferents. Hippocampus 7: 146-183.

Jensen P, Farago AF, Awatramani RB, Scott MM, Deneris ES, Dymecki SM. (2008). Redefining the serotonergic system by genetic lineage. Nature Neurosci 11: 417-419.

Jones BE. (1995). Reticular formation: Cytoarchitecture, transmitters, and projections. In G. Paxinos (Ed.), The Rat Nervous System, 2nd ed. Academic Press, San Diego.

Jones EG. (2007). The Thalamus. 2nd ed., Cambridge University Press, Cambridge.

Kanjhan R, Lipski J, Kruszewska B, and Rong W. (1995). A comparative study of presympathetic and Bötzinger neurons in the rostral ventrolateral medulla (RVLM) of the rat. Brain Res. 699: 19-32.

Keay KA, and Bandler R. (2014). Periaqueductal gray. In G. Paxinos (Ed.), The Rat Nervous System, 4th ed. Elsevier Academic Press, San Diego.

Koelle GG, and Friedenwald JS. (1949). A histochemical method for localizing Cholinesterase activity. Proc. Soc. Exp. Biol. Med. 70: 617-622.

Krettek JE, and Price JL. (1977). Projections from the amygdaloid complex and adjacent olfactory structures to the entorhinal cortex and to the subiculum in the rat and cat. J. Comp Neurol. 172: 723-752.

Larsell O. (1952). The morphogenesis and adult pattern of the lobules and fissures of the cerebellum of the white rat. J. Comp. Neurol. 97: 281-356.

Larsell O. (1970). The Comparative Anatomy and Histology of the Cerebellum from Monotremes through Apes. University of Minnesota Press, Minneapolis, MN.

Larsen PJ. (1992). Distribution of substance P-immunoreactive elements in the preoptic area and the hypothalamus of the rat. J. Comp. Neurol. 316: 287-313.

LeDoux JE, Ruggiero DA, and Reis DJ. (1985). Projections to the subcortical forebrain from anatomically defined regions of the medial geniculate body in the rat. J. Comp. Neurol. 242: 182-213.

Lein ES, Callaway EM, Albright TD, and Gage FH. (2005). Redefining the boundaries of the hippocampal CA2 subfield in the mouse using gene expression and 3-dimensional reconstruction. J Comp Neurol 485: 1-10.

Lewis PR. (1961). The effect of varying the conditions in the Koelle method. Biblitheca Anat. Vol. 2, Karger, Basel, 11-20.

Liang H, Paxinos G, and Watson C. (2011). Projections from the brain to the spinal cord in the mouse. Brain Struct Funct 215: 159-186.

Lundy RF, and Norgren R. (2014). Gustatory System. In G. Paxinos (Ed.), The Rat Nervous System, 4th ed. Elsevier Academic Press, San Diego.

Magoul R, Ciofi P, and Tramu G. (1994). Visualization of an efferent projection route of the hypothalamic rat arcuate nucleus through the stria terminalis after labeling with carbocyanine dye (DiI) or proopiomelanocortin-immunohistochemistry. Neurosci. Lett. 172: 134-138.

Mai JK, Majtanic M, and Paxinos G. (2016). Atlas of the Human Brain, 4th ed. Academic Press, San Diego.

Malmierca MS. (2014). Auditory System. In G. Paxinos (Ed.), The Rat Nervous System, 2nd ed. Academic Press, San Diego.

Marfurt C, and Rajchert DM. (1991). Trigeminal primary afferent projections to "Non-Trigeminal" areas of the rat central nervous system. J. Comp. Neurol. 303: 489-511.

Marin O. (2014). Tangential migration in the telencephalon. In G. Paxinos (Ed.), The Rat Nervous System, 4th ed. Elsevier Academic Press, San Diego.

Martin PR, Harvey AR, and Sefton A. (2014). Visual system. In G. Paxinos (Ed.), The Rat Nervous System, 4th ed. Elsevier Academic Press, San Diego.

McRitchie DA, Hardman CD, and Halliday GM. (1996). Cytoarchitectural distribution of calcium binding proteins in midbrain dopaminergic regions of rats and humans. J. Comp. Neurol. 364: 121-150.

McRitchie DA. (1992). Cytoarchitecture and chemical neuroanatomy of the nucleus of the solitary tract: Comparative and experimental studies in the human and the rat. Unpublished Ph.D. thesis, Univ. of New South Wales.

Morin LP, and Blanchard J. (1995). Organization of the hamster intergeniculate leaflet: NPY and ENK projections to the suprachiasmatic nucleus, intergeniculate leaflet and posterior limitans nucleus. Visual Neurosci. 12: 57-67.

Mulders WHAM, West MJ, and Slomianka I. (1997). Neuron numbers in the presubiculum, parasubiculum, and the entorhinal area of the rat. J Comp. Neurol. 385: 83-94.

Neafsey EJ, Bold EL, Haas G, Hurley-Gius KM, Quirk G, Sievert CF, and Terreberry RR. (1986). The organization of the rat motor cortex: A microstimulation mapping study. Brain Res. Rev. 11, 77-96.

Oldfield BJ, and McKinley MJ. (2014). Circumventricular organs. In G. Paxinos (Ed.), The Rat Nervous System, 4th ed. Elsevier Academic Press, San Diego.

Olmos J, Hardy H, and Heimer L. (1978). The afferent connections of the main and the accessory olfactory bulb formation in the rat: An experimental HRP study. J. Comp. Neurol. 181: 213-244.

Olszelwski J, and Baxter D. (1954). Cytoarchitecture of the Human brain Stem. Karger, Basel.

Olucha-Bordonau FE, Fortes-Marco L, Otero-García M, Lanuza E, and Martínez-García F. (2014). Amygdala and extension of the amygdala. In G. Paxinos (Ed.), The Rat Nervous System, 4th ed. Elsevier Academic Press, San Diego.

Palomero-Gallagher N, and Zilles K. (2004). Isocortex. In G. Paxinos (Ed.), The Rat Nervous System, 3rd ed. Elsevier Academic Press, San Diego.

Palomero-Gallagher N, and Zilles K. (2014). Isocortex. In G. Paxinos (Ed.), The Rat Nervous System, 4th ed. Elsevier Academic Press, San Diego.

Paxinos G. (2014). The Rat Nervous System, 4th ed. Elsevier Academic Press, San Diego.

Paxinos G, and Watson C. (2014). Paxinos and Watson's The Rat Brain in Stereotaxic Coordinates, 7th ed. Elsevier Academic Press, San Diego.

Paxinos G, Watson C, Calabrese E, Badea A, and Johnson GA. (2015). MRI/DTI Atlas of the Rat Brain. Academic Press/Elsevier, San Diego.

Paxinos G, Watson C, Carrive P, Kirkcaldie M, and Ashwell K. (2009). Chemoarchitectonic atlas of the rat brain, 2nd ed. Elsevier Academic Press, San Diego.

Paxinos G, Watson C, Petrides M, Rosa M, and Tokuno H. (2012). The Marmoset Brain in Stereotaxic Coordinates. Elsevier Academic Press, SanDiego.

Paxinos G, and Huang X-F. (1995). Atlas of the Human Brainstem. Academic Press, San Diego.

Paxinos G, and Watson C. (1982). The Rat Brain in Stereotaxic Coordinates. Academic Press, Sydney.

Paxinos G, and Watson C. (1986). The Rat Brain in Stereotaxic Coordinates, 2nd ed. Academic Press, San Diego.

Paxinos G, and Watson C. (1997). The Rat Brain in Stereotaxic Coordinates, Compact 3rd ed. CD-Rom Academic Press, San Diego.

Paxinos G, and Watson C. (2005). The Rat Brain in Stereotaxic Coordinates: the new coronal set, 5th ed. Elsevier Academic Press, San Diego.

Paxinos G, and Watson C. (2007). The Rat Brain in Stereotaxic Coordinates, 6th ed. Elsevier Academic Press, San Diego.

Paxinos G, Ashwell KW, and Tork I. (1994). Atlas of the Developing Rat Nervous System, 2nd ed. Academic Press, San Diego.

Paxinos G, Carrive P, Wang H, and Wang P-Y. (1999b). Chemoarchitecture of the Rat Brainstem. Academic Press, San Diego.

Paxinos G, Halliday G, Watson C, Koutcherov Y, and Wang H. (2007). Atlas of the Developing Mouse Brain: E17.5, P0, P6. Elsevier Academic Press, San Diego.

Paxinos G, Huang X-F, and Toga AW. (2000). The Rhesus Monkey Brain in Stereotaxic Coordinates. Academic Press, San Diego.

Paxinos G, Kus L, Ashwell K, and Watson C. (1999a). Chemoarchitecture of the Rat Forebrain. Academic Press, San Diego.

Paxinos G, Tork I, Halliday G, and Mehler WR. (1990). Human homologs to brainstem nuclei identified in other animals as revealed by acetylcholinesterase. In G. Paxinos (Ed.), The Human Nervous System, Academic Press, San Diego.

Paxinos G, Watson C, Pennisi M, and Topple A. (1985). Bregma, lambda and the interaural midpoint in stereotaxic surgery with rats of different sex, strain and weight. J. Neurosci. Meth. 13: 139-143.

Paxinos G. (Ed.). (2004). The Rat Nervous System, 3rd ed. Elsevier Academic Press, San Diego.

Paxinos G. (Ed.). (2014). The Rat Nervous System, 4th ed. Elsevier Academic Press, San Diego.

Paxinos G, and Franklin K. (2013). Paxinos and Franklin's The Mouse Brain in Stereotaxic Coordinates, 4th ed. Elsevier Academic Press, San Diego.

Puelles E, Puelles L, and Watson C. (2012). Midbrain. In Watson C, Paxinos G, and Puelles L (Eds.), The Mouse Nervous System. Elsevier Academic Press, San Diego.

Puelles L, Martinez-de-la-Torre M, Bardet S, and Rubenstein JLR. (2012a). Hypothalamus. In Watson C, Paxinos G, and Puelles L (Eds.), The Mouse Nervous System. Elsevier Academic Press, San Diego.

Puelles L, Martinez-de-la-Torre M, Ferran J-L, and Watson C. (2012b). Diencephalon. In Watson C, Paxinos G, & L. Puelles (Eds.), The Mouse Nervous System. Elsevier Academic Press, San Diego.

Puelles L, Martinez-de-la-Torre M, Paxinos G, Watson C, and Martinez S. (2007). The Chick Brain in Stereotaxic Coordinates: An atlas featuring neuromeres and mammalian homologies. Elsevier Academic Press, San Diego.

Puelles L, Martinez-de-la-Torre M, Martinez S, Watson C, and Paxinos G. (2007). The Chick Brain in Stereotaxic Coordinates: An Atlas Correlating Avian and Mammalian Anatomy. Elsevier Academic Press, San Diego.

Puelles L, Martinez S, Martinez-de-la-Torre M, and Rubenstein JLR. (2014). Gene maps and related histogenetic domains in the forebrain and midbrain. In G. Paxinos (Ed.), The Rat Nervous System, 4th ed. Elsevier Academic Press, San Diego.

Puelles L, Paxinos G and Ashwell K. (2014). Neuromeric landmarks in the rat. In G. Paxinos (Ed.), The Rat Nervous System, 4th ed. Elsevier Academic Press, San Diego.

Ray JP, and Price JL. (1992). The organization of the thalamocortical connections of the mediodorsal thalamic nucleus in the rat, related to the ventral forebrain-prefrontal cortex topography. J. Comp. Neurol. 323: 167-197.

Ribeiro-da-Silva A. (2014). Substantia gelatinosa of the spinal cord. In G. Paxinos (Ed.), The Rat Nervous System, 4th ed. Elsevier Academic Press, San Diego.

Ruigrok TJH, Sillitoe R, and Voogd J. (2014). Cerebellum and cerebellar connections. In G. Paxinos (Ed.), The Rat Nervous System, 4th ed. Elsevier Academic Press, San Diego.

Saper CB, and Stornetta RL. (2014). Central Autonomic System. In G. Paxinos (Ed.), The Rat Nervous System, 4th ed. Elsevier Academic Press, San Diego.

Scremin OU. (2014). Cerebral vascular system. In G. Paxinos (Ed.), The Rat Nervous System, 4th ed. Elsevier Academic Press, San Diego.

Sengul G, Watson C, Tanaka I, and Paxinos G. (2013). Atlas of spinal cords of the rat, mouse, marmoset, macaque, and human. Elsevier Academic Press, San Diego.

Sengul G, and Watson C. (2014). Ascending and descending pathways in the spinal cord. In G. Paxinos (Ed.), The Rat Nervous System, 4th ed. Elsevier Academic Press, San Diego.

Sengul G. (2014a). Primary afferent projections to the spinal cord. In G. Paxinos (Ed.), The Rat Nervous System, 4th ed. Elsevier Academic Press, San Diego.

Sengul G. (2014b). Spinal cord cyto- and chemoarchitecture. In G. Paxinos (Ed.), The Rat Nervous System, 4th ed. Elsevier Academic Press, San Diego.

Sherin JE, Shiromani PJ, McCarley RW, and Saper CB. (1996). Activation of ventrolateral preoptic neurons during sleep. Science 271: 216-219.

Shipley MT, Ennis M, and Puche A. (2004). Olfactory system. In G. Paxinos (Ed.), The Rat Nervous System, 3rd ed. Elsevier Academic Press, San Diego.

Simerly RB, Swanson LW, and Gorski RA. (1984). Demonstration of a sexual dimorphism in the distribution of serotonin-immunoreactive fibers in the medial preoptic nucleus of the rat. J. Comp. Neurol. 225: 151-166.

Simerly RB. (2014). Autonomic sub Hy. In G. Paxinos (Ed.), The Rat Nervous System, 4th ed. Academic Press, San Diego.

Slotnick BM, and Brown DL. (1980). Variability in the stereotaxic position of cerebral points in the albino rat. Brain Res. Bull. 5: 135-139.

Swanson L. (2004). Brain maps: structure of the rat brain, 3rd ed. Elsevier, Amsterdam.

Travers JB. (2014). Oromotor nuclei. In G. Paxinos (Ed.), The Rat Nervous System, 4th ed. Elsevier Academic Press, San Diego.

Tsuneoka Y, Maruyama T, Yoshida, Nishimori K, Kato T, Numan M, and Kuroda KO. (2013). Functional, anatomical, and neurochemical differentiation of medial preoptic area subregions in relation to maternal behavior in the mouse. J Comp Neurol. 521: 1633-1663.

Van Groen T, and Wyss JM. (1992). Connections of the retrosplenial dysgranular cortex in the rat. J. Comp. Neurol. 315: 200-216.

Vertes R. (2014). Thalamus. In G. Paxinos (Ed.), The Rat Nervous System, 4th ed. Elsevier Academic Press, San Diego.

Vidal PP, Cullen K, Curthoys IS, Du Lac S, Hostein G, Idoux E, Lysakowsky L, Peussner K, and Smith P. (2014). Vestibular system. In G. Paxinos (Ed.), The Rat Nervous System, 4th ed. Elsevier Academic Press, San Diego.

Vogt BA. (2009). Architecture, cytology and comparative organization of primate cingulate cortex. In: BA Vogt (Ed.), Cingulate Neurobiology and Disease. Oxford University Press, London. pp 65-93.

Vogt BA, Hof P, Zilles K, Vogt LJ, Herold C, and Palomero-Gallagher N. (2013). Anterior cingulate area 32 duality in mouse, rat, macaque and human: Cytoarchitecture and receptor architecture. J Comp Neurol (in press).

Vogt BA, and Paxinos G. (2012). Cytoarchitecture of Mouse and Rat Cingulate Cortex with Human Homologies. Brain Struc. Fune. DOI. https://doi.org/10.1007/s00429-012-0493-3.

Vogt BA, Vogt L, and Farber NB. (2004). Cingulate cortex and disease models. In G. Paxinos (Ed.), The Rat Nervous System, 3rd ed. Elsevier Academic Press, San Diego.

Vogt BA. (2014). Cingulate cortex and pain architecture. In G. Paxinos (Ed.), The Rat Nervous System, 4th ed. Academic Press, San Diego.

Wang PY, and Zhang FC. (1995). Outlines and Atlas of Learning Rat Brain Slides. Westnorth University Press, China.

Watson C. (2012) Hindbrain. The Mouse Nervous System. C Watson, G Paxinos, and L Puelles (Eds.). Elsevier Academic Press, San Diego.

Watson C, and Harrison M. (2012). The location of the major ascending and descending spinal cord tracts in all spinal segments in the mouse - actual and extrapolated. Anat Rec 295: 1692-1697.

Watson C, and Paxinos G. (2010). Chemoarchitectonic Atlas of the Mouse Brain. Elsevier Academic Press, San Diego.

Watson C, Kirkcaldie M, and Paxinos G. (2010). The Brain: An Introduction to Functional Neuroanatomy. Elsevier Academic Press, San Diego.

Watson C, Mitchelle A, and Puelles L. (2017). A new mammalian brain ontology based on developmental gene expression. In J. Kaas, Evolution of Nervous Systems, 2 ed. Elsevier, Oxford, Vol2: (pp. 53–75). ISBN 9780128040423. https://doi.org/10.1016/B978-0-12-804042-3.00138-X.

Watson C, Paxinos G, & Puelles L. (Eds.). (2012). The Mouse Nervous System. Elsevier Academic Press, San Diego.

Watson C, Paxinos, G, and Kayalioglu G (Eds.). (2009). The Spinal Cord. A Christopher and Dana Reeve Foundation Text and Atlas. Elsevier Academic Press, San Diego.

Westlund, KN, and Willis WD. (2014). Pain System. In G. Paxinos (Ed.), The Rat Nervous System, 4th ed. Elsevier Academic Press, San Diego.

Whishaw IQ, Cioe JDD, Previsich N, and Kolb B. (1977). The variability of the interaural line vs the stability of bregma in rat stereotaxic surgery. Physiol. Behav. 19: 719-722.

Whitehead MC. (1990). Sibdivisions and neruon types of the nucleus of the solitary tract in the hamster. J. Comp. Neurol. 310, 554-574.

Witter MP, and Amaral DG. (2004). Hippocampal formation. In G. Paxinos (Ed.), The Rat Nervous System, 3rd ed. Elsevier Academic Press, San Diego.

Yasui Y, Saper C, Cechetto D. (1989). Calcitonin gene-related peptide immunoreactivity in the visceral sensory cortex, thalamus, and related pathways in the rat. J. Comp. Neurol. 290: 487-501.

Zaborsky L. (2014). Basal forebrain. In G. Paxinos (Ed.), The Rat Nervous System, 4th ed. Elsevier Academic Press, San Diego.

Zaborszky L, Alheid GF, Beinfeld MC, Eidens LE, Heimer L, and Palkovits M. (1985). Colecystokinin innervation of the ventral striatum: A morphological and radioimmunological study. Neurosci. 14: 427-453.

Zilles K. (1985). The Cortex of the Rat: A Stereotaxic Atlas. Springer-Verlag, Berlin.

List of Structures

Names of the structures are listed in alphabetical order.
Each name is followed by abbreviation of the structure.

3rd ventricle   3V
4th ventricle   4V

A

abducens nerve   6n
abducens nucleus   6N
abducens nucleus, retractor bulbi part   6RB
accessory nerve   11n
accessory nerve nucleus   11N
accessory neurosecretory nuclei   ANS
accessory olfactory bulb   AOB
accessory olfactory tract   aot
accessory olfactory tract   aot
accumbens nucleus   Acb
accumbens nucleus, core region   AcbC
accumbens nucleus, shell region   AcbSh
acoustic stria   as
Ad1 adrenalin cells   Ad1
Ad1 adrenalin cells and NA1 noradrenalin cells   Ad1/NA1
Ad2 adrenalin cells   Ad2
Ad3 adrenalin cells   Ad3
agranular insular cortex   AI
agranular insular cortex, dorsal part   AID
agranular insular cortex, posterior part   AIP
agranular insular cortex, ventral part   AIV
alveus of the hippocampus   alv
ambiguus nucleus   Amb
ambiguus nucleus, compact part   AmbC
ambiguus nucleus, loose part   AmbL
ambiguus nucleus, semicompact part   AmbSC
amygdalohippocampal area   AHi
amygdalohippocampal area, anterolateral part   AHiAL
amygdalohippocampal area, posterolateral   AHiPL
amygdalohippocampal area, posteromedial part   AHiPM
amygdaloid fissure   af
amygdaloid intramedullary gray   IMG
amygdalopiriform transition area   APir
amygdalostriatal transition area   ASt
angular thalamic nucleus   Ang
ansa lenticularis   al

ansoparamedian fissure   apmf
anterior amygdaloid area   AA
anterior cerebral artery   acer
anterior commissural nucleus   AC
anterior commissure   ac
anterior commissure, anterior part   aca
anterior commissure, intrabulbar part   aci
anterior commissure, posterior limb   acp
anterior cortical amygdaloid nucleus   ACo
anterior hypothalamic area, anterior part   AHA
anterior hypothalamic area, central part   AHC
anterior hypothalamic area, posterior part   AHP
anterior hypothalamic nucleus   AH
anterior lobe of the pituitary   APit
anterior olfactory area external part   AOE
anterior olfactory area posterior part   AOP
anterior olfactory area, dorsal part   AOD
anterior olfactory area, ventral part   AOV
anterior olfactory area, ventral posterior part   AOVP
anterior olfactory nucleus, lateral part   AOL
anterior olfactory nucleus, medial part   AOM
anterior perifornical nucleus   APF
anterior pretectal nucleus   APT
anterior pretectal nucleus, dorsal part   APTD
anterior pretectal nucleus, ventral part   APTV
anterior spinal artery   asp
anterior tegmental nucleus   ATg
anterodorsal thalamic nucleus   AD
anteromedial preoptic nucleus   AMPO
anteromedial thalamic nucleus   AM
anteromedial thalamic nucleus, ventral part   AMV
anteroventral periventricular nucleus   AVPe
anteroventral thalamic nucleus   AV
anteroventral thalamic nucleus, dorsomedial part   AVDM
anteroventral thalamic nucleus, ventrolateral part   AVVL
aqueduct   Aq
arcuate hypothalamic nucleus   Arc
arcuate hypothalamic nucleus, dorsal part   ArcD
arcuate hypothalamic nucleus, lateral part   ArcL
arcuate hypothalamic nucleus, lateroposterior part   ArcLP
arcuate hypothalamic nucleus, medial part   ArcM
arcuate hypothalamic nucleus, medial posterior part   ArcMP
area postrema   AP
area subpostrema   SubP
artery   a

ascending fibers of the facial nerve   asc7
azygous anterior cerebral artery   azac
azygous pericallosal artery   azp

B

Barrington's nucleus   Bar
basal interstitial nucleus   BI
basal nucleus (Meynert)   B
basilar artery   bas
basolateral amygdaloid nucleus   BL
basolateral amygdaloid nucleus, anterior part   BLA
basolateral amygdaloid nucleus, posterior part   BLP
basolateral amygdaloid nucleus, ventral part   BLV
basomedial amygdaloid nucleus, anterior part   BMA
basomedial amygdaloid nucleus, posterior part   BMP
bed nucleus of stria terminalis, fusiform part   Fu
bed nucleus of stria terminalis, dorsal division   STD
bed nucleus of stria terminalis, lateral division   STL
bed nucleus of stria terminalis, medial division   STM
bed nucleus of stria terminalis, supracapsular division   STS
bed nucleus of stria terminalis, supracapsular division, lateral part   STSL
bed nucleus of stria terminalis, supracapsular division, medial part   STSM
bed nucleus of the accessory olfactory tract   BAOT
bed nucleus of the anterior commissure   BAC
bed nucleus of the stria terminalis, intraamygdaloid division   STIA
bed nucleus of the stria terminalis, lateral division, dorsal part   STLD
bed nucleus of the stria terminalis, lateral division, intermediate part   STLI
bed nucleus of the stria terminalis, lateral division, juxtacapsular part   STLJ
bed nucleus of the stria terminalis, lateral division, posterior part   STLP
bed nucleus of the stria terminalis, lateral division, ventral part   STLV
bed nucleus of the stria terminalis, medial division, anterior part   STMA
bed nucleus of the stria terminalis, medial division, anterolateral part   STMAL
bed nucleus of the stria terminalis, medial division, anteromedial part   STMAM
bed nucleus of the stria terminalis, medial division, posterior part   STMP
bed nucleus of the stria terminalis, medial division, posterointermediate part   STMPI
bed nucleus of the stria terminalis, medial division, posterolateral part   STMPL
bed nucleus of the stria terminalis, medial division, posteromedial part   STMPM
bed nucleus of the stria terminalis, medial division, ventral part   STMV
Botzinger complex   Bo
brachium of the inferior colliculus   bic
brachium of the superior colliculus   bsc

C

caudal interstitial nucleus of the medial longitudinal fasciculus	CI
caudal linear nucleus of the raphe	CLi
caudal periolivary nucleus	CPO
caudate putamen (striatum)	CPu
caudomedial entorhinal cortex	CEnt
caudoventrolateral reticular nucleus	CVL
cell bridges of the ventral striatum	CB
central amygdaloid nucleus	Ce
central amygdaloid nucleus, capsular part	CeC
central amygdaloid nucleus, lateral part	CeL
central amygdaloid nucleus, medial part	CeM
central canal	CC
central cervical nucleus	CeCv
central gray	CG
central gray, alpha part	CGA
central gray, beta part	CGB
central gray, gamma	CGG
central gray, nucleus O	CGO
central medial thalamic nucleus	CM
central mesencephalic nucleus	CeMe
central nucleus of the inferior colliculus	CIC
central tegmental tract	ctg
centrolateral thalamic nucleus	CL
cerebellar commissure	cbc
cerebellar white matter	cbw
cerebral peduncle	cp
choroid plexus	chp
cingulate cortex, area 24'	A24'
cingulate cortex, area 24a	A24a
cingulate cortex, area 24b	A24b
cingulate cortex, area 24b'	A24b'
cingulate cortex, area 25	A25
cingulate cortex, area 29a	A29a
cingulate cortex, area 29b	A29b
cingulate cortex, area 29c	A29c
cingulate cortex, area 30	A30
cingulate cortex, area 3 D	A32D
cingulate cortex, area 3 V	A32V
cingulate cortex, area 33	A33
cingulum	cg
circular nucleus	Cir
claustrum	Cl
cochlear root of the vestibulocochlear nerve	8cn
commissural nucleus of the inferior colliculus	Com
commissural stria terminalis	cst
commissure of the inferior colliculus	cic
commissure of the lateral lemniscus	cll
commissure of the superior colliculus	csc
conterminal nucleus	Ct
copula of the pyramis	Cop
corpus callosum	cc
cortex amygdala transition zone, layer 1	CxA1
cortex amygdala transition zone, layer 3	CxA3
cortex-amygdala transition zone	CxA
corticostriate artery	cost
crus 1 of the ansiform lobule	Crus1
crus 2 of the ansiform lobule	Crus2
cuneate fasciculus	cu
cuneate nucleus	Cu
cuneate nucleus, rotundus part	CuR
cuneiform nucleus	CnF
cuneiform nucleus, dorsal part	CnFD
cuneiform nucleus, intermediate part	CnFI
cuneiform nucleus, ventral part	CnFV

D

DA11 dopamine cells	DA11
DA12 dopamine cells	DA12
DA13 dopamine cells	DA13
DA14 dopamine cells	DA14
DA8 dopamine cells	DA8
decussation of the inferior cerebellar peduncle	xicp
decussation of the superior cerebellar peduncle	xscp
decussation of the trapezoid body	tzx
decussation of the trochlear nerve	4x
decussation of the uncinate fasciculus of the cerebellum	unx
deep cerebral white matter	dcw
deep gray layer of the superior colliculus	DpG
deep white layer of the superior colliculus	DpWh
dentate gyrus	DG
dorsal 3rd ventricle	D3V
dorsal acoustic stria	das
dorsal claustrum	DCl
dorsal cochlear nucleus	DC
dorsal cochlear nucleus, deep layer	DCDp
dorsal cochlear nucleus, fusiform layer	DCFu
dorsal cochlear nucleus, granular layer	DCGr
dorsal cochlear nucleus, molecular layer	DCMo
dorsal corticospinal tract	dcs
dorsal cortex of the inferior colliculus	DCIC
dorsal cortex of the inferior colliculus, layer 1	DCIC1
dorsal cortex of the inferior colliculus, layer 2	DCIC2
dorsal cortex of the inferior colliculus, layer 3	DCIC3
dorsal endopiriform nucleus	DEn
dorsal fornix	df
dorsal hippocampal commissure	dhc
dorsal hypothalamic area	DA
dorsal intermediate entorhinal cortex	DIEnt
dorsal lateral geniculate nucleus	DLG
dorsal lateral olfactory tract	dlo
dorsal nucleus of the lateral lemniscus	DLL
dorsal paragigantocellular nucleus	DPGi
dorsal peduncular cortex	DP
dorsal peduncular pontine nucleus	DPPn
dorsal periolivary region	DPO
dorsal raphe nucleus	DR
dorsal raphe nucleus, caudal part	DRC
dorsal raphe nucleus, dorsal part	DRD
dorsal raphe nucleus, lateral part	DRL
dorsal raphe nucleus, ventral part	DRV
dorsal spinocerebellar tract	dsc
dorsal spinocerebellar tract and olivocerebellar tract	dsc/oc
dorsal subiculum	DS
dorsal taenia tecta layer 1	DTT1
dorsal taenia tecta layer 2	DTT2
dorsal tegmental decussation	dtgx
dorsal tegmental nucleus, central part	DTgC
dorsal tegmental nucleus, pericentral part	DTgP
dorsal tenia tecta	DTT
dorsal terminal nucleus	DT
dorsal transition zone	DTr
dorsal tuberomammillary nucleus	DTM
dorsolateral entorhinal cortex	DLEnt
dorsolateral orbital cortex	DLO
dorsolateral periaqueductal gray	DLPAG
dorsomedial hypothalamic nucleus	DM
dorsomedial hypothalamic nucleus, compact part	DMC
dorsomedial hypothalamic nucleus, dorsal part	DMD
dorsomedial hypothalamic nucleus, ventral part	DMV
dorsomedial periaqueductal gray	DMPAG
dorsomedial spinal trigeminal nucleus	DMSp5
dorsomedial tegmental area	DMTg
dysgranular insular cortex	DI

E

ectorhinal cortex   Ect
ectotrigeminal nucleus   E5
Edinger-Westphal nucleus   EW
entopeduncular nucleus   EP
ependymal and subendymal layer/olfactory ventricle   E/OV
ependymal and subependymal layer   E
epifascicular nucleus   EF
epilemniscal nucleus   ELm
epipeduncular nucleus   EpP
epirubrospinal nucleus   ERS
episupraoptic nucleus   ESO
ethmoid thalamic nucleus   Eth
extension of the amygdala   EA
external capsule   ec
external cortex of the inferior colliculus   ECIC
external cortex of the inferior colliculus, layer 1   ECIC1
external cortex of the inferior colliculus, layer 2   ECIC2
external cortex of the inferior colliculus, layer 3   ECIC3
external cuneate nucleus   ECu
external medullary lamina   eml
external part of globus pallidus   EGP
external plexiform layer of the accessory olfactory b   EPlA
external plexiform layer of the olfactory bulb   EPl
extreme capsule   ex

F

F cell group of the vestibular complex   FVe
facial motor nucleus, stylohyoid part   7SH
facial nerve   7n
facial nucleus   7N
facial nucleus, dorsal intermediate subnucleus   7DI
facial nucleus, dorsal subnucleus   7D
facial nucleus, dorsolateral subnucleus   7DL
facial nucleus, dorsomedial subnucleus   7DM
facial nucleus, lateral subnucleus   7L
facial nucleus, ventral intermediate subnucleus   7VI
facial nucleus, ventromedial subnucleus   7VM
fasciculus retroflexus   fr
fasciola cinereum   FC
field CA1 of the hippocampus   CA1
field CA2 of the hippocampus   CA2
field CA3 of the hippocampus   CA3

fimbria of the hippocampus   fi
flocculus   Fl
forceps major of the corpus callosum   fmj
forceps minor of the corpus callosum   fmi
fornix   f
frontal association cortex   FrA
frontal cortex, area 3   Fr3

G

gelatinous layer of the caudal spinal trigeminal nucleus   Ge5
gemini hypothalamic nucleus   Gem
genu of the corpus callosum   gcc
genu of the facial nerve   g7
gigantocellular reticular nucleus   Gi
gigantocellular reticular nucleus, alpha part   GiA
gigantocellular reticular nucleus, ventral part   GiV
glomerular layer of the accessory olfactory bulb   GlA
glomerular layer of the olfactory bulb   Gl
glossopharyngeal nerve   9n
gracile fasciculus   gr
gracile nucleus   Gr
granular insular cortex   GI
granule cell layer of cerebellum   GrCb
granule cell layer of the accessory olfactory bulb   GrA
granule cell layer of the cochlear nuclei   GrC
granule cell layer of the dentate gyrus   GrDG
granule cell layer of the olfactory bulb   GrO

H

h2 field of Forel   h2
habenular commissure   hbc
hilus of the dentate gyrus   Hil
hippocampal fissure   hif
hypoglossal nerve   12n
hypoglossal nucleus   12N
hypoglossal nucleus, geniohyoid part   12GH

I

indusium griseum   IG
inferior cerebellar peduncle icp inferior colliculus   IC
inferior olivary nucleus   IO
inferior olive, beta subnucleus of the medial nucleus   IOBe

inferior olive, cap of Kooy of the medial nucleus   IOK
inferior olive, dorsal nucleus   IOD
inferior olive, dorsomedial cell group   IODM
inferior olive, medial nucleus   IOM
inferior olive, principal nucleus   IOPr
inferior olive, subnucleus A of medial nucleus   IOA
inferior olive, subnucleus B of medial nucleus   IOB
inferior olive, subnucleus C of medial nucleus   IOC
inferior olive, ventrolateral protrusion   IOVL
inferior salivatory nucleus   IS
infundibular recess   IRe
infundibular stem   InfS
interanterodorsal thalamic nucleus   IAD
interanteromedial thalamic nucleus   IAM
intercalated amygdaloid nucleus, main part   IM
intercalated nuclei of the amygdala   I
intercalated nucleus   In
intercrural fissure   icf
interfascicular nucleus   IF
interfascicular trigeminal nucleus   IF5
intergeniculate leaflet   IGL
intermediate endopiriform nucleus   IEn
intermediate gray layer of the superior colliculus   InG
intermediate interstitial nucleus of the medial longitudinal fasciculus   II
intermediate lobe of the pituitary   IPit
intermediate nucleus of the lateral lemniscus   ILL
intermediate reticular nucleus   IRt
intermediate reticular nucleus, alpha part   IRtA
intermediate white layer of the superior colliculus   InWh
intermediodorsal thalamic nucleus   IMD
intermedioventral thalamic commissure   imvc
intermedius nucleus of the medulla   InM
internal acoustic stria   ias
internal arcuate fibers   ia
internal capsule   ic
internal carotid artery   ictd
internal medullary lamina   iml
internal part of globus pallidus   IGP
internal plexiform layer of the olfactory bulb   IPl
interpeduncular fossa   ipf
interpeduncular nucleus   IP
interpeduncular nucleus, apical subnucleus   IPA
interpeduncular nucleus, caudal subnucleus   IPC
interpeduncular nucleus, dorsolateral subnucleus   IPDL
interpeduncular nucleus, dorsomedial subnucleu   IPDM

interpeduncular nucleus, intermediate subnucleus   IPI
interpeduncular nucleus, lateral subnucleus   IPL
interpeduncular nucleus, rostral subnucleus   IPR
interposed cerebellar nucleus, anterior part   IntA
interposed cerebellar nucleus, dorsolateral hump   IntDL
interposed cerebellar nucleus, dorsomedial crest   IntDM
interposed cerebellar nucleus, posterior part   IntP
interposed cerebellar nucleus, posterior parvicellula   IntPPC
interstitial basal nucleus of the medulla   IB
interstitial nucleus of Cajal   InC
interstitial nucleus of Cajal, shell region   InCSh
interstitial nucleus of the decussation of the superior cerebellar peduncle   ID
interstitial nucleus of the posterior limb of the anterior commissure   IPAC
interstitial nucleus of the posterior limb of the anterior commissure, lateral
    part   IPACL
interstitial nucleus of the posterior limb of the anterior commissure, medial
    part   IPACM
interstitial nucleus of the vestibular part of the 8th nerve   I8
interventricular foramen   IVF
intramedullary thalamic area   IMA
island of Calleja   ICj
island of Calleja, major island   ICjM
isthmic reticular formation   isRt

J

juxtaolivary nucleus   JxO
juxtaparaventricular part of the lateral hypothalamus   JPLH

K

Kolliker-Fuse nucleus   KF

L

lacunosum moleculare layer of the hippocampus   LMol
lambdoid septal zone   Ld
lamina 2 of the spinal gray   2Sp
lamina 3 of the spinal gray   3Sp
lamina 9 of the spinal gray   9Sp
lamina terminalis   LTer
lateral (dentate) cerebellar nucleus   Lat
lateral accumbens, shell region   LAcbSh
lateral amygdaloid nucleus, dorsolateral part   LaDL
lateral amygdaloid nucleus, ventral part   LaV

lateral amygdaloid nucleus, ventrolateral part   LaVL
lateral amygdaloid nucleus, ventromedial part   LaVM
lateral cerebellar nucleus, parvicellular part   LatPC
lateral cervical nucleus   LatC
lateral entorhinal cortex   LEnt
lateral habenular nucleus   LHb
lateral habenular nucleus, lateral part   LHbL
lateral habenular nucleus, medial part   LHbM
lateral lemniscus   ll
lateral mamillary nucleus   LM
lateral nucleus of the diagonal band   LDB
lateral olfactory tract   lo
lateral orbital cortex   LO
lateral orbitofrontal artery   lofr
lateral parabrachial nucleus   LPB
lateral parabrachial nucleus, central part   LPBC
lateral parabrachial nucleus, crescent part   LPBCr
lateral parabrachial nucleus, dorsal part   LPBD
lateral parabrachial nucleus, external part   LPBE
lateral parabrachial nucleus, internal part   LPBI
lateral parabrachial nucleus, superior part   LPBS
lateral parabrachial nucleus, ventral part   LPBV
lateral paragigantocellular nucleus   LPGi
lateral paragigantocellular nucleus, alpha part   LPGiA
lateral paragigantocellular nucleus, external par   LPGiE
lateral parietal association cortex   LPtA
lateral periaqueductal gray   LPAG
lateral posterior thalamic nucleus   LP
lateral posterior thalamic nucleus, laterocaudal par   LPLC
lateral posterior thalamic nucleus, laterorostral par   LPLR
lateral posterior thalamic nucleus, mediocaudal par   LPMC
lateral posterior thalamic nucleus, mediorostral par   LPMR
lateral preoptic area   LPO
lateral recess of the 4th ventricle   LR4V
lateral reticular nucleus   LRt
lateral reticular nucleus, parvicellular part   LRtPC
lateral reticular nucleus, subtrigeminal part   LRtS5
lateral septal nucleus, dorsal part   LSD
lateral septal nucleus, intermediate part   LSI
lateral septal nucleus, ventral part   LSV
lateral spinal nucleus   LSp
lateral stripe of the striatum   LSS
lateral superior olive   LSO
lateral terminal nucleus of the accessory optic tract   LT
lateral tuberal nucleus   LTu

lateral ventricle   LV
lateral vestibular nucleus   LVe
lateroanterior hypothalamic nucleus   LA
laterodorsal tegmental nucleus   LDTg
laterodorsal tegmental nucleus, ventral part   LDTgV
laterodorsal thalamic nucleus   LD
laterodorsal thalamic nucleus, dorsomedial par   LDDM
laterodorsal thalamic nucleus, ventrolateral part   LDVL
lateroventral periolivary nucleus   LVPO
layer   5a
layer   6a
layer   6b
layer   1
layer   2
layer   3
layer   4
layer   5
layer   6
linear nucleus of the hindbrain   Li
lithoid nucleus   Lth
lobule 1 of cerebellar   vermis
1Cb lobule 10 of cerebellar vermis   10Cb
lobule 2 of the cerebellar vermis   2Cb
lobule 2b of the cerebellar vermis   2bCb
lobule 3 of the cerebellar vermis   3Cb
lobule 4 of the cerebellar vermis   4Cb
lobule 5 of the cerebellar vermis   5Cb
lobule 6 of the cerebellar vermis   6Cb
lobule 6a of the cerebellar vermis   6aCb
lobule 6a,b of the cerebellar vermis   6a,bCb
lobule 6b of the cerebellar vermis   6bCb
lobule 6c of the cerebellar vermis   6cCb
lobule 7 of cerebellar vermis   7Cb
lobule 8 of cerebellar vermis   8Cb
lobule 9 of cerebellar vermis   9Cb
lobule 9a of the cerebellar vermis   9aCb
lobule 9a,b of the cerebellar vermis   9a,bCb
lobule 9b of the cerebellar vermis   9bCb
lobule 9c of the cerebellar vermis   9cCb
lobules 1 and 2 of the cerebellar vermis   1/2Cb
lobules 2 and 3 of the cerebellar vermis   2/3Cb
lobules 3 and 4 of the cerebellar vermis   3/4Cb
lobules 4 and 5 of the cerebellar vermis   4/5Cb
locus coeruleus   LC
longitudinal fasciculus of the pons   lfp

The Rat Brain in Stereotaxic Coordinates   Compact 7th Edition   Paxinos & Watson

M

magnocellular nucleus of the lateral hypothalamu   MCLH
magnocellular nucleus of the posterior commissure   MCPC
mamillary peduncle   mp
mamillary recess of the 3rd ventricle   MRe
mamillotegmental tract   mtg
mamillothalamic tract   mt
marginal zone of the medial geniculate   MZMG
matrix region of the medulla   Mx
medial accessory oculomotor nucleus   MA3
medial amygdaloid nucleus, anterior part   MeA
medial amygdaloid nucleus, anterodorsal   MeAD
medial amygdaloid nucleus, anteroventral part   MeAV
medial amygdaloid nucleus, posterodorsal part   MePD
medial amygdaloid nucleus, posteroventral part   MePV
medial cerebellar nucleus   Med
medial cerebellar nucleus, dorsolateral protuberance   MedDL
medial cerebellar nucleus, lateral part   MedL
medial corticohypothalamic tract   mch
medial eminence, external layer   MEE
medial eminence, internal layer   MEI
medial entorhinal cortex   MEnt
medial entorhinal cortex, rostral part   MEntR
medial forebrain bundle   mfb
medial forebrain bundle, 'a' component   mfba
medial forebrain bundle, 'a' component and ventral pallidum   mfba/VP
medial forebrain bundle, 'b' component   mfbb
medial geniculate nucleus   MG
medial geniculate nucleus, dorsal part   MGD
medial geniculate nucleus, medial part   MGM
medial geniculate nucleus, medial part   MGV
medial habenular nucleus   MHb
medial lemniscus   ml
medial lemniscus decussation   mlx
medial longitudinal fasciculus   mlf
medial mamillary nucleus, lateral part   ML
medial mamillary nucleus, medial part   MM
medial mamillary nucleus, median part   MnM
medial orbital cortex   MO
medial orbitofrontal artery mofr medial parabrachial nucleus   MPB
medial parabrachial nucleus, external part   MPBE
medial paralemniscial nucleus   MPL
medial parietal association cortex   MPtA
medial preoptic area   MPA
medial preoptic nucleus   MPO
medial preoptic nucleus, central part   MPOC
medial preoptic nucleus, lateral part   MPOL
medial preoptic nucleus, medial part   MPOM
medial pretectal area   MPT
medial septal nucleus   MS
medial superior olive   MSO
medial terminal nucleus   MT
medial tuberal nucleus   MTu
medial vestibular nucleus   MVe
medial vestibular nucleus, magnocellular part   MVeMC
medial vestibular nucleus, parvicellular part   MVePC
median accessory nucleus of the medulla   MnA
median eminence   ME
median preoptic nucleus   MnPO
median raphe nucleus   MnR
mediodorsal thalamic nucleus   MD
mediodorsal thalamic nucleus, central part   MDC
mediodorsal thalamic nucleus, lateral part   MDL
mediodorsal thalamic nucleus, medial part   MDM
medioventral periolivary nucleus   MVPO
medullary reticular nucleus, dorsal part   MdD
medullary reticular nucleus, ventral part   MdV
mesencephalic part of the dorsal raphe   mDR
mesencephalic reticular formation   mRt
mesencephalic trigeminal nucleus   Me5
mesencephalic trigeminal tract   me5
microcellular tegmental nucleus   MiTg
middle cerebellar peduncle   mcp
middle cerebral artery   mcer
mitral cell layer of the accessory olfactory bulb   MiA
mitral cell layer of the olfactory bulb   Mi
molecular layer of the cerebellum   MoCb
molecular layer of the dentate gyrus   MoDG
molecular layer of the dorsal cochlear nucleus   MoDC
molecular layer of the subiculum   MoS
motor root of the trigeminal nerve   m5
motor trigeminal nucleus, mylohyoid part   5MHy
motor trigeminal nucleus   5N
motor trigeminal nucleus, anterior digastric part   5ADi
motor trigeminal nucleus, masseter part   5Ma
motor trigeminal nucleus, pterygoid part   5Pt
motor trigeminal nucleus, temporalis part   5Te

N

NA1 noradrenalin cells   NA1
NA2 noradrenalin cells   NA2
NA5 noradrenalin cells   NA5
NA7 noradrenalin cells   NA7
navicular nucleus of the basal forebrain   Nv
nervus intermedius component of facial nerve   7ni
nigrostriatal bundle   ns
nucleus of Darkschewitsch   Dk
nucleus of origin of efferents of the vestibular nerv   EVe
nucleus of Roller   Ro
nucleus of stria medullaris   SM
nucleus of the ansa lenticularis   AL
nucleus of the brachium of the inferior colliculus   BIC
nucleus of the central acoustic tract   CAT
nucleus of the commissural stria terminalis   CST
nucleus of the fields of Forel   F
nucleus of the horizontal limb of the diagonal band   HDB
nucleus of the lateral olfactory tract   LOT
nucleus of the lateral olfactory tract, layer 1   LOT1
nucleus of the lateral olfactory tract, layer 2   LOT2
nucleus of the posterior commissure   PCom
nucleus of the trapezoid body   Tz
nucleus of the vertical limb of the diagonal band   VDB
nucleus X   X
nucleus Y of the vestibular complex   Y
nucleus Z   Z

O

obex   Obex
oculomotor nerve   3n
oculomotor nucleus   3N
oculomotor nucleus, parvicellular part   3PC
olfactory artery   olfa
olfactory bulb   OB
olfactory nerve layer   ON
olfactory tubercle   Tu
olfactory tubercle, layer 1   Tu1
olfactory tubercle, layer 3   Tu3
olfactory ventricle   OV
olivary pretectal nucleus   OPT
olivocerebellar tract   oc
olivocerebellar tract and dorsal spinocerebellar tract   oc/dsc
olivocochlear bundle   ocb
optic chiasm   och
optic nerve   2n
optic nerve layer of the superior colliculus   Op
optic tract   opt

oriens layer of the hippocampus   Or
oval paracentral thalamic nucleus   OPC

P

paraabducens nucleus   Pa6
parabigeminal nucleus   PBG
parabrachial pigmented nucleus of the ventral tegmentalarea   PBP
paracentral thalamic nucleus   PC
paracochlear glial substance   PCGS
paradiagonal zone   PDZ
parafascicular thalamic nucleus   PaF
parafloccular sulcus pfs paraflocculus   PFl
parainterfascicular nucleus of the ventral tegmental area   PIF
paralambdoid septal nucleus   PLd
paralemniscal nucleus   PL
paramedian lobule   PM
paramedian raphe nucleus   PMnR
paramedian reticular nucleus   PMn
paramedian sulcus   pms
paranigral nucleus of the ventral tegmental area   PN
parapyramidal nucleus of the raphe   PPy
pararubral nucleus   PaR
parasolitary nucleus   PSol
parastrial nucleus   PS
parasubiculum   PaS
parasubthalamic nucleus   PSTh
paratenial thalamic nucleus   PT
paraterete nucleus   PTe
paratrigeminal nucleus   Pa5
paratrochlear nucleus   Pa4
paraventricular hypothalamic nucleus   Pa
paraventricular hypothalamic nucleus, anterior parvic   PaAP
paraventricular hypothalamic nucleus, dorsal cap   PaDC
paraventricular hypothalamic nucleus, lateral magnocellular part   PaLM
paraventricular hypothalamic nucleus, medial magnocellular part   PaMM
paraventricular hypothalamic nucleus, medial parvicellular part   PaMP
paraventricular hypothalamic nucleus, posterior part   PaPo
paraventricular hypothalamic nucleus, ventral part   PaV
paraventricular thalamic nucleus   PV
paraventricular thalamic nucleus, anterior part   PVA
paraventricular thalamic nucleus, posterior part   PVP
paraxiphoid nucleus of thalamus   PaXi
parietal cortex, posterior area, caudal part   PtPC
parietal cortex, posterior area, dorsal part   PtPD

parietal cortex, posterior area, rostral part   PtPR
parvicellular reticular nucleus   PCRt
parvicellular reticular nucleus, alpha part   PCRtA
peduncular lateral hypothalamus   PLH
pedunculotegmental nucleus   PTg
periaqueductal gray   PAG
perifacial zone   P7
perifornical nucleus   PeF
perifornical part of lateral hypothalamus   PeFLH
perilemniscal nucleus, ventral part   PLV
periolivary horn   POH
peripeduncular area   PPA
peripeduncular nucleus   PP
perirhinal cortex   PRh
peritrigeminal zone   P5
periventricular gray   PVG
periventricular hypothalamic nucleus   Pe
pineal gland   Pi
pineal recess   PiRe
pineal stalk   PiSt
piriform cortex   Pir
piriform cortex, layer 1   Pir1
piriform cortex, layer 1a   Pir1a
piriform cortex, layer 2   Pir2
piriform cortex, layer 3   Pir3
pleioglia periaqueductal gray   PlPAG
polymorph layer of the dentate gyrus   PoDG
pontine nuclei   Pn
pontine raphe nucleus   PnR
pontine reticular nucleus, caudal part   PnC
pontine reticular nucleus, oral part   PnO
pontine reticular nucleus, ventral part   PnV
post superior fissure   psf
posterior cerebral artery   pcer
posterior commissure   pc
posterior hypothalamic area   PHA
posterior hypothalamic area, dorsal part   PHD
posterior hypothalamic nucleus   PH
posterior intralaminar thalamic nucleus   PIL
posterior limitans thalamic nucleus   PLi
posterior lobe of the pituitary   PPit
posterior thalamic nuclear group   Po
posterior thalamic nuclear group, triangular part   PoT
posterodorsal preoptic nucleus   PDPO
posterodorsal raphe nucleus   PDR

posterodorsal tegmental nucleus   PDTg
posterolateral cortical amygdaloid area   PLCo
posterolateral cortical amygdaloid area, layer 1   PLCo1
posterolateral fissure   plf
posteromedial cortical amygdaloid area   PMCo
posteromedian thalamic nucleus   PoMn
postsubiculum   Post
pre-Botzinger complex   PrBo
pre-Edinger-Westphal nucleus   PrEW
precommissural nucleus   PrC
preculminate fissure   pcuf
precuneiform area   PrCnF
pregeniculate nucleus of the prethalamus   PrG
premamillary nucleus, dorsal part   PMD
premamillary nucleus, ventral part   PMV
prepositus nucleus   Pr
prepositus nucleus, magnocellular part   PrMC
prepyramidal fissure   ppf
prerubral field   PR
presubiculum   PrS
prethalamic eminence   PrThE
primary auditory cortex   Au1
primary fissure   prf
primary motor cortex   M1
primary somatosensory cortex   S1
primary somatosensory cortex, barrel field   S1BF
primary somatosensory cortex, dysgranular zon   S1DZ
primary somatosensory cortex, dysgranular zone, oralregion   S1DZO
primary somatosensory cortex, hindlimb region   S1HL
primary somatosensory cortex, jaw region   S1J
primary somatosensory cortex, shoulder region   S1Sh
primary somatosensory cortex, trunk region   S1Tr
primary somatosensory cortex, upper lip regio   S1ULp
primary visual cortex   V1
primary visual cortex, binocular area   V1B
primary visual cortex, monocular area   V1M
principal mamillary tract   pm
principal sensory trigeminal nucleus   Pr5
principal sensory trigeminal nucleus, dorsomedial par   Pr5DM
principal sensory trigeminal nucleus, ventrolateral par   Pr5VL
prosomere 1   p1
prosomere 1 periaqueductal gray   p1PAG
prosomere 1 reticular formation   p1Rt
prosomere 2   p2
prosomere 3   p3

The Rat Brain in Stereotaxic Coordinates   Compact 7th Edition   Paxinos & Watson

prosubiculum   ProS
Purkinje cell layer of the cerebellum   Pk
pyramidal cell layer of the hippocampus   Py
pyramidal decussation   pyx
pyramidal tract   py

R

radiatum layer of the hippocampus   Rad
raphe interpositus nucleus   RIP
raphe magnus nucleus   RMg
raphe obscurus nucleus   ROb
raphe pallidus nucleus   RPa
recess of the inferior colliculus   ReIC
red nucleus   R
red nucleus, magnocellular part   RMC
red nucleus, parvicellular part   RPC
region where VA and VL overlap   VA/VL
reticular nucleus (prethalamus)   Rt
reticulostrial nucleus   RtSt
reticulotegmental nucleus of the pons   RtTg
reticulotegmental nucleus of the pons, lateral par   RtTgL
reticulotegmental nucleus of the pons, pericentral part   RtTgP
retroambiguus nucleus   RAmb
retrochiasmatic area   RCh
retrochiasmatic area, lateral part   RChL
retroendopirifom nucleus   REn
retroethmoid nucleus   REth
retroisthmic nucleus   RIs
retrolemniscal nucleus   RL
retromamillary decussation rmx retromamillary nucleus   RM
retromamillary nucleus, lateral part   RML
retromamillary nucleus, medial part   RMM
retroparafascicular nucleus   RPF
retroreuniens nucleus   RRe
retrorubral field   RRF
retrotrapezoid nucleus   RTz
reuniens thalamic nucleus   Re
rhabdoid nucleus   Rbd
rhinal fissure   rf
rhinal incisure   ri
rhomboid thalamic nucleus   Rh
rhombomere 1   r1
rhombomere 1 reticular formation   r1Rt
rhombomere 10 r10
rhombomere 11   r11

rhombomere 2   r2
rhombomere 2 reticular formation   r2Rt
rhombomere 3   r3
rhombomere 3 reticular formation   r3Rt
rhombomere 4   r4
rhombomere 5   r5
rhombomere 6   r6
rhombomere 7   r7
rhombomere 8   r8
rhombomere 9   r9
rostral amygdalopiriform area   RAPir
rostral interstitial nucleus of the medial longitudinal fasciculus   RI
rostral linear nucleus (midbrain)   RLi
rostral migratory stream   RMS
rostral migratory stream/olfactory ventricle   RMS/OV
rostral ventral respiratory group   RVRG
rostroventrolateral reticular nucleus   RVL
rostrum of the corpus callosum   rcc
rubrospinal tract   rs

S

sagulum nucleus   Sag
scaphoid thalamic nucleus   Sc
secondary auditory cortex, dorsal area   AuD
secondary auditory cortex, ventral area   AuV
secondary fissure   sf
secondary motor cortex   M2
secondary somatosensory cortex   S2
secondary visual cortex, lateral area   V2L
secondary visual cortex, mediolateral area   V2ML
secondary visual cortex, mediomedial area   V2MM
sensory root of the trigeminal nerve   s5
septofimbrial nucleus   SFi
septohippocampal nucleus   SHi
septohypothalamic nucleus   SHy
simple lobule   Sim
simple lobule A   SimA
simple lobule B   SimB
simplex fissure   simf
solitary nucleus   Sol
solitary nucleus, central part   SolCe
solitary nucleus, commissural part   SolC
solitary nucleus, dorsal part   SolD
solitary nucleus, dorsolateral part   SolDL
solitary nucleus, dorsomedial part   SolDM

solitary nucleus, gelatinous part   SolG
solitary nucleus, intermediate part   SolIM
solitary nucleus, interstitial part   SolI
solitary nucleus, lateral part   SolL
solitary nucleus, medial part   SolM
solitary nucleus, retrocentral part   SolRC
solitary nucleus, rostrolateral part   SolRL
solitary nucleus, ventral intermediate part   SolVI
solitary nucleus, ventral part   SolV
solitary nucleus, ventrolateral part   SolVL
solitary tract   sol
sphenoid nucleus   Sph
spinal trigeminal nucleus, caudal part   Sp5C
spinal trigeminal nucleus, interpolar part   Sp5I
spinal trigeminal nucleus, oral part   Sp5O
spinal trigeminal tract   sp5
spinal vestibular nucleus   SpVe
splenium of the corpus callosum   scc
stigmoid hypothalamic nucleus   Stg
stratum lucidum of the hippocampus   SLu
stria medullaris   sm
stria terminalis   st
strial part of the preoptic area   StA
striohypothalamic nucleus   StHy
subbrachial nucleus   SubB
subcoeruleus nucleus, alpha part   SubCA
subcoeruleus nucleus, dorsal part   SubCD
subcoeruleus nucleus, ventral part   SubCV
subcommissural organ   SCO
subfornical organ   SFO
subgeniculate nucleus of prethalamus   SubG
subiculum   S
subiculum, transition area   STr
subincertal nucleus   SubI
submamillothalamic nucleus   SMT
submedius thalamic nucleus   Sub
submedius thalamic nucleus, dorsal part   SubD
submedius thalamic nucleus, ventral part   SubV
subparafascicular thalamic nucleus   SPF
subparafascicular thalamic nucleus, parvicellular part   SPFPC
subparaventricular zone of the hypothalamus   SPa
subpeduncular tegmental nucleus   SPTg
substantia innominata, basal part   SIB
substantia nigra   SN
substantia nigra, compact part   SNC

substantia nigra, compact part, dorsal tier   SNCD
substantia nigra, compact part, medial tier   SNCM
substantia nigra, lateral part   SNL
substantia nigra, reticular part   SNR
subtantia nigra, compact part, ventral tier   SNCV
subthalamic nucleus   STh
sulcus limitans   sl
superficial gray and zonal layers of the superior colliculus   SuG/Zo
superficial gray layer of the superior colliculus   SuG
superior cerebellar peduncle   scp
superior cerebellar peduncle, descending limb   scpd
superior colliculus   SC
superior medullary velum   SMV
superior medullary velum and trochlear nerve   SMV/4n
superior olive   SOl
superior paraolivary nucleus   SPO
superior salivatory nucleus   SuS
superior thalamic radiation   str
superior vestibular nucleus   SuVe
suprachiasmatic nucleus   SCh
suprachiasmatic nucleus, dorsolateral part   SChDL
suprachiasmatic nucleus, ventromedial part   SChVM
suprageniculate thalamic nucleus   SG
supragenual nucleus of the raphe   SGe
supralemniscal nucleus   SuL
supraoculomotor cap   Su3C
supraoculomotor periaqueductal gray   Su3
supraoptic decussation   sox
supraoptic nucleus   SO
supraoptic nucleus, retrochiasmatic part   SOR
supratrigeminal nucleus   Su5

T

tectal gray   TG
tectospinal tract   ts
temporal association cortex   TeA
terete hypothalamic nucleus   Te
transverse fibers of the pons   tfp
trapezoid body   tz
triangular nucleus of the lateral lemniscus   TrLL
triangular septal nucleus   TS

trigeminal ganglion   5Gn
trigeminal nerve   5n
trigeminal transition zone   5Tr
trigeminal-solitary transition zone   5Sol
trigeminothalamic tract tth trochlear nerve   4n
trochlear nerve/mesencephalic trigeminal tract   4n/me5
trochlear nerve/superior medullary velum   4n/SMV
trochlear nucleus   4N
trochlear nucleus shell region   4Sh
tuberal region of lateral hypothalamus   TuLH
tuberomamillary nucleus   TuM

U

uncinate fasciculus of the cerebellum   un

V

vagus nerve   10n
vagus nerve nucleus   10N
vascular organ of the lamina terminalis   VOLT
vein   v
ventral anterior thalamic nucleus   VA
ventral claustrum   VCl
ventral cochlear nucleus, anterior part   VCA
ventral cochlear nucleus, posterior part   VCP
ventral cochlear nucleus, posterior part, octopus cell area   VCPO
ventral entopiriform nucleus   VEn
ventral hippocampal commissure vhc ventral horn   VH
ventral intermediate entorhinal cortex   VIEnt
ventral linear nucleus of the thalamus   VLi
ventral nucleus of the lateral lemniscus   VLL
ventral orbital cortex   VO
ventral pallidum   VP
ventral posterior nucleus of the thalamus, parvicellular   VPPC
ventral posterolateral thalamic nucleus   VPL
ventral posteromedial thalamic nucleus   VPM
ventral reuniens thalamic nucleus   VRe
ventral spinocerebellar tract   vsc
ventral subiculum   VS
ventral taenia tecta, layer 1   VTT1

ventral taenia tecta, layer 2   VTT2
ventral tegmental area   VTA
ventral tegmental area, rostral part   VTAR
ventral tegmental decussation   vtgx
ventral tegmental nucleus   VTg
ventral tenia tecta   VTT
ventral tuberomamillary nucleus   VTM
ventrolateral hypothalamic nucleus   VLH
ventrolateral hypothalamic tract   vlh
ventrolateral periaqueductal gray   VLPAG
ventrolateral preoptic nucleus   VLPO
ventrolateral thalamic nucleus   VL
ventromedial hypothalamic nucleus   VMH
ventromedial hypothalamic nucleus, central par   VMHC
ventromedial hypothalamic nucleus, dorsomedial par   VMHDM
ventromedial hypothalamic nucleus, shell regio   VMHSh
ventromedial hypothalamic nucleus, ventrolateral par   VMHVL
ventromedial preoptic nucleus   VMPO
ventromedial thalamic nucleus   VM
vertebral artery   vert
vestibular root of the vestibulocochlear nerve   8vn
vestibulocerebellar nucleus   VeCb
vestibulocochlear nerve   8n
vestibulomesencephalic tract   veme
vestibulospinal tract   vesp

X

xiphoid thalamic nucleus   Xi

Z

zona incerta   ZI
zona incerta, caudal part   ZIC
zona incerta, dorsal part   ZID
zona incerta, rostral part   ZIR
zona incerta, ventral part   ZIV
zona limitans   ZL
zonal layer and superficial gray layer of superior colliculus   Zo/SuG
zonal layer of superior colliculus   Zo

The Rat Brain in Stereotaxic Coordinates   Compact 7th Edition   Paxinos & Watson

Index of Abbreviations

The abbreviations are listed in alphabetical order. Each abbreviation is followed by the structure name and the number of the figures on which the abbreviation appears. Numbers 1-161 refer to coronal figures herein. Numbers 162-207 refer to the sagittal and horizontal figures in the full edition of the atlas.

1   layer   1-166, 168-207
1Cb   lobule 1 of cerebellar vermis   112-128, 162-166, 190-195
2   layer   7-74, 94, 99-109, 163, 168, 170-172, 174-195, 199-204, 207
2/3Cb lobules 2 and 3 of the cerebellar vermis   106-109, 172-175, 192-195
2bCb   lobule 2b of the cerebellar vermis   110
2Cb   lobule 2 of the cerebellar vermis   103-121, 162-171, 192-199
2n   optic nerve   163, 183
3   layer   8-74, 100, 104-109, 168, 170-179, 185, 189-191, 194-195, 199-204, 207
3/4Cb lobules 3 and 4 of the cerebellar vermis   127-129, 196
3Cb   lobule 3 of the cerebellar vermis   106-126, 162-171, 197-206
3N   oculomotor nucleus   85-91, 162-164, 193-194
3n   oculomotor nerve   79-82, 165-169, 181-190
3PC   oculomotor nucleus, parvicellular part   84-91, 162-164, 195
3V   3rd ventricle   29-73, 162-164, 181-199
4   layer   8-10, 30-33, 52, 194
4/5Cb lobules 4 and 5 of the cerebellar vermis   106-109, 176-178, 194-203
4Cb   lobule 4 of the cerebellar vermis   108-126, 162-175, 197-198, 200-202, 204-207
4N   trochlear nucleus   92-95, 162, 164, 192-194
4n   trochlear nerve   94-112, 166-170, 173, 175, 182-183, 185, 187, 192-197
4n/me5 trochlear nerve/mesencephalic trigeminal tract   195
4n/SMV trochlear nerve/superior medullary velum   167
4Sh   trochlear nucleus shell region   92-95, 163, 195
4V   4th ventricle   104-146, 162-165, 167, 187-195
4x   decussation of the trochlear nerve   162, 164, 171
5   layer   52
5a   layer   52, 94
5ADi   motor trigeminal nucleus, anterior digastric part   113-116, 168, 184-186
5Cb   lobule 5 of the cerebellar vermis   109-131, 162-175, 200-202, 204-207
5Gn   trigeminal ganglion   182
5Ma   motor trigeminal nucleus, masseter part   109-115
5MHy   motor trigeminal nucleus, mylohyoid part   112-113
5N   motor trigeminal nucleus   108, 169-171, 186-188
5Pt   motor trigeminal nucleus, pterygoid part   109-111
5Sol   trigeminal-solitary transition zone   122-148, 169-170, 187
5Te   motor trigeminal nucleus, temporalis part   109-115, 189

5Tr   trigeminal transition zone   113-117, 171, 187-189
6   layer   94, 186
6a   layer   52
6aCb   lobule 6a of the cerebellar vermis   121-127, 129-133, 162-167, 170
6b   layer   52
6bCb   lobule 6b of the cerebellar vermis   134-138, 162-167, 170
6Cb   lobule 6 of the cerebellar vermis   128, 133, 203-206
6cCb   lobule 6c of the cerebellar vermis   134-143, 162-166, 168
6N   abducens nucleus   117-119, 164, 185-186
6n   abducens nerve   115-116, 165
6RB   abducens nucleus, retractor bulbi part   117-120, 182, 184
7Cb   lobule 7 of cerebellar vermis   137-150, 162-167, 200, 202, 205
7DI   facial nucleus, dorsal intermediate subnucleus   123-131
7DL   facial nucleus, dorsolateral subnucleus   122-132
7DM   facial nucleus, dorsomedial subnucleus   121-131
7L   facial nucleus, lateral subnucleus   122-134
7N   facial nucleus   167-172, 181
7n   facial nerve   111-119, 166-175, 181-185
7ni   nervus intermedius component of facial nerve   121-123, 173, 186
7SH   facial motor nucleus, stylohyoid part   120-125, 128-132, 168-169, 172, 182-183
7VI   facial nucleus, ventral intermediate subnucleus   121-132
7VM   facial nucleus, ventromedial subnucleus   121-130
8Cb   lobule 8 of cerebellar vermis   135-155, 162-166, 197-201
8cn   cochlear root of the vestibulocochlear nerve   112-117, 121-123, 126-127, 181-185
8n   vestibulocochlear nerve   171-172, 174-175, 177, 182, 185-190
8vn   vestibular root of the vestibulocochlear nerve   115-125, 173, 176, 182, 186, 188
9aCb   lobule 9a of the cerebellar vermis   162-164, 166
9bCb   lobule 9b of the cerebellar vermis   162-164
9Cb   lobule 9 of cerebellar vermis   131-139, 164, 188-198
9cCb   lobule 9c of the cerebellar vermis   140-158, 162-167
9n   glossopharyngeal nerve   130-132, 182-183
10Cb   lobule 10 of cerebellar vermis   129-149, 162-168, 189-194
10N   vagus nerve nucleus   136-156, 163-165, 182-187
10n   vagus nerve   132-137, 139, 141-143, 147, 149-150, 172, 182-183
11N   accessory nerve nucleus   157-161
12GH   hypoglossal nucleus, geniohyoid part   147-156
12N   hypoglossal nucleus   138-157, 162-164, 182-186
12n   hypoglossal nerve   137, 141, 144, 149-155

A

a   artery   38-39, 43, 47-63, 65-74, 76-79, 81, 83-85, 87, 91-105, 107, 109, 113-114, 118, 120-126, 128, 130-131, 148-149, 163, 169-170, 176-177

A24a   cingulate cortex, area 24a   12-21, 23-36, 162-167, 203-207
A24a   prime cingulate cortex, area 24a prime   37-45
A24b   cingulate cortex, area 24b   12-36, 162-167
A24b   prime cingulate cortex, area 24b prime   37-45
A25   cingulate cortex, area   25, 10-11, 13, 163-165, 197-202
A29a   cingulate cortex, area 29a   78-89, 164-171, 207
A29b   cingulate cortex, area 29b   79-104, 162-168
A29c   cingulate cortex, area 29c   47-91, 162-164, 166-167
A30   cingulate cortex, area   30, 47-111, 162-180, 206-207
A32D   cingulate cortex, area 32D   6-11, 162-164, 207
A32V   cingulate cortex, area 32V   7-11, 162-164, 201-204
A33   cingulate cortex, area   33, 13-42, 163-164, 205-207
AA   anterior amygdaloid area   35-48, 171-176, 182-186
AC   anterior commissural nucleus   36-38, 164-165, 189-191
ac   anterior commissure   34-37, 162-168, 191-194
aca   anterior commissure, anterior part   8-33, 167-169, 171, 189-191, 196
AcbC   accumbens nucleus, core region   11-29, 166-171, 189-195
AcbSh   accumbens nucleus, shell region   10-29, 164-165, 167-170, 187-195
acer   anterior cerebral artery   16-31, 33-37, 40, 186
aci   anterior commissure, intrabulbar part   1-7, 167-170, 191-195
ACo   anterior cortical amygdaloid nucleus   37-55, 174-178, 181-183
acp   anterior commissure, posterior limb   33-43, 169-180, 187-191
AD   anterodorsal thalamic nucleus   43-52, 166-168, 199-202
Ad1   Ad1 adrenalin cells   152-156, 158-159, 161, 167, 169
Ad1/NA1   Ad1 adrenalin cells and NA1 noradrenalin cells   147-151
Ad2   Ad2 adrenalin cells   138-143
Ad3   Ad3 adrenalin cells   135-137, 162
af   amygdaloid fissure   46-47, 50, 54, 67-80, 181-185
AHA   anterior hypothalamic area, anterior part   41-45, 164-166, 185-186
AHC   anterior hypothalamic area, central part   46-50, 164-166, 184-186
AHi   amygdalohippocampal area   184-186
AHiAL   amygdalohippocampal area, anterolateral part   59-66, 171, 174-178, 181-184
AHiPL   amygdalohippocampal area, posterolateral   67-70, 177, 179
AHiPM   amygdalohippocampal area, posteromedial part   67-81, 172-177, 179-180, 182-184
AHP   anterior hypothalamic area, posterior part   47-52, 164-165, 186-187
AI   agranular insular cortex   189-192, 194-204
AID   agranular insular cortex, dorsal part   8-32, 172-180
AIP   agranular insular cortex, posterior part   33-57, 187-192
AIV   agranular insular cortex, ventral part   8-32, 172-180
AL   nucleus of the ansa lenticularis   52, 173, 187
al   ansa lenticularis   169, 186-188
alv   alveus of the hippocampus   47-96, 167-180, 187-207

AM anteromedial thalamic nucleus 42-53, 164-168, 170, 192-196

AmbC ambiguus nucleus, compact part 133-140, 169-170, 181

AmbL ambiguus nucleus, loose part 146-151, 169, 181

AmbSC ambiguus nucleus, semicompact part 141-145, 169-170, 181

AMV anteromedial thalamic nucleus, ventral part 45-49, 164, 166

Ang angular thalamic nucleus 53-55, 169, 197

ANS accessory neurosecretory nuclei 43-50, 166, 185-186, 189

AOB accessory olfactory bulb 165, 167, 169

AOD anterior olfactory area, dorsal part 4-7, 168-169, 193, 196-200

AOE anterior olfactory area external part 3-5, 166-167

AOL anterior olfactory nucleus, lateral part 3-8, 168-171, 191-193, 195

AOM anterior olfactory nucleus, medial part 4-7, 164-167, 193, 195

AOP anterior olfactory area posterior part 9-11, 164-168, 188-194

aot accessory olfactory tract 46

AOV anterior olfactory area ventral part 4-5, 165-167, 191-192, 194

AOVP anterior olfactory area, ventral posterior part 6-9, 165-166, 191-192

AP area postrema 147-152, 162-163, 187-188

APF anterior perifornical nucleus 41

APir amygdalopiriform transition area 64-88, 178, 180-186

APit anterior lobe of the pituitary 162-170, 181

apmf ansoparamedian fissure 139-150, 168-172, 174, 196-198

APT anterior pretectal nucleus 80-82, 168, 197-200

APTD anterior pretectal nucleus, dorsal part 69-79, 167, 169-172, 201-202

APTV anterior pretectal nucleus, ventral part 70-79, 169-172, 194-196

Aq aqueduct 74-103, 162-164, 195-199

Arc arcuate hypothalamic nucleus 47-48, 162-164, 181

ArcD arcuate hypothalamic nucleus, dorsal part 49-61

ArcL arcuate hypothalamic nucleus, lateral part 49-61

ArcLP arcuate hypothalamic nucleus, lateroposterior part 62-68

ArcM arcuate hypothalamic nucleus, medial part 49-61

ArcMP arcuate hypothalamic nucleus, medial posterior part 62-69, 162

as acoustic stria 169

asc7 ascending fibers of the facial nerve 121-126, 165, 167, 184, 186

asp anterior spinal artery 137, 139-142, 144-145

ASt amygdalostriatal transition area 46-50, 54-61, 178-180, 187-188, 190

ATg anterior tegmental nucleus 96-98, 164, 189-190

Au1 primary auditory cortex 60-90, 197-206

AuD secondary auditory cortex, dorsal area 58-90, 195-207

AuV secondary auditory cortex, ventral area 60-90, 194-202

AV anteroventral thalamic nucleus 42

AVDM anterovent thalamic nucleus, dorsomedial part 43-53, 167-169, 197-201

AVPe anteroventral periventricular nucleus 32-34, 162, 185-187

AVVL anteroventral thalamic nucleus, ventrolateral part 43-51, 167-169, 195-201

azac azygous anterior cerebral artery 10-15

azp azygous pericallosal artery 11-18

B

B basal nucleus (Meynert) 36-55, 57-59, 171-180, 187-195

BAC bed nucleus of the anterior commissure 37-40, 165, 193-194

BAOT bed nucleus of the accessory olfactory tract 46-51, 171, 181-182

Bar Barrington's nucleus 108-112, 166, 189-191

bas basilar artery 85-96, 98-135, 162

BIC nucleus of the brachium of the inferior colliculus 88-98, 173-174, 195-200, 202

bic brachium of the inferior colliculus 82-102, 173-175, 193-205

BL basolateral amygdaloid nucleus 183

BLA basolateral amygdaloid nucleus, anterior part 46-61, 177-180, 184-188

BLP basolateral amygdaloid nucleus, posterior part 53-72, 178-180, 183-186

BLV basolateral amygdaloid nucleus, ventral part 49-61, 179-183

BMA basomedial amygdaloid nucleus, anterior part 43-56, 175-178, 182-184

BMP basomedial amygdaloid nucleus, posterior part 54-67, 176-186

Bo Botzinger complex 133-137, 169-170

bsc brachium of the superior colliculus 69-86, 165, 167, 169, 171-177, 197-204

C

CA1 field CA1 of the hippocampus 52-66, 68-90, 162-180, 185-207

CA2 field CA2 of the hippocampus 48-77, 163-165, 167-180, 186-207

CA3 field CA3 of the hippocampus 47-52, 54-81, 83-84, 163-165, 167-180, 184-207

CAT nucleus of the central acoustic tract 105-109, 168-169, 181

CB cell bridges of the ventral striatum 24-34, 167, 169, 185-187

cbc cerebellar commissure 120-122, 124

cbw cerebellar white matter 108-156, 162-167, 169-172, 174-180, 200

CC central canal 147-162, 182-186

cc corpus callosum 14-74, 162-180, 204-207

Ce central amygdaloid nucleus 189

CeC central amygdaloid nucleus, capsular part 46-61, 176-179, 185-188

CeCv central cervical nucleus 149-161

CeL central amygdaloid nucleus, lateral part 49-60, 177-178, 188

CeM central amygdaloid nucleus, medial part 45-57, 174-177, 185-188

CeMe central mesencephalic nucleus 84-87, 168-170, 195-196

CEnt caudomedial entorhinal cortex 89-112, 176-180, 186-205

CG central gray 106-108, 112-114, 162, 189, 192

cg cingulum 11-83, 166-170, 198-207

CGA central gray, alpha part 109-118, 162-164, 188-189

CGB central gray, beta part 109-115, 164

CGG central gray, gamma 115-118, 165, 188

CGO central gray, nucleus O 110-114, 188-190

chp choroid plexus 34-76, 118-119, 124-143, 145-146, 163, 166, 168, 170, 189-190, 204-205

CI caudal interstitial nucleus of the medial longitudinal fasciculus 122-128, 131-133, 162

CIC central nucleus of the inferior colliculus 98-109, 165-172, 196-207

cic commissure of the inferior colliculus 98-102, 162-165, 202-206

Cir circular nucleus 45, 185

CL centrolateral thalamic nucleus 51-69, 167-168, 197-201

Cl claustrum 8-14, 167-168, 171-180, 189, 194-203

CLi caudal linear nucleus of the raphe 85-95, 162-164, 187-191

cll commissure of the lateral lemniscus 100-103

CM central medial thalamic nucleus 44-66, 162-164, 192-197

CnF cuneiform nucleus 171-172, 195-197

CnFD cuneiform nucleus, dorsal part 101-106, 167-170, 198

CnFI cuneiform nucleus, intermediate part 98-105, 168-170

CnFV cuneiform nucleus, ventral part 98-105, 168-170, 194

Com commissural nucleus of the inferior colliculus 99-102, 162-163, 206-207

Cop copula of the pyramis 135-154, 167-176, 190-199

cp cerebral peduncle 61-93, 165, 167-173, 175-176, 184-191

CPO caudal periolivary nucleus 121, 168

CPu caudate putamen (striatum) 12-66, 166-171, 173-180, 187-207

Crus1 crus 1 of the ansiform lobule 117-141, 169-180, 195-207

Crus2 crus 2 of the ansiform lobule 132-150, 168-180, 195-203, 205

csc commissure of the superior colliculus 76-86, 162-164, 200-201

CST nucleus of the commissural stria terminalis 35, 194

cst commissural stria terminalis 39, 41-64, 175-177, 184-188, 192

Ct conterminal nucleus 152-155, 168

ctg central tegmetal tract 167 Cu cuneate nucleus 140-161, 165-169, 185, 187-190

cu cuneate fasciculus 143-161, 165-168, 170, 185, 187-189

CuR cuneate nucleus, rotundus part 147-154

CVL caudoventrolateral reticular nucleus 138-145, 150-151, 170-171

CxA cortex-amygdala transition zone 174, 176, 181-184

CxA1 cortex amygdala transition zone, layer 1, 34-47, 177-178

CxA3 cortex amygdala transition zone, layer 3, 175, 179, 185

D

D3V dorsal 3rd ventricle 42-76, 162-164, 198-203

DA dorsal hypothalamic area 52-60, 163-164, 187

DA11 DA11 dopamine cells 58-70, 163, 187-188, 191-192

DA13 DA13 dopamine cells 50-57, 165-166, 188

DA14 DA14 dopamine cells 36-37, 163, 190-191

das dorsal acoustic stria 126-129, 168

DC dorsal cochlear nucleus 169-171, 176-177, 187-189, 191

DCDp dorsal cochlear nucleus, deep layer 121-130, 173-175, 189-191

DCFu dorsal cochlear nucleus, fusiform layer 121-131, 173-175, 189-191

The Rat Brain in Stereotaxic Coordinates   Compact 7th Edition   Paxinos & Watson

DCIC dorsal cortex of the inferior colliculus 101-110, 162-169, 201-207
DCl dorsal claustrum 15-49, 192-193
DCMo dorsal cochlear nucleus, molecular layer 121-131, 173-175, 189-191
dcs dorsal corticospinal tract 160-161
dcw deep cerebral white matter 55-59, 61-72, 74-98, 184-186, 198-207
DEn dorsal endopiriform nucleus 8-70, 165, 167-180, 184-195
df dorsal fornix 40-68, 162, 164, 166, 205-207
DG dentate gyrus 66, 164-165, 169, 171, 177, 183, 186-187
dhc dorsal hippocampal commissure 49-92, 162-168, 170, 203-207
DI dysgranular insular cortex 10-57, 174-180, 191-196
DIEnt dorsal intermediate entorhinal cortex 85-95, 183, 185-192, 195
Dk nucleus of Darkschewitsch 72-82, 163-164, 192-195
DLEnt dorsolateral entorhinal cortex 59-107, 183-184, 186-199
DLG dorsal lateral geniculate nucleus 60-78, 174-178, 197-203
DLL dorsal nucleus of the lateral lemniscus 100-106, 172-174, 192-194
DLO dorsolateral orbital cortex 6-7, 198-204
dlo dorsal lateral olfactory tract 3-4
DLPAG dorsolateral periaqueductal gray 82-101, 164-165, 200-201
DM dorsomedial hypothalamic nucleus 63-64, 164-165, 182-184
DMC dorsomedial hypothalamic nucleus, compact part 58-62, 163-164, 185-186
DMD dorsomedial hypothalamic nucleus, dorsal part 52-62, 163-164, 185-186
DMPAG dorsomedial periaqueductal gray 77-106, 162-164, 200-206
DMSp5 dorsomedial spinal trigeminal nucleus 120-140, 171, 187
DMTg dorsomedial tegmental area 105-115, 163-166, 187-189
DMV dorsomedial hypothalamic nucleus, ventral part 59-62, 163
DP dorsal peduncular cortex 9-18, 162-165, 195-198
DpG deep gray layer of the superior colliculus 78-100, 163-170, 196-204
DPGi dorsal paragigantocellular nucleus 120-132, 163-165, 184-186
DPO dorsal periolivary region 113-116, 118-120, 169
DPPn dorsal peduncular pontine nucleus 91-92, 165
DpWh deep white layer of the superior colliculus 79-100, 164-168, 197-203
DR dorsal raphe nucleus 111, 192-194
DRC dorsal raphe nucleus, caudal part 104-110, 162, 164, 191
DRD dorsal raphe nucleus, dorsal part 93-103, 162-163, 195-197
DRL dorsal raphe nucleus, lateral part 93-100
DRV dorsal raphe nucleus, ventral part 93-103, 162-164
DS dorsal subiculum 74-90, 167-180, 202-207
dsc dorsal spinocerebellar tract 150, 152-161, 182
dsc/oc dorsal spinocerebellar tract and olivocerebellar track 135, 137-149, 151, 185-186, 188-189
DT dorsal terminal nucleus 83-87, 173-174
DTgC dorsal tegmental nucleus, central part 107 110, 162 164, 191
DTgP dorsal tegmental nucleus, pericentral part 104-110, 162-164, 191-192
dtgx dorsal tegmental decussation 84-89, 162-164, 191, 193-194

DTM dorsal tuberomamillary nucleus 63-65, 163, 181, 183
DTr dorsal transition zone 8-9, 164-166, 196
DTT dorsal tenia tecta 9-14, 164-166, 192-193, 195-199
DTT1 dorsal taenia tecta layer 1, 8, 194
DTT2 dorsal taenia tecta layer 2, 163

E

E ependyma and subependymal layer 14-39, 163, 166-168, 170
E/OV ependymal and subendymal layer/olfactory ventricle 196-198
E5 ectotrigeminal nucleus 143-144, 175
EA extension of the amygdala 39-48, 169-175, 185-191
ec external capsule 13-66, 173, 175-180, 188-206
ECIC external cortex of the inferior colliculus 91-98, 100, 104-110, 165-174, 196-198, 200-206
ECIC1 external cortex of the inferior colliculus, layer 1 119
ECIC3 external cortex of the inferior colliculus, layer 3 99, 101-103, 171
Ect ectorhinal cortex 58-111, 178-180, 193-207
ECu external cuneate nucleus 135-151, 168-172, 188-191
EF epifascicular nucleus 132-134
EGP external part of globus pallidus 35-39, 41-59, 176-180, 189-199
ELm epilemniscal nucleus 82, 190
eml external medullary lamina 47-65, 171-173, 175, 177, 193-199
EP entopeduncular nucleus 49-56, 171-175, 188-191
EPl external plexiform layer of the olfactory bulb 1-5, 164
EPlA external plexiform layer of the accessory olfactory b 3-4
EpP epipeduncular nucleus 78-80
ERS epirubrospinal nucleus 98-99, 169-170, 188
ESO episupraoptic nucleus 39-45, 167-169, 183
Eth ethmoid thalamic nucleus 70-73, 167-172, 192-196
EVe nucleus of origin of efferents of the vestibular nerve 116-125, 166, 187-188
EW Edinger-Westphal nucleus 84-90, 163, 194-195

F

F nucleus of the fields of Forel 67-71, 167-169, 189-191
f fornix 34-72, 162-166, 170, 183-201
FC fasciola cinereum 57-75, 77, 205-207
fi fimbria of the hippocampus 31-68, 164-165, 169, 171-180, 194-206
Fl flocculus 107-125, 176-179, 189-193
fmi forceps minor of the corpus callosum 8-13, 166-176, 179, 198-207
fmj forceps major of the corpus callosum 77-92, 94, 96-101, 166-173, 175-180, 204-207
fr fasciculus retroflexus 58-80, 164-166, 185-199
Fr3 frontal cortex, area 3, 7-12, 163, 175-180, 202-203, 205-207
FrA frontal association cortex 4-5, 164-174, 204-207

Fu bed nucleus of stria terminalis, fusiform part 32-36, 167-168
FVe F cell group of the vestibular complex 137-140, 169, 188

G

g7 genu of the facial nerve 118-124, 164-165, 187
gcc genu of the corpus callosum 14-18, 162-165, 201-203
Ge5 gelatinous layer of the caudal spinal trigeminal nucleus 158-161, 169-170, 182-183
Gem gemini hypothalamic nucleus 66-70, 165, 187-188
GI granular insular cortex 12-57, 177-180, 193-199
Gi gigantocellular reticular nucleus 119-148, 162-168, 181-186
GiA gigantocellular reticular nucleus, alpha part 119-133, 162-164
GiV gigantocellular reticular nucleus, ventral part 134-143, 162-165
Gl glomerular layer of the olfactory bulb 1-5, 163
GlA glomerular layer of the accessory olfactory bulb 4-5, 167-169
Gr gracile nucleus 144-166, 188-189
gr gracile fasciculus 153-162, 164, 166
GrA granule cell layer of the accessory olfactory bulb 1-5, 167
GrC granule cell layer of the cochlear nuclei 109-128, 175, 188-190
GrCb granule cell layer of cerebellum 120, 122, 125, 131, 134, 162-167, 169-180, 189, 197, 199-201
GrDG granule cell layer of the dentate gyrus 47-89, 162-163, 166-168, 170-180, 184-185, 188-207
GrO granule cell layer of the olfactory bulb 1-6, 164

H

h2 h2 field of Forel 61-62, 64-65
hbc habenular commissure 67-72, 162-163, 202
HDB nucleus of the horizontal limb of the diagonal band 24-46, 165-172, 182-187
hif hippocampal fissure 54-85, 163, 165, 167-178, 185-207
Hil hilus of the dentate gyrus 169, 173, 175, 177, 189, 197, 200-201, 205-206

I

intercalated nuclei of the amygdala 39-40, 42-48, 52-60, 174-179, 183-186
I8 interstitial nucleus of the vestibular part of the 8th nerve 118, 121-123, 176-177, 186
ia internal arcuate fibers 151-156
IAD interanterodorsal thalamic nucleus 44-48, 164-165, 196-198
IAM interanteromedial thalamic nucleus 47-51, 162-163, 193-195
IB interstitial basal nucleus of the medulla 156-161
IC inferior colliculus 111
ic internal capsule 33-68, 70-71, 169, 171-180, 187-201
icf intercrural fissure 132-140, 169-180, 195-202

ICj   island of Calleja   10, 12-32, 163-165, 167-169, 171, 173, 182-187

ICjM   island of Calleja, major island   15-16, 18-24, 164, 190-194

icp   inferior cerebellar peduncle   116-118, 120-143, 168-169, 171-176, 185-192, 194-195

ictd   internal carotid artery   41-45, 47, 181-184

ID   interstitial nucleus of the decussation of the superior cerebellar peduncle   89-92

IEn   intermediate endopiriform nucleus   8-39, 169-178, 185-192

IF   interfascicular nucleus   75-87, 162-163

IF5   interfascicular trigeminar nucleus   108-111, 171-172, 184-186

IG   indusium griseum   14-76, 162-164, 200-207

IGL   intergeniculate leaflet   65-75, 176-178, 195-199

IGP   internal part of globus pallidus   41-46, 171-175, 189-199

II   intermediate interstitial nucleus of the medial longitudinal fasciculus   115-116, 163

ILL   intermediate nucleus of the lateral lemniscus   99-105, 173-174, 190-191

IM   intercalated amygdaloid nucleus, main part   49-51, 177, 184

IMA   intramedullary thalamic area   65-78, 175, 199-203

IMD   intermediodorsal thalamic nucleus   53-64, 162-163, 195-199

IMG   amygdaloid intramedullary gray   54-56, 189-190

iml   internal medullary lamina   44, 50, 53, 65, 167, 196, 198

imvc   intermedioventral thalamic commissure   60, 194

In   intercalated nucleus   140-141

InC   interstitial nucleus of Cajal   75-90, 164, 192-194

InCSh   interstitial nucleus of Cajal, shell region   75-84, 165, 193-195

InfS   infundibular stem   62-65, 162

InG   intermediate gray layer of the superior colliculus   77-99, 164-173, 198-205

InM   intermedius nucleus of the medulla   142-147

IntA   interposed cerebellar nucleus, anterior part   122-129, 167-173, 194-197

IntDL   interposed cerebellar nucleus, dorsolateral hump   123-132, 173-174, 196-197

IntDM   interposed cerebellar nucleus, dorsomedial crest   124-127, 130-131

IntP   interposed cerebellar nucleus, posterior part   128-133, 167-172, 192-196

IntPPC   interposed cerebellar nucleus, posterior parvicellular   128-131, 168, 171

InWh   intermediate white layer of the superior colliculus   76-99, 164-172, 198-205

IOA   inferior olive, subnucleus A of medial nucleus   148-154, 165-166

IOB   inferior olive, subnucleus B of medial nucleus   146-154

IOBe   inferior olive, beta subnucleus of the medial nucleus   146-153, 162-163

IOC   inferior olive, subnucleus C of medial nucleus   146-154, 162-164

IOD   inferior olive, dorsal nucleus   134-149, 162-168

IODM   inferior olive, dorsomedial cell group   142-143

IOK   inferior olive, cap of Kooy of the medial nucleus   147-151, 162

IOM   inferior olive, medial nucleus   134-145, 155-156, 162-163

IOPr   inferior olive, principal nucleus   133-147, 163-166

IOVL   inferior olive, ventrolateral protrusion   144-146

IP   interpeduncular nucleus   164

IPA   interpeduncular nucleus, apical subnucleus   89-92, 162-163, 186

IPAC   interstitial nucleus of the posterior limb of the anterior commissure   28-34, 44-45, 169-171, 173-177, 187-191

IPACL   interstitial nucleus of the posterior limb of the anterior commissure, lateral part   35-43

IPACM   interstitial nucleus of the posterior limb of the anterior commissure, medial part   35-43

IPC   interpeduncular nucleus, caudal subnucleus   80-93, 162-163, 184-186

IPDL   interpeduncular nucleus, dorsolateral subnucleus   85-90

IPDM   interpeduncular nucleus, dorsomedial subnucleus   84-87

ipf   interpeduncular fossa   75-80, 162-165, 185-186

IPI   interpeduncular nucleus, intermediate subnucleus   87-91, 162-163, 184-186

IPit   intermediate lobe of the pituitary   162-164, 181

IPL   interpeduncular nucleus, lateral subnucleus   80-93, 184-186

IPl   internal plexiform layer of the olfactory bulb   1-5, 164

IPR   interpeduncular nucleus, rostral subnucleus   79-88, 162-163, 184-186

IRe   infundibular recess   162

IRt   intermediate reticular nucleus   117-119, 133-161, 165-172, 181-186

IRtA   intermediate reticular nucleus, alpha part   120-132, 167-169, 181-183, 186

IS   inferior salivatory nucleus   126-135, 162-163, 165-169, 171, 173

isRt   isthmic reticular formation   92-103, 166-169, 191-194

IVF   interventricular foramen   39-42, 165, 167-168, 197, 199

J

JPLH   juxtaparaventricular part of the lateral hypothalamus   45-48, 165, 188

JxO   juxtaolivary nucleus   134-141, 165-166

K

KF   Kolliker-Fuse nucleus   105-110, 172-173, 188-190

L

LA   lateroanterior hypothalamic nucleus   40-45, 164-166, 183-185

LAcbSh   lateral accumbens, shell region   15-26, 171-173, 187-189

LaDL   lateral amygdaloid nucleus, dorsolateral part   50-67, 180, 189-191

Lat   lateral (dentate) cerebellar nucleus   122-130, 173-175, 192-197

LatPC   lateral cerebellar nucleus, parvicellular part   124-128, 172-175, 192-193

LaV   lateral amygdaloid nucleus, ventral part   186

LaVL   lateral amygdaloid nucleus, ventrolateral part   55-61, 180

LaVM   lateral amygdaloid nucleus, ventromedial part   55-67, 180

LC   locus coeruleus   112-119, 167, 189-192

LD   laterodorsal thalamic nucleus   169, 203

Ld   lambdoid septal zone   23-30, 162-164, 194-199

LDB   lateral nucleus of the diagonal band   32-41, 43-46, 171-173, 183-186

LDDM   laterodorsal thalamic nucleus, dorsomedial part   50-59, 167-168, 199-202

LDTg   laterodorsal tegmental nucleus   101-112, 165-166, 190-195

LDTgV   laterodorsal tegmental nucleus, ventral part   104-108, 165-168, 190-192

LDVL   laterodorsal thalamic nucleus, ventrolateral part   48-61, 170-175, 200-202

lfp   longitudinal fasciculus of the pons   94-107, 163-165, 167-169, 181-184

LHb   lateral habenular nucleus   52-56, 69, 164, 166

LHbL   lateral habenular nucleus, lateral part   57-68, 165, 200-201

LHbM   lateral habenular nucleus, medial part   57-68, 200-201

Li   linear nucleus of the hindbrain   135-143, 168, 170, 181, 183

ll   lateral lemniscus   92-108, 167-174, 181-196

LM   lateral mamillary nucleus   68-73, 165, 167-168, 182-183

LMol   lacunosum moleculare layer of the hippocampus   54-89, 166-171, 173-180, 189-207

LO   lateral orbital cortex   5-13, 167-175, 177, 193, 195-205

lo   lateral olfactory tract   3-44, 165-178, 180, 182-201

lofr   lateral orbitofrontal artery   5, 7-15

LOT   nucleus of the lateral olfactory tract   37, 181-184

LOT1   nucleus of the lateral olfactory tract, layer   1, 38-45, 171

LOT2   nucleus of the lateral olfactory tract, layer   2, 172-174

LP   lateral posterior thalamic nucleus   171, 175, 177, 199, 203

LPAG   lateral periaqueductal gray   77-105, 162-166, 196-197

LPB   lateral parabrachial nucleus   105-108, 111-114, 167, 169-171, 191-195

LPBC   lateral parabrachial nucleus, central part   106-110

LPBCr   lateral parabrachial nucleus, crescent part   107-110

LPBD   lateral parabrachial nucleus, dorsal part   107-110

LPBE   lateral parabrachial nucleus, external part   106-111, 171-172

LPBI   lateral parabrachial nucleus, internal part   107-113, 168, 194-195

LPBS   lateral parabrachial nucleus, superior part   105-106, 171, 194

LPBV   lateral parabrachial nucleus, ventral part   107-113, 168

LPGi   lateral paragigantocellular nucleus   119-120, 132-142, 165-169

LPGiA   lateral paragigantocellular nucleus, alpha part   121-131, 166

LPGiE   lateral paragigantocellular nucleus, external part   121-136

LPLC   lateral posterior thalamic nucleus, laterocaudal part   71-78, 173-174, 200-202

LPLR   lateral posterior thalamic nucleus, laterorostral part   62-71, 172-174, 200-202

LPMC   lateral posterior thalamic nucleus, mediocaudal part   70-81, 173-175, 200-202

LPMR   lateral posterior thalamic nucleus, mediorostral part   59-72, 167-170, 172, 199-202

LPO   lateral preoptic area   29-42, 165-169, 183-191

LPtA   lateral parietal association cortex   59-61, 63-67, 173-180

LR4V   lateral recess of the 4th ventricle   113-140, 168-177, 188-192

LRt   lateral reticular nucleus   142-157, 167-172

LRtPC   lateral reticular nucleus, parvicellular part   148-157, 171-172

LRtS5   lateral reticular nucleus, subtrigeminal part   143-150, 173

LSD   lateral septal nucleus, dorsal part   16-41, 164-166, 199-205
LSI   lateral septal nucleus, intermediate part   14-37, 162-167, 192-204
LSO   lateral superior olive   112-122, 169-171, 181
LSS   lateral stripe of the striatum   14-38, 171, 173-178, 188-189
LSV   lateral septal nucleus, ventral part   16-36, 165-168, 193-198
LT   lateral terminal nucleus of the accessory optic tract   71-78, 176-177, 192
LTer   lamina terminalis   35-37, 162, 184, 195-196
Lth   lithoid nucleus   71-76, 164, 193-196
LV   lateral ventricle   13-97, 166-180, 185-193, 195-207
LVe   lateral vestibular nucleus   122-127, 169-171, 188-194
LVPO   lateroventral periolivary nucleus   109-119, 169, 171, 181

M

M1   primary motor cortex   7-59, 167-180, 204, 207
M2   secondary motor cortex   6-60, 163-174
m5   motor root of the trigeminal nerve   88-115, 171-173, 181-188
MA3   medial accessory oculomotor nucleus   75-83, 162-164, 192-194
mcer   middle cerebral artery   23-38, 40, 165-166
mch   medial corticohypothalamic tract   38, 40-41, 189-192
MCLH   magnocellular nucleus of the lateral hypothalamus   54-59, 169-171, 185-186
mcp   middle cerebellar peduncle   92-118, 170-177, 181, 183-195
MCPC   magnocellular nucleus of the posterior commissure   71-78, 165-166, 196-197
MD   mediodorsal thalamic nucleus   47-50, 194-195
MDC   mediodorsal thalamic nucleus, central part   55-61, 165-166, 196-199
MdD   medullary reticular nucleus, dorsal part   149-161, 166-172, 182-186
MDL   mediodorsal thalamic nucleus, lateral part   50-66, 165-168, 196-201
MDM   mediodorsal thalamic nucleus, medial part   51-58, 60-66, 164, 196-200
mDR   mesencephalic part of the dorsal raphe   90-92, 163, 196-197
MdV   medullary reticular nucleus, ventral part   149-161, 165-168, 181-184
ME   median eminence   48, 162
Me5   mesencephalic trigeminal nucleus   86-107, 109-116, 166-167, 169, 190-199
me5   mesencephalic trigeminal tract   86-115, 165-169, 188-194, 196-199
MeAD   medial amygdaloid nucleus, anterodorsal   42-55, 171-175, 182-187
MeAV   medial amygdaloid nucleus, anteroventral part   47-54, 171-174, 181-182
Med   medial cerebellar nucleus   124-133, 164-167, 192-197
MedDL   medial cerebellar nucleus, dorsolateral protuberance   127-133, 166-168, 196-199
MedL   medial cerebellar nucleus, lateral part   127-131, 166, 168, 196
MEE   medial eminence, external layer   49-61
MEI   medial eminence, internal layer   49-61
MEnt   medial entorhinal cortex   87-105, 175, 177-180, 182-199
MEntR   medial entorhinal cortex, rostral part   82-86

MePD   medial amygdaloid nucleus, posterodorsal part   53-63, 172-176, 184-188
MePV   medial amygdaloid nucleus, posteroventral part   55-63, 171-174, 181-184
mfb   medial forebrain bundle   12-31, 40-70, 167-169, 171, 183-188
mfba   medial forebrain bundle, 'a' component   32-39, 165, 167, 169, 171
mfba/VP   medial forebrain bundle, 'a' component and ventral pallidum   173
mfbb   medial forebrain bundle, 'b' component   32-39, 165
MG   medial geniculate nucleus   178
MGD   medial geniculate nucleus, dorsal part   73-87, 174-177, 197-199
MGM   medial geniculate nucleus, medial part   74-87, 173-175, 193-196
MGV   medial geniculate nucleus, ventral part   73-88, 174-177, 192-197
MHb   medial habenular nucleus   48-70, 162, 164, 200-203
Mi   mitral cell layer of the olfactory bulb   1-5, 164
MiA   mitral cell layer of the accessory olfactory bulb   3-4, 167
MiTg   microcellular tegmental nucleus   90-101, 171-174, 191-195
ML   medial mamillary nucleus, lateral part   68-78, 162-166, 182-184
ml   medial lemniscus   55-153, 162-171, 173, 181-193
mlf   medial longitudinal fasciculus   78-165, 182-194
mlx   medial lemniscus decussation   148-155
MM   medial mamillary nucleus, medial part   68-73, 162-164, 182-185
MnA   median accessory nucleus of the medulla   155-161
MnM   medial mammillary nucleus, median part   68-70, 162, 182-184
MnPO   median preoptic nucleus   29-35, 162-164, 187-194
MnR   median raphe nucleus   93-108, 162, 185-190
MO   medial orbital cortex   5-8, 163-164, 166, 197-205
MoCb   molecular layer of the cerebellum   120, 122, 125, 131, 134, 162, 165-167, 169-180, 197-201
MoDG   molecular layer of the dentate gyrus   46-53, 55-90, 162, 167-180, 189-191, 197-198, 201-207
mofr   medial orbitofrontal artery   5, 7-9
MoS   molecular layer of the subiculum   90-92
mp   mamillary peduncle   71-83, 164-166, 184-186
MPA   medial preoptic area   29-45, 163-165, 183-188
MPB   medial parabrachial nucleus   105-116, 167-171, 190-192
MPBE   medial parabrachial nucleus, external part   108-111, 190
MPL   medial paralemniscal nucleus   99-105, 171, 189-192
MPO   medial preoptic nucleus   44, 163-164, 184-188
MPOC   medial preoptic nucleus, central part   37-39, 163-164, 187
MPOL   medial preoptic nucleus, lateral part   34-40
MPOM   medial preoptic nucleus, medial part   35-43
MPT   medial pretectal area   70-77, 164-166, 200-201
MPtA   medial parietal association cortex   60-67, 167-172
MRe   mamillary recess of the 3rd ventricle   66-70, 162-163
mRt   mesencephalic reticular formation   81-91, 167-172, 192-197
MS   medial septal nucleus   20-33, 162-164, 191-198

MSO   medial superior olive   109-120, 169-170, 181
MT   medial terminal nucleus   74-79, 167-168, 184-188
mt   mamillothalamic tract   47-68, 164-166, 185-195
mtg   mamillotegmental tract   69-83, 164, 188, 190
MTu   medial tuberal nucleus   55-62, 165-169, 181-182
MVe   medial vestibular nucleus   139-142, 164-165, 170, 191
MVeMC   medial vestibular nucleus, magnocellular part   115-138, 165-169, 171, 187-190
MVePC   medial vestibular nucleus, parvicellular part   117-138, 163, 165-167, 188-190, 192
MVPO   medioventral periolivary nucleus   104-121, 167-170
Mx   matrix region of the medulla   129-158, 163-164, 168-172, 187
MZMG   marginal zone of the medial geniculate   73-88, 175, 177

N

NA1   NA1 noradrenalin cells   133-146, 157, 160, 169, 171
NA2   NA2 noradrenalin cells   148-161, 163
NA5   NA5 noradrenalin cells   108-125, 127, 169, 171-172, 181-184
NA7   NA7 noradrenalin cells   102-108, 168-172, 186-188
ns   nigrostriatal bundle   47-72, 168-169, 187-188
Nv   navicular nucleus of the basal forebrain   10-17, 162-168, 189-194

O

OB   olfactory bulb   165-166, 201
Obex   obex   153, 162
oc   olivocerebellar tract   129-134, 136, 182
oc/dsc   olivocerebellar tract and dorsal spinocerebellar tract   181, 183-184
ocb   olivocochlear bundle   115-121, 162-163, 168-171, 173, 176-177, 182-184, 186-187
och   optic chiasm   29-40, 162-164, 182-183
olfa   olfactory artery   19-20
ON   olfactory nerve layer   1-4, 163
Op   optic nerve layer of the superior colliculus   77-98, 163-173, 200-207
OPC   oval paracentral thalamic nucleus   60-66, 165-168, 193-196
OPT   olivary pretectal nucleus   70-74, 165-166, 200, 202
opt   optic tract   41-79, 165-180, 182-201
Or   oriens layer of the hippocampus   47-89, 166-180, 189-207
OV   olfactory ventricle   1-5

P

p1   prosomere 1, 162-163, 165, 167, 169, 171, 173, 175
p1PAG   prosomere 1 periaqueductal gray   74-76
p1Rt   prosomere 1 reticular formation   74-80, 167-168, 192-197
p2   prosomere 2, 162-163, 165, 167, 169, 171, 173, 175
p3   prosomere 3, 165, 167, 169, 171, 173, 175

P5   peritrigeminal zone   107-116, 168-169, 185-189

P7   perifacial zone   120-131, 133-134, 167-172, 181-182

Pa   paraventricular hypothalamic nucleus   187

Pa4   paratrochlear nucleus   94-98, 165, 192-193

Pa5   paratrigeminal nucleus   141-150, 171-173, 188-189

Pa6   paraabducens nucleus   117-120, 164, 186

PaAP   paraventricular hypothalamic nucleus, anterior parvic   38-44, 163-164, 189-191

PaDC   paraventricular hypothalamic nucleus, dorsal cap   47-49, 163

PaF   parafascicular thalamic nucleus   64-70, 164-167, 193-199

PAG   periaqueductal gray   164, 194, 198-199

PaLM   paraventricular hypothalamic nucleus, lateral magnocellular part   47-49, 164, 188

PaMM   paraventricular hypothalamic nucleus, medial magnocellular part   45-46

PaMP   paraventricular hypothalamic nucleus, medial parvicellular part   45-50, 163, 188-189

PaPo   paraventricular hypothalamic nucleus, posterior part   50-52, 164-166, 188

PaR   pararubral nucleus   79-88, 167-170, 191-193

PaS   parasubiculum   87-111, 174-175, 177, 179, 187-205

PaV   paraventricular hypothalamic nucleus, ventral part   44-49, 185-186

PaXi   paraxiphoid nucleus of thalamus   46-56, 162-164, 188

PBG   parabigeminal nucleus   91-99, 174, 191-194

PBP   parabrachial pigmented nucleus of the ventral tegmental area   71-87, 164, 166-169, 186-190

PC   paracentral thalamic nucleus   44-66, 165-168, 193-195

pc   posterior commissure   71-79, 162-165, 197-201

pcer   posterior cerebral artery   85-87, 183-184

PCGS   paracochlear glial substance   111-119, 172-173, 189-191

PCom   nucleus of the posterior commissure   73-78, 165-166, 198-199

PCRt   parvicellular reticular nucleus   128-148, 166-172, 181-186

PCRtA   parvicellular reticular nucleus, alpha part   115-127, 169-171, 182-186

pcuf   preculminate fissure   109-125, 162-175, 197-206

PDPO   posterodorsal preoptic nucleus   38-39, 189

PDR   posterodorsal raphe nucleus   93-101

PDTg   posterodorsal tegmental nucleus   111-115, 162-164, 189-190

PDZ   paradiagonal zone   18-19, 21, 23-28, 189-191

Pe   periventricular hypothalamic nucleus   35-62, 162, 182-189

PeF   perifornical nucleus   55-63, 166-168, 184-185

PeFLH   perifornical part of lateral hypothalamus   51-65, 165-168, 183-187

PFl   paraflocculus   113-137, 176-180, 185-194

pfs   paraflocular sulcus   119-131, 133-135, 177, 179-180

PH   posterior hypothalamic nucleus   61-69, 162-164, 185-189

PHA   posterior hypothalamic area   70-72, 162-164, 187-189

PHD   posterior hypothalamic area, dorsal part   57-64, 162-163, 187-189

PIF   parainterfascicular nucleus of the ventral tegmental area   79-88, 164, 186

PIL   posterior intralaminar thalamic nucleus   73-84, 171, 173-174, 192

Pir   piriform cortex   61-65, 67, 69, 167-169, 171, 173, 175, 181, 195-198

Pir1   piriform cortex, layer   1, 7, 9-60, 63-66, 68, 70-74, 170, 172, 174, 176-179, 185, 189-191

Pir1a   piriform cortex, layer 1a   8

Pir3   piriform cortex, layer   3, 178, 180-184, 186-188, 192-194

PiRe   pineal recess   162, 204-206

PiSt   pineal stalk   92-93, 207

Pk   Purkinje cell layer of the cerebellum   120, 122, 125, 131, 134, 162, 164-167, 169-175, 177-178, 189

PL   paralemniscal nucleus   97-99, 172-173, 187-188

PLCo   posterolateral cortical amygdaloid area   51-53, 68-74, 176-177, 179-182

PLCo1   posterolateral cortical amygdaloid area, layer   1, 54-67, 178

PLd   paralambdoid septal nucleus   22-29, 195

plf   posterolateral fissure   115-121, 123-124, 132-148, 162, 164-167, 177, 179

PLH   peduncular lateral hypothalamus   43-71, 165-171, 184-189

PLi   posterior limitans thalamic nucleus   70-81, 171, 194-201

PlPAG   pleioglia periaqueductal gray   77-80, 199

PLV   perilemniscal nucleus, ventral part   103-106, 169, 181

PM   paramedian lobule   132-153, 168-178, 180, 192-196, 198-199

pm   principal mammillary tract   69-72

PMCo   posteromedial cortical amygdaloid area   56-80, 172-182

PMD   premamillary nucleus, dorsal part   66-67, 162-165, 182-184

PMn   paramedian reticular nucleus   139-144

PMnR   paramedian raphe nucleus   92-106, 162-164, 185-190

pms   paramedian sulcus   136-142, 144-146, 148

PMV   premamillary nucleus, ventral part   63-66, 165-167, 181-184

PN   paranigral nucleus of the ventral tegmental area   76-87, 164, 186

Pn   pontine nuclei   89-105, 162-170, 181-184

PnC   pontine reticular nucleus, caudal part   108-118, 162-168, 181-186

PnO   pontine reticular nucleus, oral part   93-107, 164-169, 182-189

PnR   pontine raphe nucleus   107-109, 162, 188

PnV   pontine reticular nucleus, ventral part   113-118, 162-164, 181

Po   posterior thalamic nuclear group   54-79, 81, 168-175, 194-200

PoDG   polymorph layer of the dentate gyrus   51-87, 164, 167-168, 170-180, 188-207

POH   periolivary horn   112-116, 171

PoMn   posteromedian thalamic nucleus   65-66, 162-163, 195-197

Post   postsubiculum   83-103, 169, 171-178, 202-207

PoT   posterior thalamic nuclear group, triangular part   74-84, 171, 173, 193-195

PP   peripeduncular nucleus   75-82, 174-176, 192

PPA   peripeduncular area   73-77, 173-174, 190-191

ppf   prepyramidal fissure   137-150, 162-167, 169-174, 176, 193-194, 196-198, 200

PPit   posterior lobe of the pituitary   162-164, 181

PPy   parapyramidal nucleus if the raphe   122-132

PR   prerubral field   65-75, 165-168, 189-191

Pr   prepositus nucleus   119-135, 140, 163-164, 185-189

Pr5   principal sensory trigeminal nucleus   182

Pr5DM   principal sensory trigeminal nucleus, dorsomedial part   110-119, 172-173, 186-190

Pr5VL   principal sensory trigeminal nucleus, ventrolateral part   107-120, 172-174, 183-188

PrBo   pre-Botzinger complex   138-140, 169-170

PrC   precommissural nucleus   67-72, 164-165, 197-199

PrCnF   precuneiform area   91-100, 167-172, 195-198

PrEW   pre-Edinger-Westphal nucleus   75-83, 163, 192-195

prf   primary fissure   111-126, 128-131, 162-178, 195-207

PrG   pregeniculate nucleus of the prethalamus   60-77, 176-179, 193-199

PRh   perirhinal cortex   58-111, 179-180, 187-206

PrMC   prepositus nucleus, magnocellular part   136-139, 163

PrS   presubiculum   87-95, 176-180, 189-201

PrThE   prethalamic eminence   163

PS   parastrial nucleus   32-35, 165-166, 189-190

psf   post superior fissure   115-135, 139-143, 162-167, 169, 171-180, 195-198, 200-204, 206

PSol   parasolitary nucleus   142-147, 166

PSTh   parasubthalamic nucleus   64-66, 169

PT   paratenial thalamic nucleus   41-51, 164-165, 195-199

PTe   paraterete nucleus   51-59, 185

PTg   pedunculotegmental nucleus   89-106, 167-171, 173, 188-195

PtPC   parietal cortex, posterior area, caudal part   83-84

PtPD   parietal cortex, posterior area, dorsal part   65-75, 177-180

PtPR   parietal cortex, posterior area, rostral part   68-76

PV   paraventricular thalamic nucleus   51-57, 162-163, 199

PVA   paraventricular thalamic nucleus, anterior part   40-50, 162-164, 192-201

PVG   periventricular gray   68-72, 162-163, 190-196

PVP   paraventricular thalamic nucleus, posterior part   58-67, 162-164, 196-199

Py   pyramidal cell layer of the hippocampus   49-89, 166-171, 173-180, 190-207

py   pyramidal tract   108-157, 162, 164-165, 167

pyx   pyramidal decussation   155-162, 164

R

R   red nucleus   163, 165, 167

r1   rhombomere   1, 162-163, 165, 167, 169, 171, 173

r10   rhombomere   10, 165, 167, 171, 173

r11   rhombomere   11, 165, 173

r1Rt   rhombomere 1 reticular formation   185-187

r2   rhombomere   2, 162-163, 165, 167, 169, 171, 173, 175

r2Rt rhombomere 2 reticular formation 188
r3 rhombomere 3, 162-163, 165, 167, 169, 171, 173, 175
r3Rt rhombomere 3 reticular formation 188
r4 rhombomere 4, 162-163, 165, 167, 169, 171, 173, 175
r5 rhombomere 5, 162-163, 165, 167, 169, 171, 173, 175
r6 rhombomere 6, 162-163, 165, 167, 169, 171, 173
r7 rhombomere 7, 162-163, 165, 167, 169, 171, 173
r8 rhombomere 8, 165, 167, 169, 171, 173
r9 rhombomere 9, 165, 167, 171, 173
Rad radiatum layer of the hippocampus 49-89, 166-180, 189-207
RAmb retroambiguus nucleus 152-160, 181
RAPir rostral amygdalopiriform area 57-63, 180-183
Rbd rhabdoid nucleus 91-103, 162-163, 187-191
rcc rostrum of the corpus callosum 16-18, 163, 200
RCh retrochiasmatic area 42-47, 162-164, 181-183
RChL retrochiasmatic area, lateral part 45-54, 165-167
Re reuniens thalamic nucleus 42-62, 162, 164, 189-192
ReIC recess of the inferior colliculus 104-108, 162, 200-206
REn retroendopirifom nucleus 73-75, 185
REth retroethmoid nucleus 72-77, 171-172, 192-194
rf rhinal fissure 4-111, 164-165, 167, 169, 171, 173, 175, 177-178, 180, 191-201
Rh rhomboid thalamic nucleus 48-60, 162-164, 192-194
RI rostral interstitial nucleus of the medial longitudinal fasciculus 69-72, 164-165, 189-192
ri rhinal incisure 4-9
RIP raphe interpositus nucleus 112-119, 162, 182-183
RIs retroisthmic nucleus 93-97, 171-172, 188-190
RL retrolemniscal nucleus 105-107, 173, 194
RLi rostral linear nucleus (midbrain) 74-84, 162-163, 187, 189-191
RM retromamillary nucleus 74, 164
RMC red nucleus, magnocellular part 78-88, 166, 189-190
RMg raphe magnus nucleus 112-134, 162-164
RML retromamillary nucleus, lateral part 68-72, 166, 184-186
RMM retromammillary nucleus, medial part 67-73, 162-163, 165, 185-186
RMS rostral migratory stream 1-5, 13-19, 166, 196-198
RMS/OV rostral migratory stream/olfactory ventricle 6-12, 165-167, 194-195
rmx retromamillary decussation 69-73, 162-163, 166, 187
Ro nucleus of Roller 135-148, 163-164, 184
ROb raphe obscurus nucleus 130-152, 162, 181-184, 186
RPa raphe pallidus nucleus 134-152, 162
RPC red nucleus, parvicellular part 76-87, 166, 190-192
RPF retroparafascicular nucleus 72-74, 165-166, 195
RRe retroreuniens nucleus 63-66, 162-164, 192
RRF retrorubral field 86-93, 165-171, 188-191
rs rubrospinal tract 86-161, 166-173, 181-183, 185-188

Rt reticular nucleus (prethalamus) 42-67 165-179, 190-201
RtSt reticulostrial nucleus 42-52, 169-173, 198-201
RtTg reticulotegmental nucleus of the pons 98-113, 162-166, 181-186
RtTgL reticulotegmental nucleus of the pons, lateral part 112-114, 162
RtTgP reticulotegmental nucleus of the pons, pericentral part 99-104, 163-166, 183
RVL rostroventrolateral reticular nucleus 133-137, 169-172
RVRG rostral ventral respiratory group 141-151, 170

S

primary somatosensory cortex 65-71, 197
S1BF primary somatosensory cortex, barrel field 36-69, 176-180, 203-207
S1DZ primary somatosensory cortex, dysgranular zone 13-61, 176-180
S1DZO primary somatosensory cortex, dysgranular zone, oral region 14-23, 177-178, 180
S1HL primary somatosensory cortex, hindlimb region 30-52, 169-174
S1J primary somatosensory cortex, jaw region 10-23, 174-180, 200-207
S1Sh primary somatosensory cortex, shoulder region 46-52, 174-177
S1Tr primary somatosensory cortex, trunk region 53-61, 169-178
S1ULp primary somatosensory cortex, upper lip region 14-59, 176-180, 198-207
S2 secondary somatosensory cortex 23-65, 180, 195-203, 206
s5 sensory root of the trigeminal nerve 88-117, 172-173, 175-176, 181-184, 188
Sag sagulum nucleus 102-105, 172-173, 195-196
SC superior colliculus 162
Sc scaphoid thalamic nucleus 70-72, 168-170, 194-196
scc splenium of the corpus callosum 75-76, 162-167
SCh suprachiasmatic nucleus 37-38, 163-164, 182-183
SChDL suprachiasmatic nucleus, dorsolateral part 39-41
SChVM suprachiasmatic nucleus, ventromedial part 39-41, 162
SCO subcommissural organ 70-75, 162-163, 199-201
scp superior cerebellar peduncle 61-82, 87, 96, 98-129, 162-174, 189-195
scpd superior cerebellar peduncle, descending limb 112-115, 189
sf secondary fissure 136-155, 162-167
SFi septofimbrial nucleus 31-42, 162-168, 196-203
SFO subfornical organ 38-48, 162-164, 198-203
SG suprageniculate thalamic nucleus 75-85, 173-174, 197-199
SGe supragenual nucleus of the raphe 116-119, 164, 188
SHi septohippocampal nucleus 14-35, 162-163, 192-203
SHy septohypothalamic nucleus 28-37, 163-166, 189-196
SIB substantia innominata, basal part 25-40, 165-170, 185-188
Sim simple lobule 111-113, 124-135, 167, 173, 175, 180, 195-207
SimA simple lobule A 114-123, 169, 171-179
SimB simple lobule B 114-123, 169, 171-179
simf simplex fissure 121-130

SLu stratum lucidum of the hippocampus 49-82, 166-169, 171-179, 187-193, 195-196, 198-206
SM nucleus of stria medullaris 42-45, 167-169, 187-191
sm stria medullaris 38-69, 163-169, 187-203
SMT submamillothalamic nucleus 66-68, 165, 185-186
SMV superior medullary velum 106, 111-112, 114, 116, 122-123, 126, 128, 162, 166, 192, 194, 197-199
SMV/4n superior medullary velum and trochlear nerve 165
SNC substantia nigra, compact part 185-186
SNCD substantia nigra, compact part, dorsal tier 71-88, 167-172, 187-191
SNCM substantia nigra, compact part, medial tier 81-88, 166-167
SNCV subtantia nigra, compact part, ventral tier 81-88, 170
SNL substantia nigra, lateral part 71-87, 172-175, 189-191
SNR substantia nigra, reticular part 69-89, 166-169, 171-174, 184-191
SO supraoptic nucleus 33-48, 165-170, 182-184
Sol nucleus of the solitary tract 125, 170, 182-188
sol solitary tract 128-135, 137-157, 165-167, 185, 187
SolC solitary nucleus, commissural part 147-162, 184-186
SolCe solitary nucleus, central part 143-147, 165
SolDL solitary nucleus, dorsolateral part 141-156
SolDM solitary nucleus, dorsomedial part 131-140
SolG solitary nucleus, gelatinous part 144-148, 165
SolI solitary nucleus, interstitial part 141-150, 167
SolIM solitary nucleus, intermediate part 126-151, 169
SolL solitary nucleus, lateral part 129-143, 168
SolM solitary nucleus, medial part 129-161, 163-168
SolRC solitary nucleus, retrocentral part 133-137
SolRL solitary nucleus, rostrolateral part 127-130
SolV solitary nucleus, ventral part 129-157, 165-166
SolVL solitary nucleus, ventrolateral part 135-157, 165, 167-169
SOR supraoptic nucleus, retrochiasmatic part 49-54, 168-169, 181
sox supraoptic decussation 38-71, 162-173, 175-177, 181-192
sp5 spinal trigeminal tract 118-161, 169-175, 181-189
Sp5C spinal trigeminal nucleus, caudal part 149-161, 168-173, 181-187
Sp5I spinal trigeminal nucleus, interpolar part 129-151, 171-175, 181-187
Sp5O spinal trigeminal nucleus, oral part 120-130, 172-175, 181-187
SPa subparaventricular zone of the hypothalamus 41-50, 163, 187
SPF subparafascicular thalamic nucleus 65-68, 164-165, 191-192
SPFPC subparafascicular thalamic nucleus, parvicellular part 68-76, 167, 169-172, 174, 192
Sph sphenoid nucleus 111-113, 162-164, 190-191
SPO superior paraolivary nucleus 110-120, 167-168, 181
SPTg subpeduncular tegmental nucleus 96-104, 165-168, 189-191
SpVe spinal vestibular nucleus 125-142, 167-172, 187-191
st stria terminalis 37-67, 167-180, 185-201

ST   bed nucleus of stria terminalis   197

StA   strial part of the preoptic area   33-35, 189-190

STD   bed nucleus of stria terminalis, dorsal division   194-199

Stg   stigmoid hypothalamic nucleus   50-52, 187

STh   subthalamic nucleus   59-68, 170-174, 186-190

StHy   striohypothalamic nucleus   38-42, 164, 166, 188-191

STIA   bed nucleus of the stria terminalis, intraamygdaloid division   52-63, 175-177, 184-188

STL   bed nucleus of stria terminalis, lateral division   167, 169-170, 191

STLD   bed nucleus of the stria terminalis, lateral division, dorsal part   31-35, 193-195

STLI   bed nucleus of the stria terminalis, lateral division, intermediate part   35-38, 166, 168, 192

STLJ   bed nucleus of the stria terminalis, lateral division, juxtacapsular part   34-36, 192-195

STLP   bed nucleus of the stria terminalis, lateral division, posterior part   30-40, 192, 194-195

STLV   bed nucleus of the stria terminalis, lateral division, ventral part   30-37, 189-190

STM   bed nucleus of stria terminalis, medial division   190

STMA   bed nucleus of the stria terminalis, medial division, anterior part   26-34, 165-168, 191-195

STMAL   bed nucleus of the stria terminalis, medial division, anterolateral part   34-37

STMAM   bed nucleus of the stria terminalis, medial division, anteromedial part   34-37

STMP   bed nucleus of the stria terminalis, medial division, posterior part   42, 165, 167, 189-190

STMPI   bed nucleus of the stria terminalis, medial division, posterointermediate part   38-42, 166, 168, 191-196

STMPL   bed nucleus of the stria terminalis, medial division, posterolateral part   38-44, 166, 168, 191-196

STMPM   bed nucleus of the stria terminalis, medial division, posteromedial part   38-45, 165-166, 168-169, 188, 191-196

STMV   bed nucleus of the stria terminalis, medial division, ventral part   30-36, 165-168, 189

STr   subiculum, transition area   91-96, 194-201

str   superior thalamic radiation   66-73, 171-177, 194, 196-200

STS   bed nucleus of stria terminalis, supracapsular division   169

STSL   bed nucleus of stria terminalis, supracapsular division, lateral part   45-46

STSM   bed nucleus of stria terminalis, supracapsular division, medial part   45-46

Su3   supraoculomotor periaqueductal gray   83-92, 162-164, 195

Su3C   supraoculomotor cap   84-92, 164, 195

Su5   supratrigeminal nucleus   106-113, 168-171, 188-189

Sub   submedius thalamic nucleus   49-51, 57-60, 162, 164, 166, 191-194

SubB   subbrachial nucleus   83-92, 173-176, 192-194

SubCA   subcoeruleus nucleus, alpha part   109-112, 166-167, 188, 190-191

SubCD   subcoeruleus nucleus, dorsal part   107-116, 166-167, 169, 187-189

SubCV   subcoeruleus nucleus, ventral part   107-116, 167-169, 181-185

SubD   submedius thalamic nucleus, dorsal part   52-56, 165

SubG   subgeniculate nucleus of prethalamus   67-72, 176-178, 191-195

SubI   subincertal nucleus   56-60, 167, 169, 187-188

SubP   area subpostrema   146-152

SubV   submedius thalamic nucleus, ventral part   52-56, 165, 190

SuG   superficial gray layer of the superior colliculus   77-98, 162-172, 203-206

SuG/Zo   superficial gray and zonal layers of the superior colliculus   202-203

SuL   supralemniscal nucleus   90-96, 165-166, 168, 186-187

SuS   superior salivatory nucleus   118-128, 169, 181, 183

SuVe   superior vestibular nucleus   114-123, 169-172, 189-192

T

Te   terete hypothalamic nucleus   58-66, 168, 181-182

TeA   temporal association cortex   71-106, 193, 195-207

tfp   transverse fibers of the pons   88-91, 93-107, 162-165, 167-169, 181-183

TG   tectal gray   72-81, 165-173, 198-203

TrLL   triangular nucleus of the lateral lemniscus   98-105

TS   triangular septal nucleus   35-45, 162-166, 196-205

ts   tectospinal tract   85-166, 181-184, 186-190, 193, 196

tth   trigeminothalamic tract   83-128, 163

Tu   olfactory tubercle   16, 35, 164-167, 169, 171, 173, 175, 188-189

Tu1   olfactory tubercle, layer   1, 10-34, 168, 174

Tu3   olfactory tubercle, layer   3, 163, 170, 172, 182-187

TuLH   tuberal region of lateral hypothalamus   47-62, 165-169, 181-183

TuM   tuberomamillary nucleus   170

Tz   nucleus of the trapezoid body   107-120, 165-167

tz   trapezoid body   105-125, 162-165, 167, 169, 171, 173-177, 181-186

tzx   decussation of the trapezoid body   118

U

un   uncinate fasciculus of the cerebellum   115-122, 168-169, 171, 194

unx   decussation of the uncinate fasciculus of the cerebellum   123-126, 128, 195

V

v   vein   92-93, 95, 98, 101-102, 105-107

V1   primary visual cortex   70-81, 177-180

V1B   primary visual cortex, binocular area   82-111, 173-180

V1M   primary visual cortex, monocular area   82-111, 165, 167-176

V2L   secondary visual cortex, lateral area   71-111, 177-180, 204-207

V2ML   secondary visual cortex, mediolateral area   68-98, 169-176

V2MM   secondary visual cortex, mediomedial area   68-107, 165-171

VA   ventral anterior thalamic nucleus   46, 48-52, 166-169, 192-193, 195-196

VA/VL   region where VA and VL overlap   47

VCA   ventral cochlear nucleus, anterior part   109-121, 175-177, 184-190

VCl   ventral claustrum   15-49, 190-192

VCP   ventral cochlear nucleus, posterior part   120-122, 125-127, 175-177, 184-188

VCPO   ventral cochlear nucleus, posterior part, octopus cell area   123-125

VDB   nucleus of the vertical limb of the diagonal band   18-29, 162-164, 188-192

VeCb   vestibulocerebellar nucleus   122-126, 166-168, 193-194

veme   vestibulomesencephalic tract   116-122, 168

VEn   ventral entopiriform nucleus   40-58, 177-180, 182-185

vert   vertebral artery   137, 139-145, 147-161

vesp   vestibulospinal tract   123-126, 168

vhc   ventral hippocampal commissure   38-48, 162-167, 199-204

VIEnt   ventral intermediate entorhinal cortex   85-100, 182-199

VL   ventrolateral thalamic nucleus   48-59, 166-172, 192-199

VLH   ventrolateral hypothalamic nucleus   39-46, 167-169, 184

vlh   ventrolateral hypothalamic tract   47-56

VLi   ventral linear nucleus of the thalamus   69-74, 169-172, 192

VLL   ventral nucleus of the lateral lemniscus   97-108, 170-173, 181-189

VLPAG   ventrolateral periaqueductal gray   88-106, 162, 165-166, 195-197

VLPO   ventrolateral preoptic nucleus   33-38, 165-167, 184

VM   ventromedial thalamic nucleus   48-64, 165-170, 189-192

VMH   ventromedial hypothalamic nucleus   47-49, 60-61, 164-165, 182

VMHC   ventromedial hypothalamic nucleus, central part   50-59

VMHDM   ventromedial hypothalamic nucleus, dorsomedial part   50-59, 163, 183

VMHSh   ventromedial hypothalamic nucleus, shell region   47-61, 164-166

VMHVL   ventromedial hypothalamic nucleus, ventrolateral part   50-59, 181

VMPO   ventromedial preoptic nucleus   31-36, 163-164, 184

VO   ventral orbital cortex   5-11, 165-171, 173-175, 194-205

VOLT   vascular organ of the lamina terminalis   28-33, 162, 186-187

VP   ventral pallidum   11-42, 165-174, 182-190

VPL   ventral posterolateral thalamic nucleus   51-70, 171-177, 191-199

VPM   ventral posteromedial thalamic nucleus   54-73, 169-175, 192-199

VPPC   ventral posterior nucleus of the thalamus, parvicellular   62-69, 165-169, 192

VRe   ventral reuniens thalamic nucleus   43-61, 164-165, 189-192

VS   ventral subiculum   71-90, 174-180, 182-193

vsc   ventral spinocerebellar tract   105-161, 167-174, 187-194

VTA   ventral tegmental area   89-90, 164-166, 186

VTAR   ventral tegmental area, rostral part   72-75, 164, 166, 168, 186-188

VTg   ventral tegmental nucleus   100-105, 164, 190-191

vtgx ventral tegmental decussation 78-86, 162-165, 187-190
VTM ventral tuberomamillary nucleus 64-74, 165-168, 181-183
VTT ventral tenia tecta 7, 164-165, 189-192, 194
VTT1 ventral taenia tecta, layer 1, 8-9
VTT2 ventral taenia tecta, layer 2, 63

X

X nucleus X 125-139, 170-171, 173, 187-190
Xi xiphoid thalamic nucleus 45-48, 51-53, 188-190
xicp decussation of the inferior cerebellar peduncle 122-124

xscp decussation of the superior cerebellar peduncle 86-98, 165, 187-191

Y

Y nucleus Y of the vestibular complex 124-127, 171-173, 192

Z

Z nucleus Z 140-142, 165
ZI zona incerta 54-55, 165, 167-169, 176, 188-189, 193

ZIC zona incerta, caudal part 70-76, 171-173, 175, 190-192
ZID zona incerta, dorsal part 56-69, 170-175, 189-192, 194
ZIR zona incerta, rostral part 49-53, 165-167, 169, 189
ZIV zona incerta, ventral part 56-69, 170-174, 189-192
ZL zona limitans 19, 30-32
Zo zonal layer of superior colliculus 77-99, 162-173, 204-206
Zo/SuG zonal layer and superficial gray layer of superior colliculus 207

Figures

Coronal sections of the brain   Figures 1-161

Plate 1

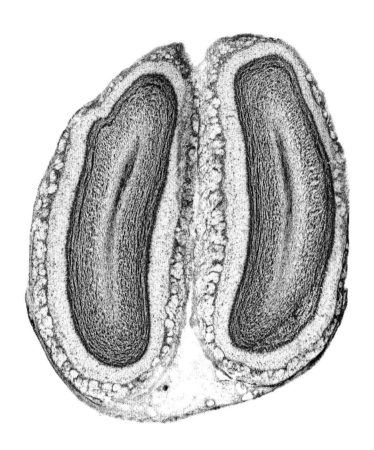

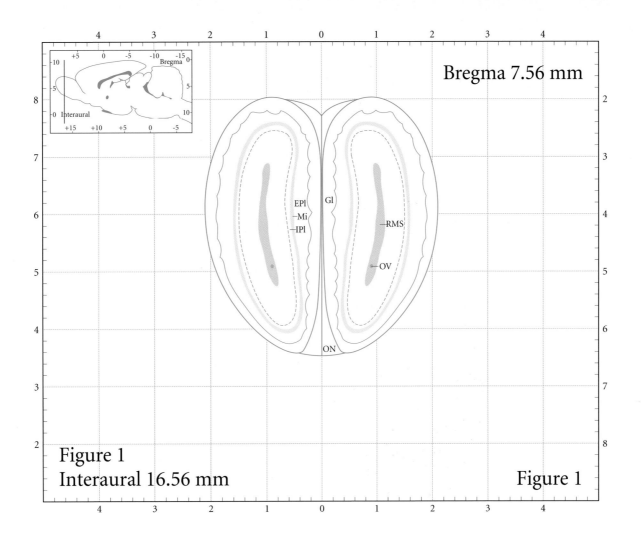

Figure 1
Interaural 16.56 mm

Bregma 7.56 mm

Figure 1

EPl external plexiform layer olf
Gl   glomerular layer of the olfactory
IPl internal plexiform layer of olf
Mi   mitral cell layer of the olf bulb
ON   olfactory nerve layer
OV   olfactory ventricle
RMS rostral migratory stream

Plate 2

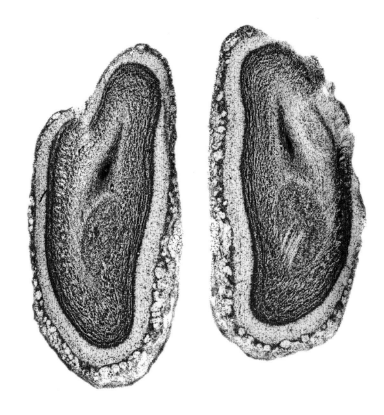

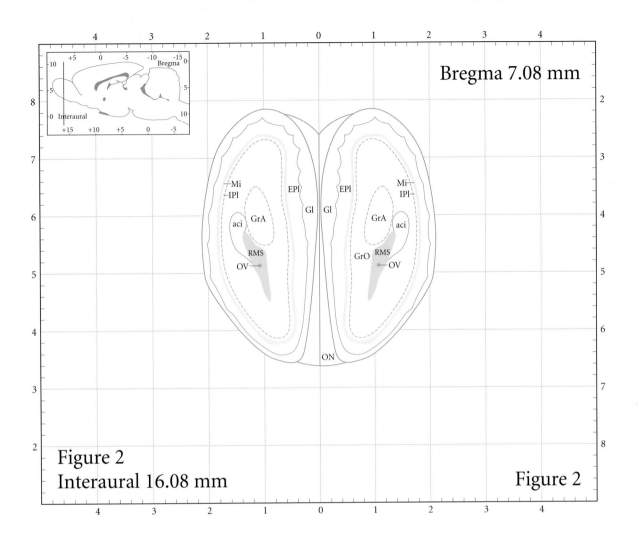

Figure 2
Interaural 16.08 mm

Bregma 7.08 mm

Figure 2

aci anterior commissure, intrabulbar
EPl external plexiform layer olf
Gl glomerular layer of the olfactory
GrA granule cell layer of access olf
GrO granule cell layer of olfactory
IPl internal plexiform layer of olf
Mi mitral cell layer of the olf bulb
ON olfactory nerve layer
OV olfactory ventricle
RMS rostral migratory stream

Plate 3

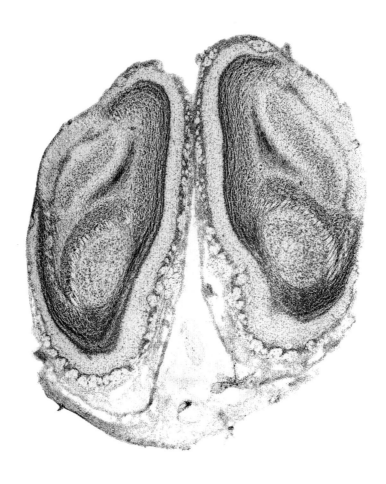

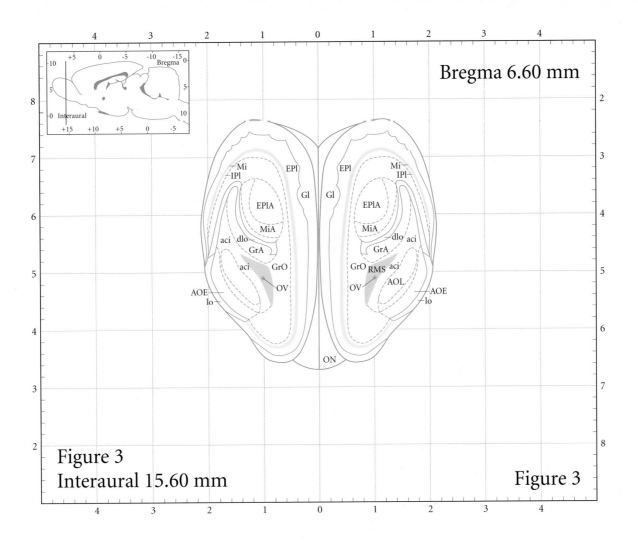

Bregma 6.60 mm

Figure 3
Interaural 15.60 mm

Figure 3

aci anterior commissure, intrabulbar	GrO granule cell layer of olfactory
AOE anterior olfactory area, external	IPl internal plexiform layer of olf
AOL anterior olfactory area, later	lo lateral olfactory tract
dlo dorsal lateral olfactory tract	Mi mitral cell layer of the olf bulb
EPl external plexiform layer olf	MiA mitral cell layer accessory olfa
EPlA external plexiform layer acc	ON olfactory nerve layer
Gl glomerular layer of the olfactory	OV olfactory ventricle
GrA granule cell layer of access olf	RMS rostral migratory stream

Plate 4

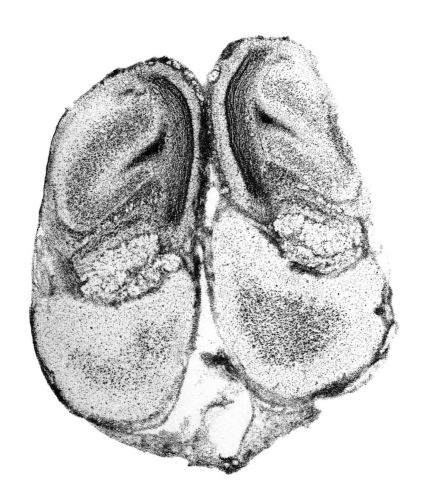

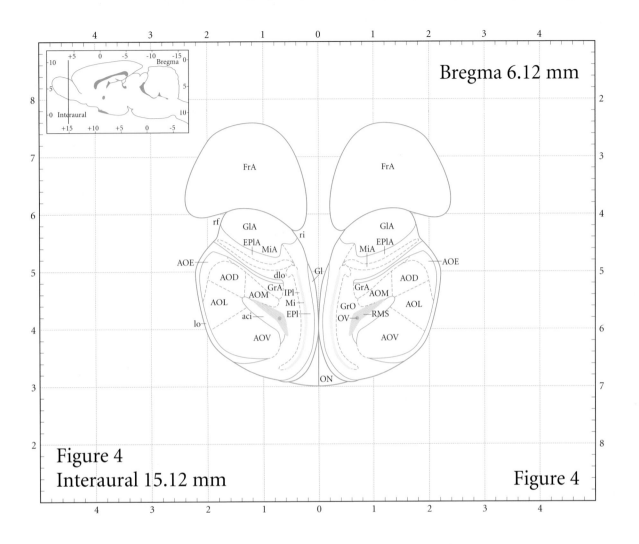

Figure 4
Interaural 15.12 mm
Bregma 6.12 mm

aci	anterior commissure, intrabulbar
AOD	anterior olfactory area, dorsal
AOE	anterior olfactory area, external
AOL	anterior olfactory area, later
AOM	anterior olfactory area, medi
AOV	anterior olfactory area, ventral
dlo	dorsal lateral olfactory tract
EPl	external plexiform layer olf
EPlA	external plexiform layer acc
FrA	frontal association cortex
Gl	glomerular layer of the olfactory
GlA	glomerular layer of the access
GrA	granule cell layer of access olf
GrO	granule cell layer of olfactory
IPl	internal plexiform layer of olf
lo	lateral olfactory tract
Mi	mitral cell layer of the olf bulb
MiA	mitral cell layer accessory olfa
ON	olfactory nerve layer
OV	olfactory ventricle
rf	rhinal fissure
ri	rhinal incisure
RMS	rostral migratory stream

Plate 5

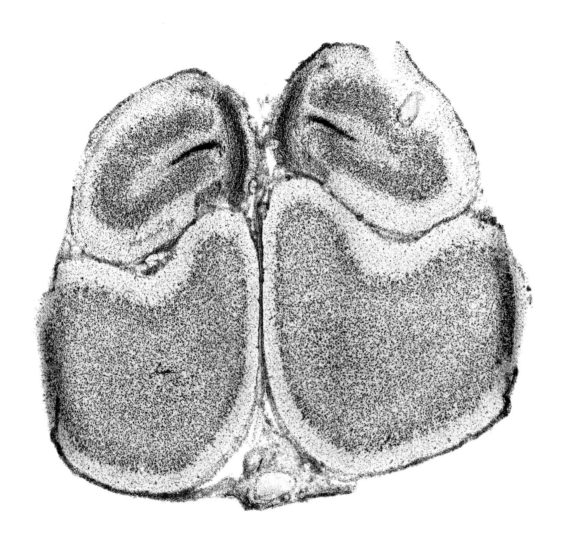

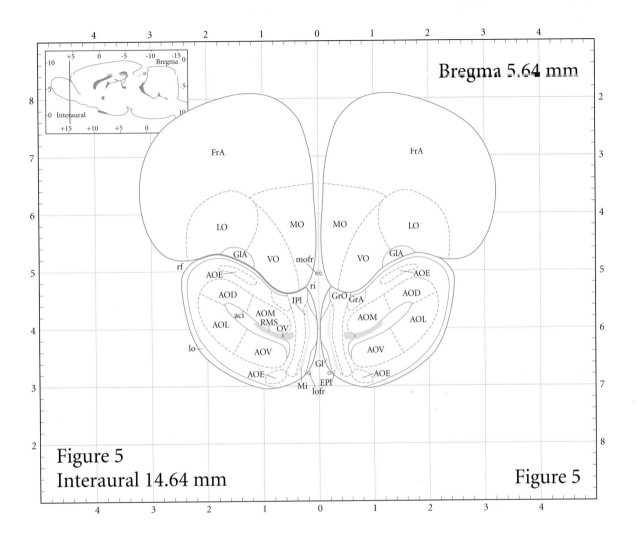

Figure 5
Interaural 14.64 mm
Bregma 5.64 mm

aci anterior commissure, intrabulbar
AOD anterior olfactory area, dorsal
AOE anterior olfactory area, external
AOL anterior olfactory area, later
AOM anterior olfactory area, medi
AOV anterior olfactory area, ventral
EPl external plexiform layer olf
FrA frontal association cortex
Gl glomerular layer of the olfactory
GlA glomerular layer of the access
GrA granule cell layer of access olf
GrO granule cell layer of olfactory
IPl internal plexiform layer of olf
LO lateral orbital cortex
lo lateral olfactory tract
lofr lateral orbitofrontal artery
Mi mitral cell layer of the olf bulb
MO medial orbital cortex
mofr medial orbitofrontal artery
OV olfactory ventricle
rf rhinal fissure
ri rhinal incisure
RMS rostral migratory stream
VO ventral orbital cortex

Plate 6

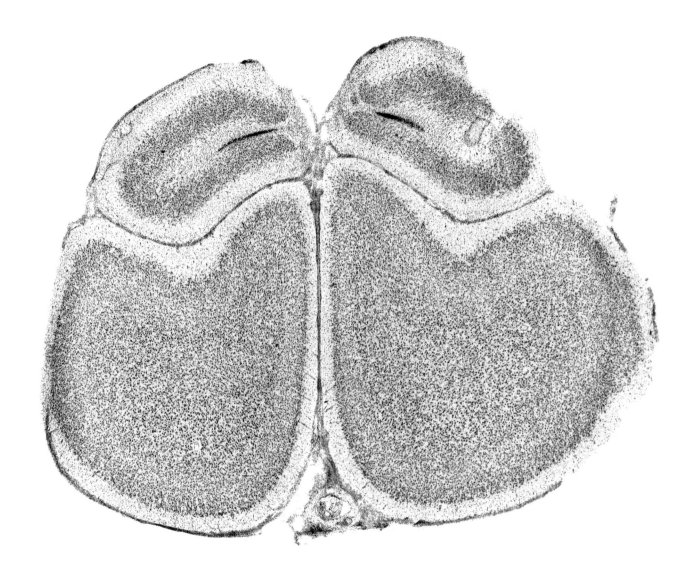

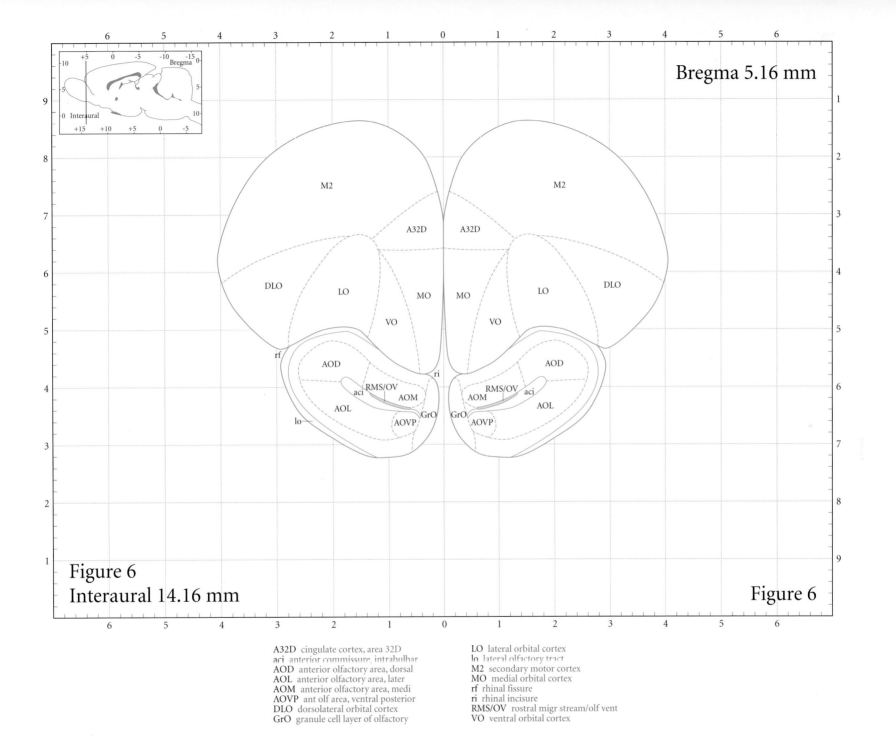

Plate 7

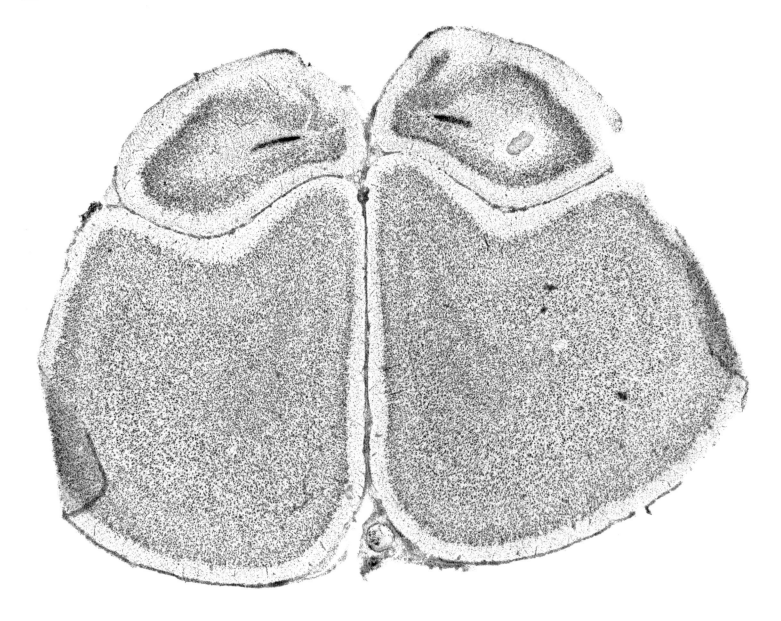

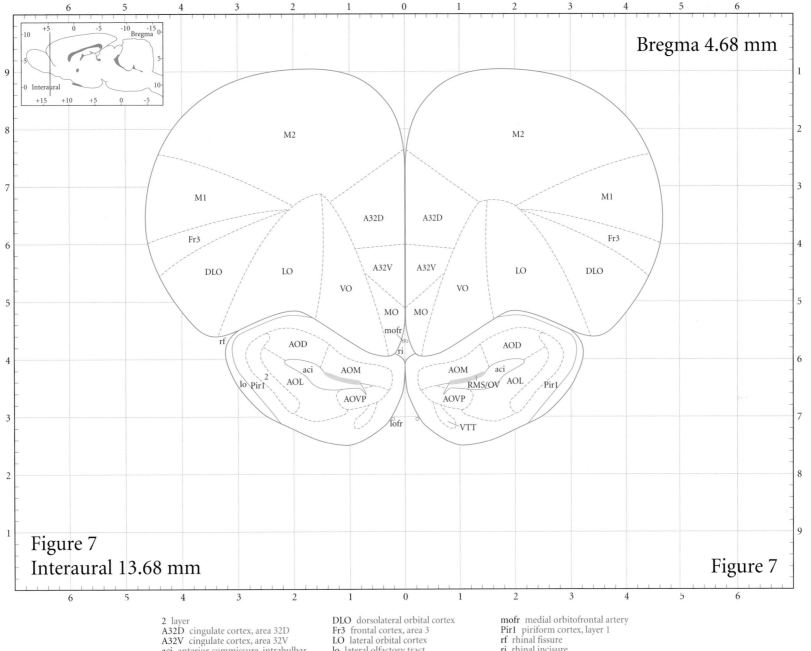

Plate 8

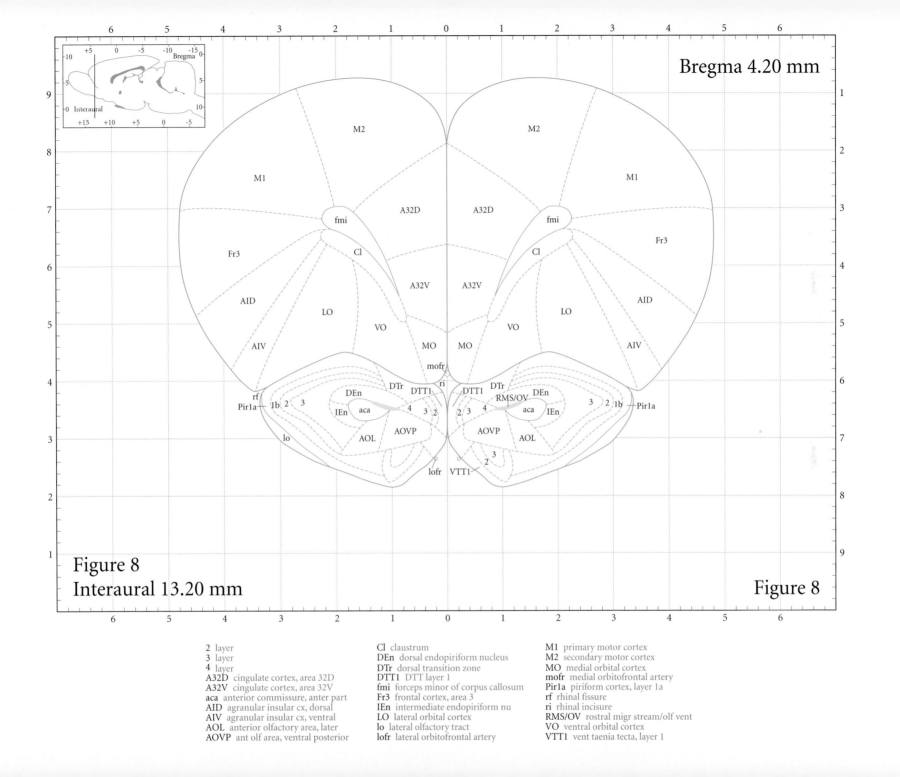

Plate 9

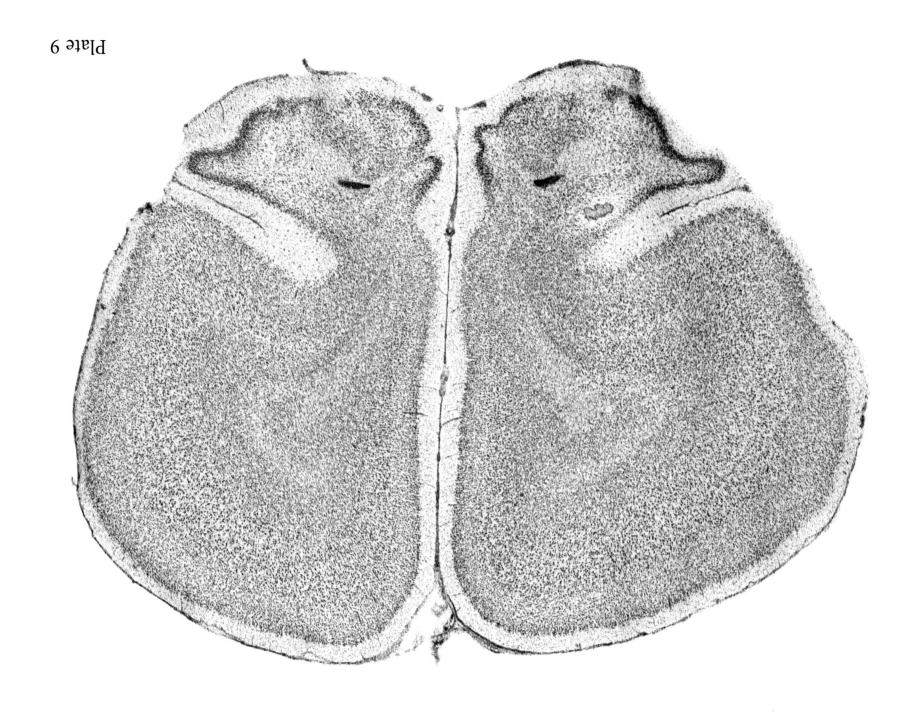

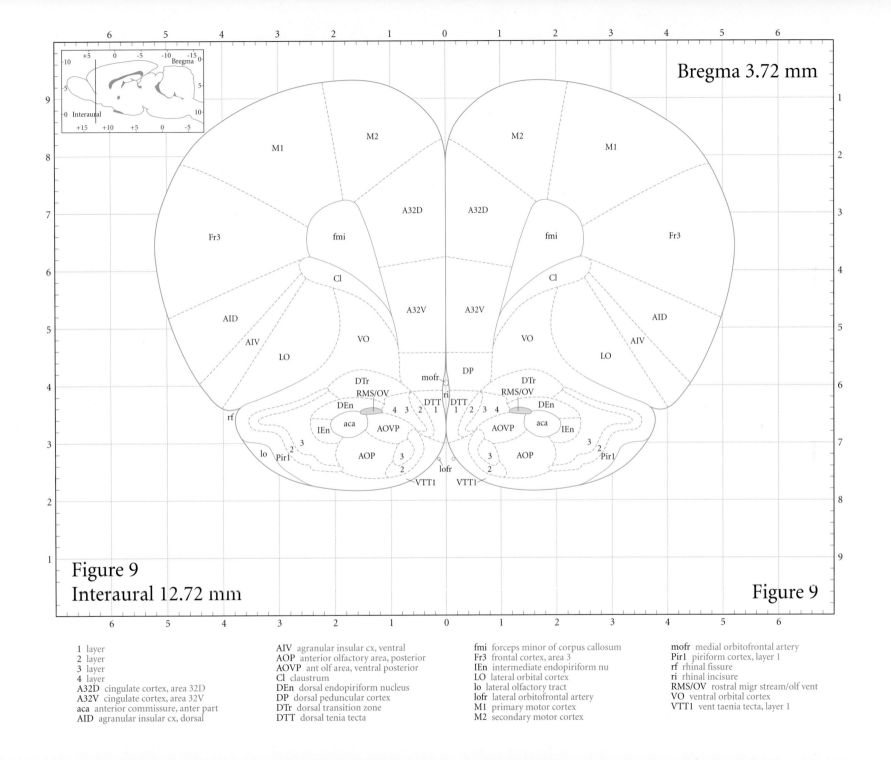

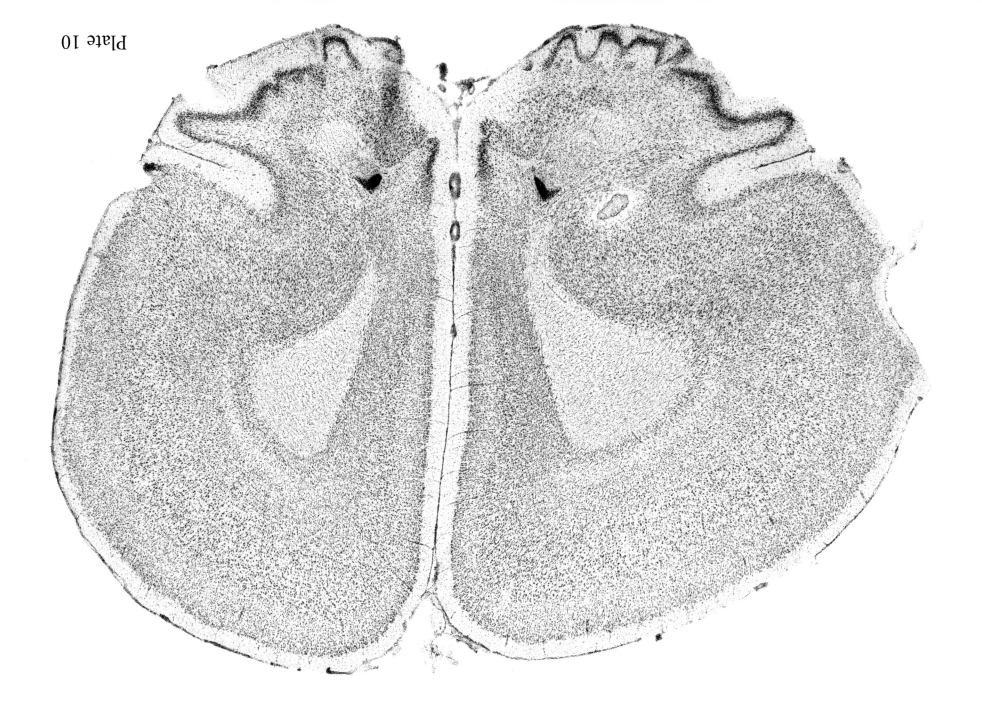

Plate 10

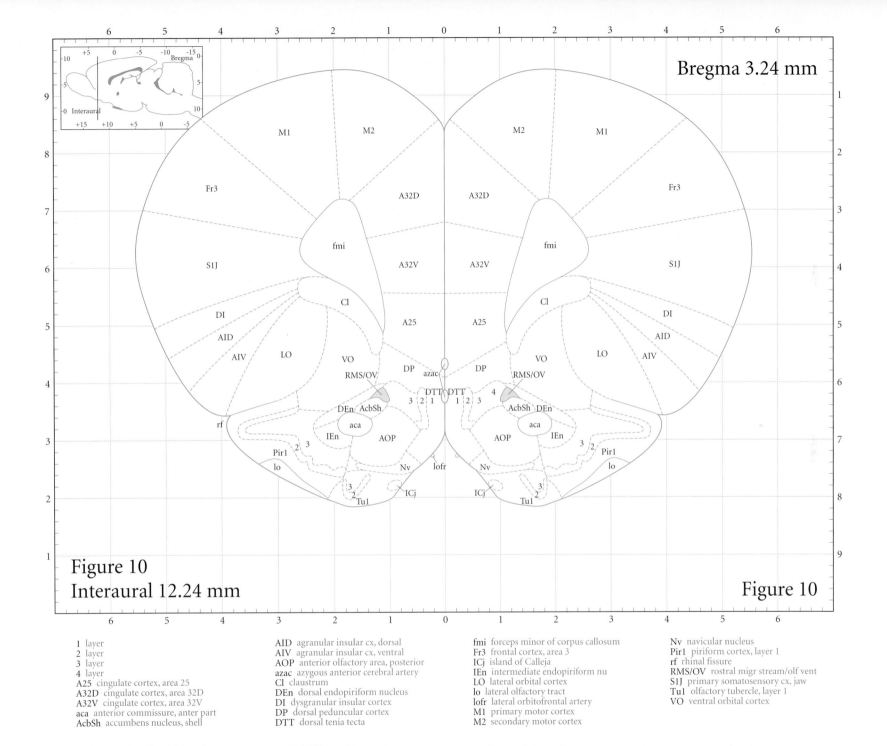

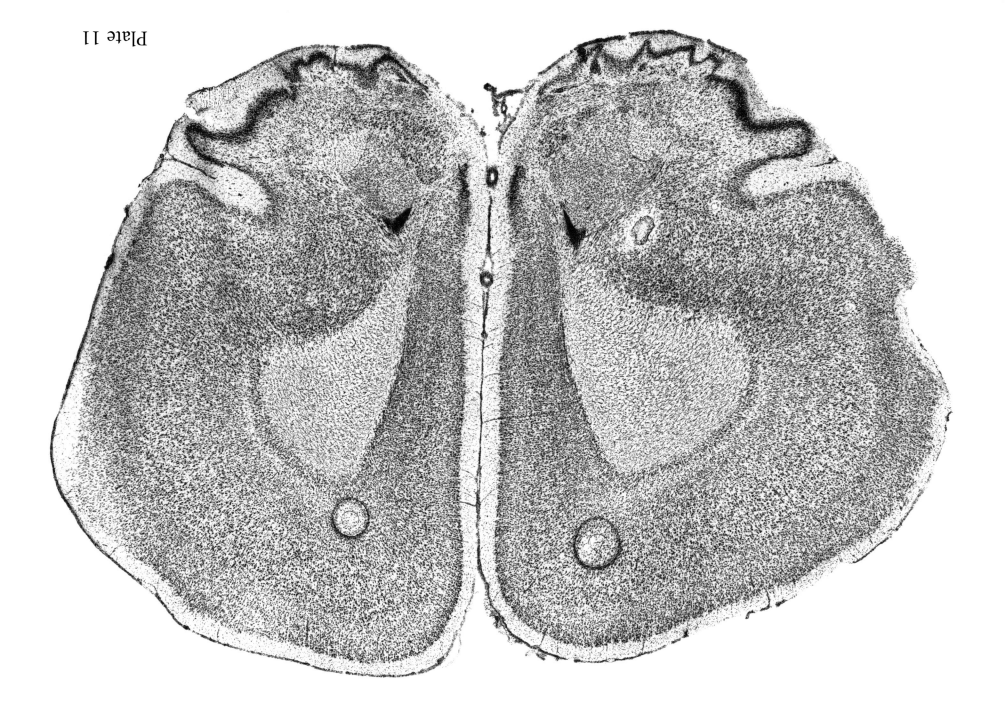

Plate 11

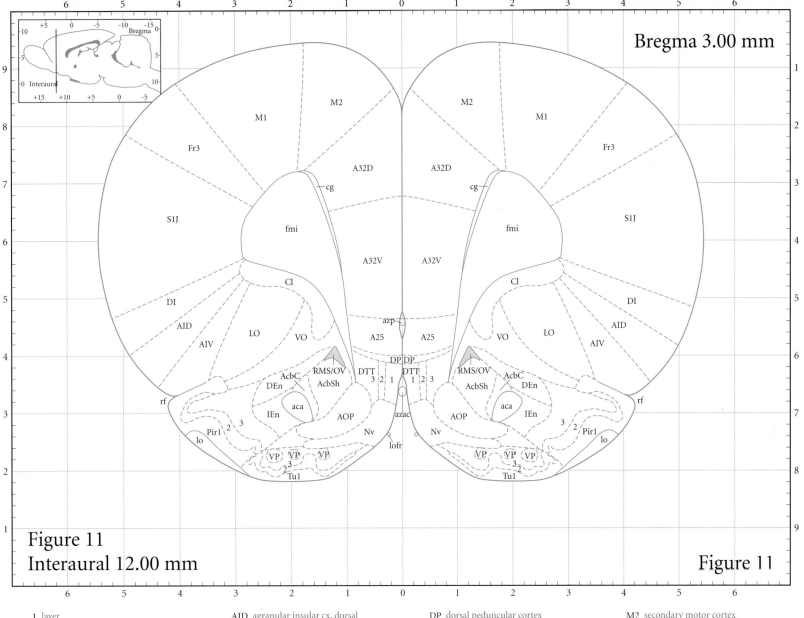

Plate 12

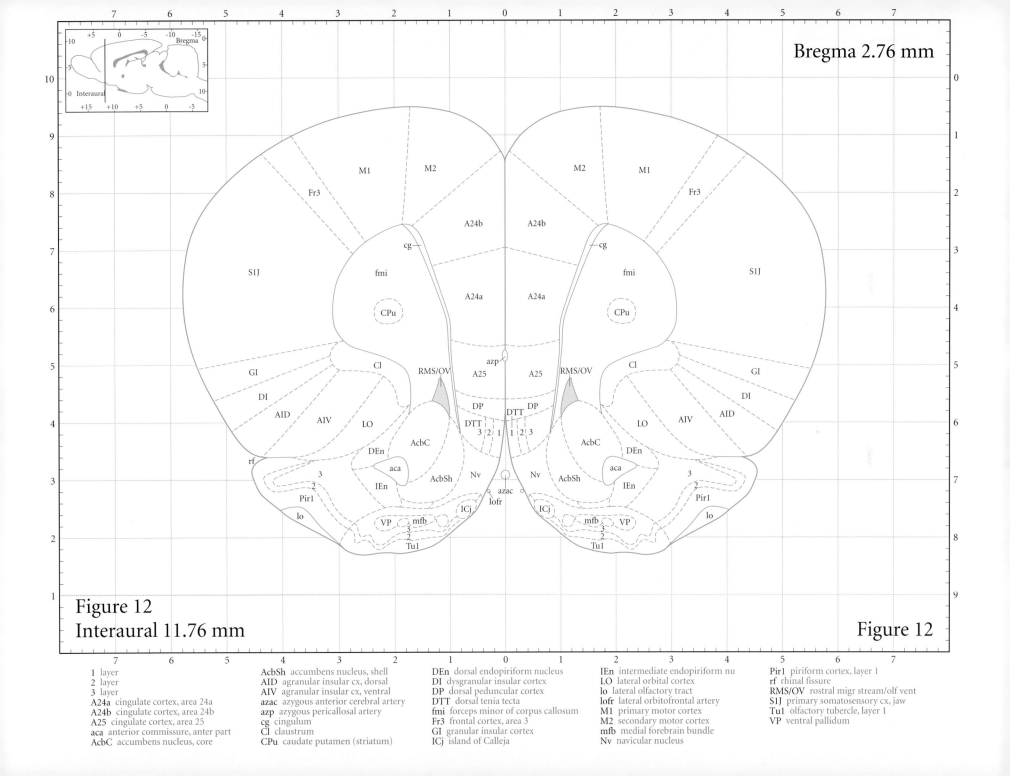

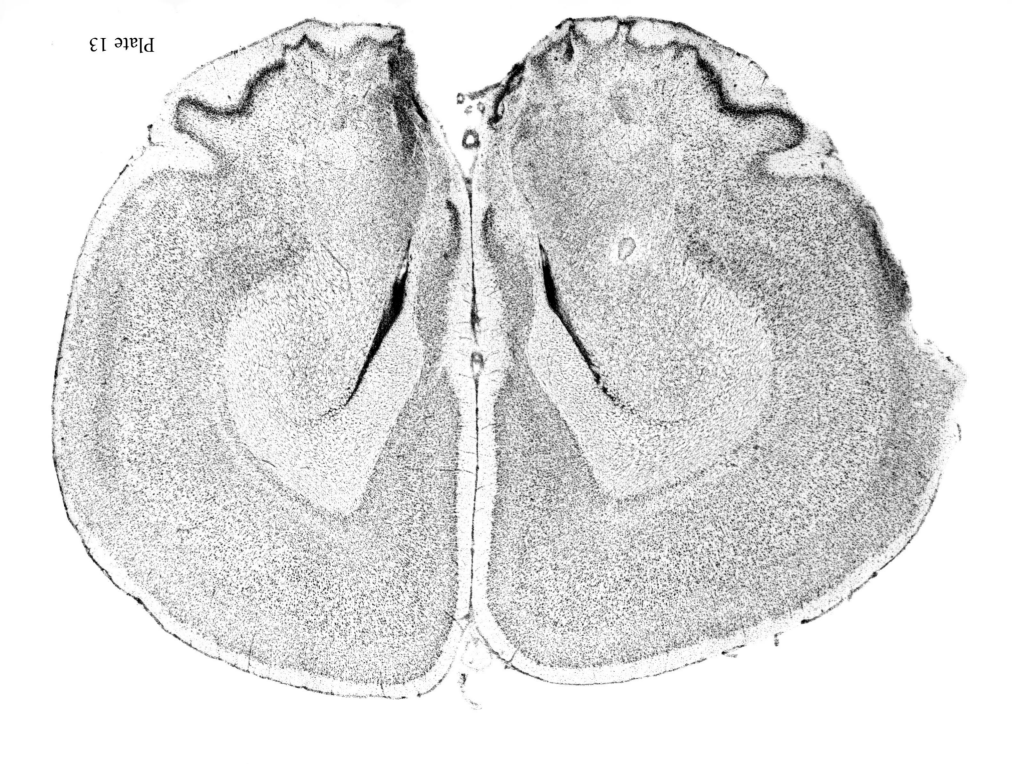

Plate 13

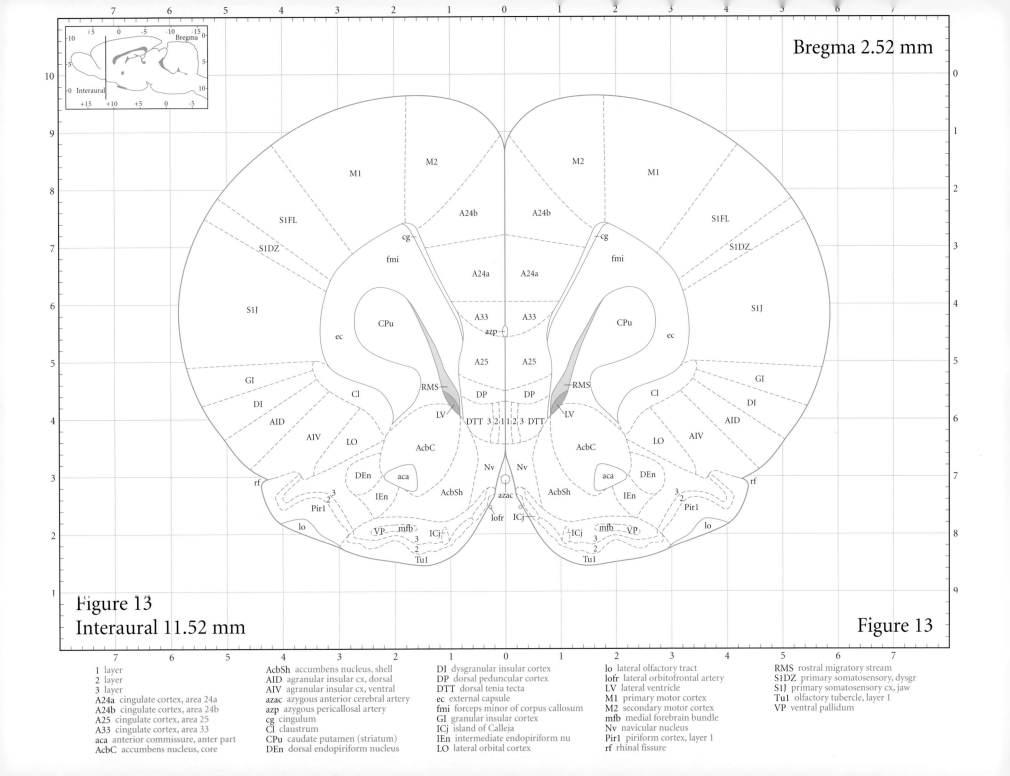

Figure 13
Interaural 11.52 mm
Bregma 2.52 mm

Plate 14

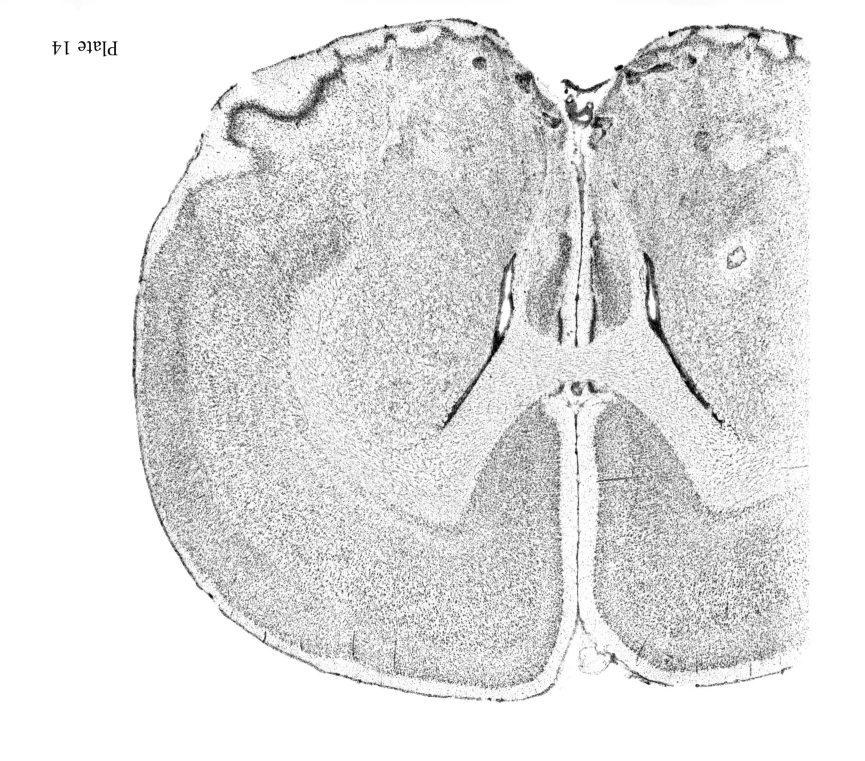

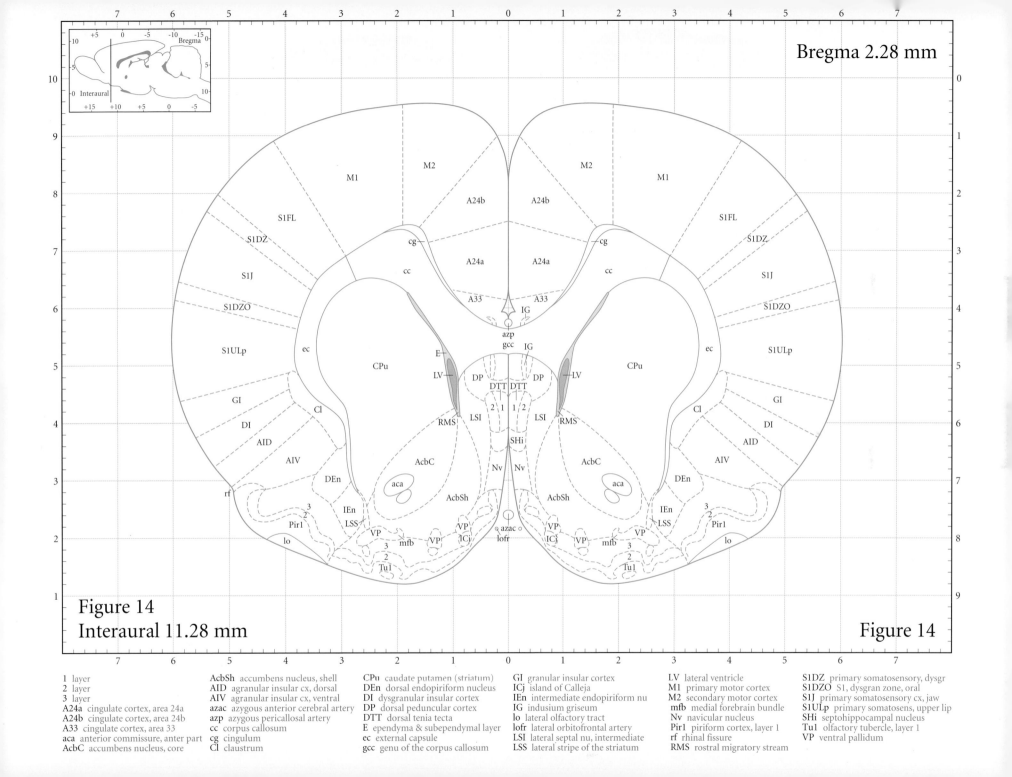

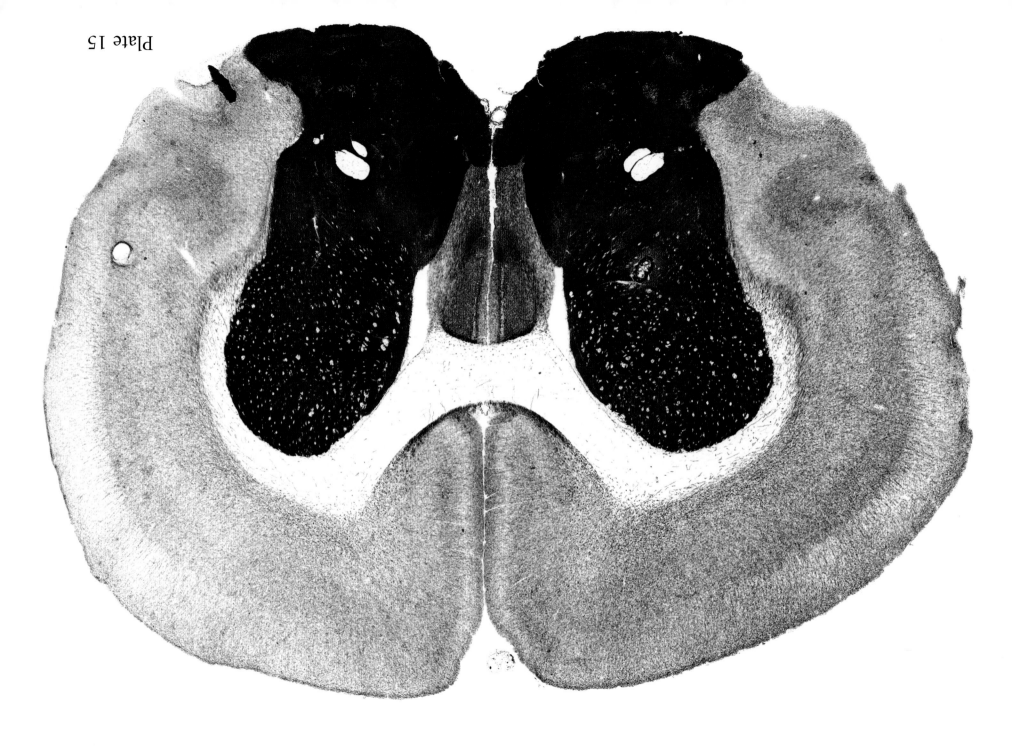

Plate 15

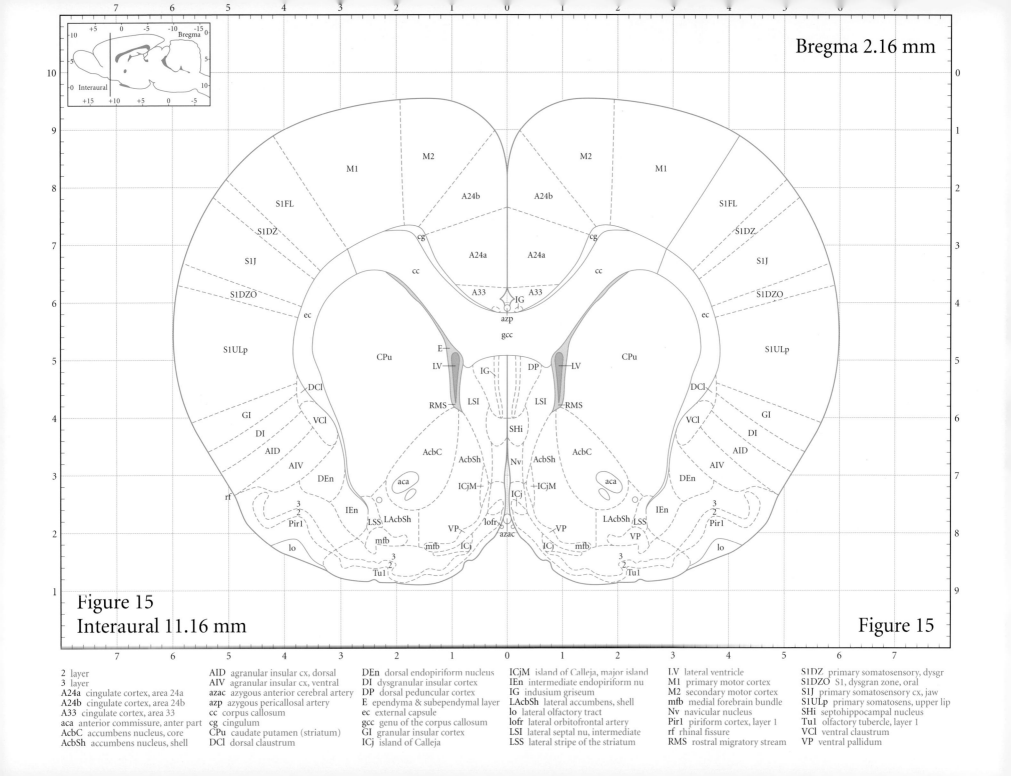

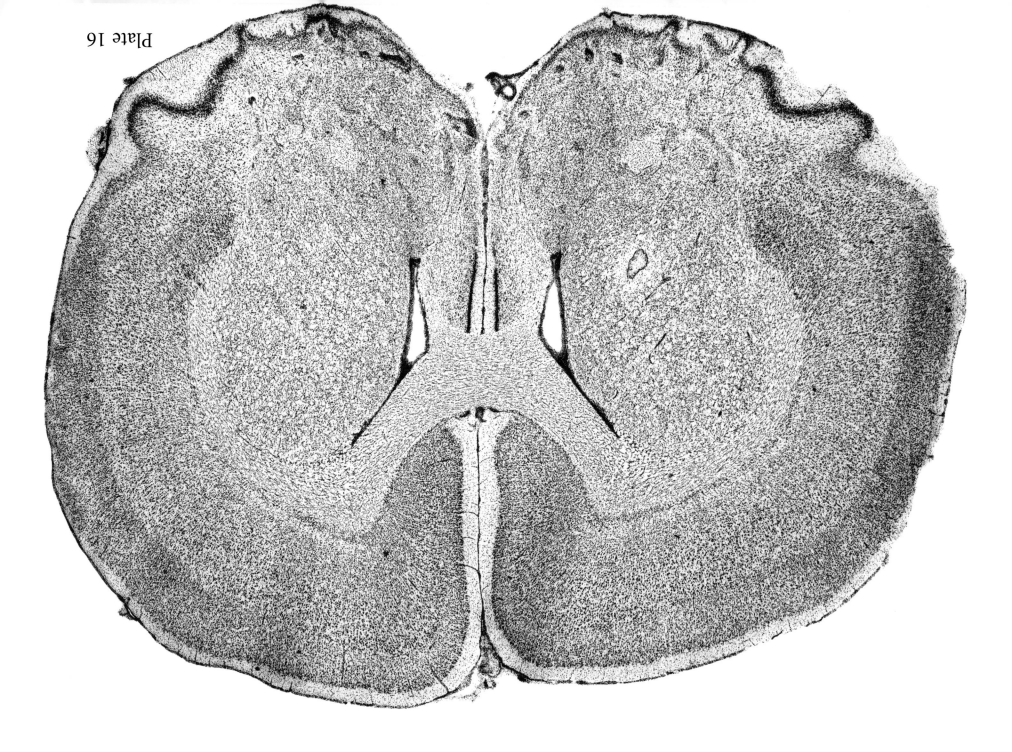

Plate 16

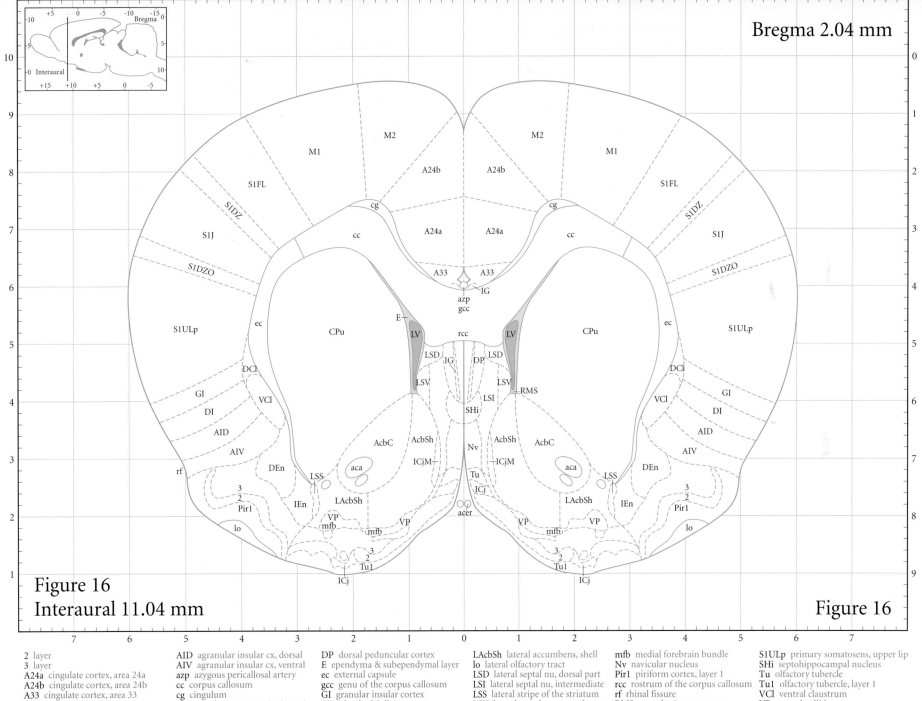

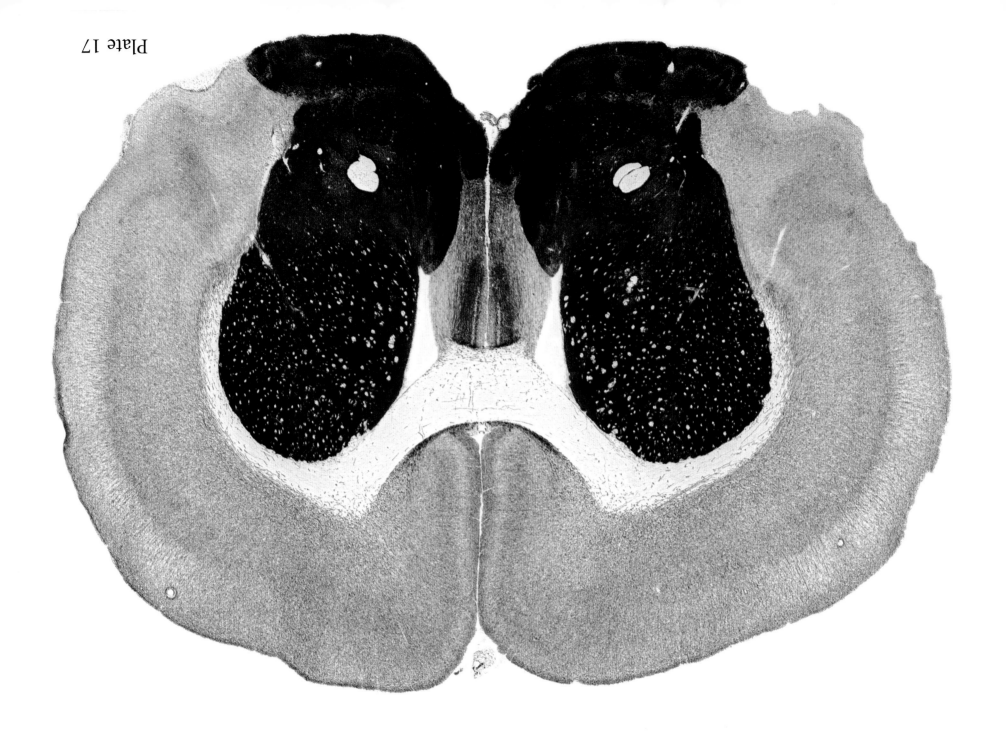

Plate 17

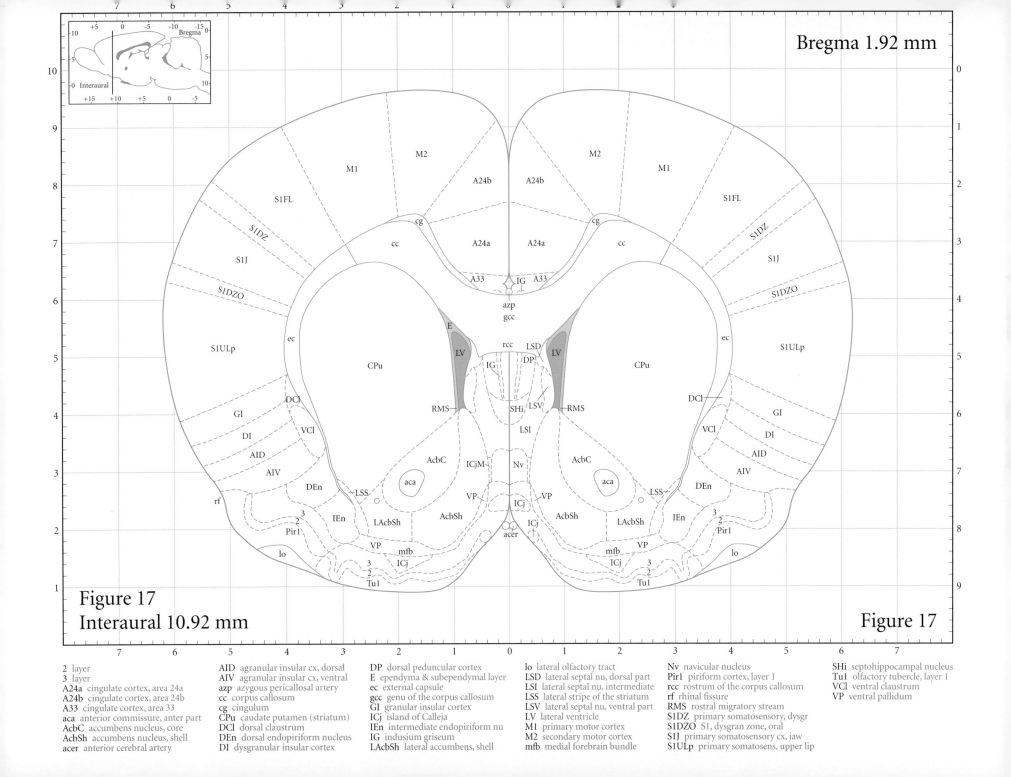

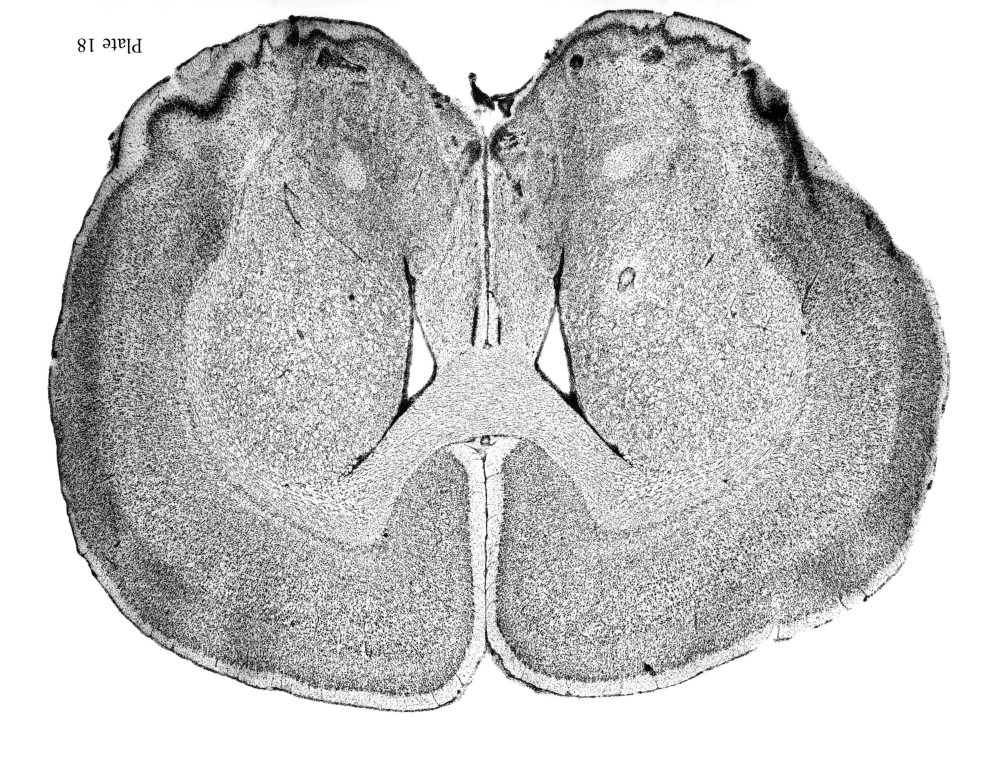

Plate 18

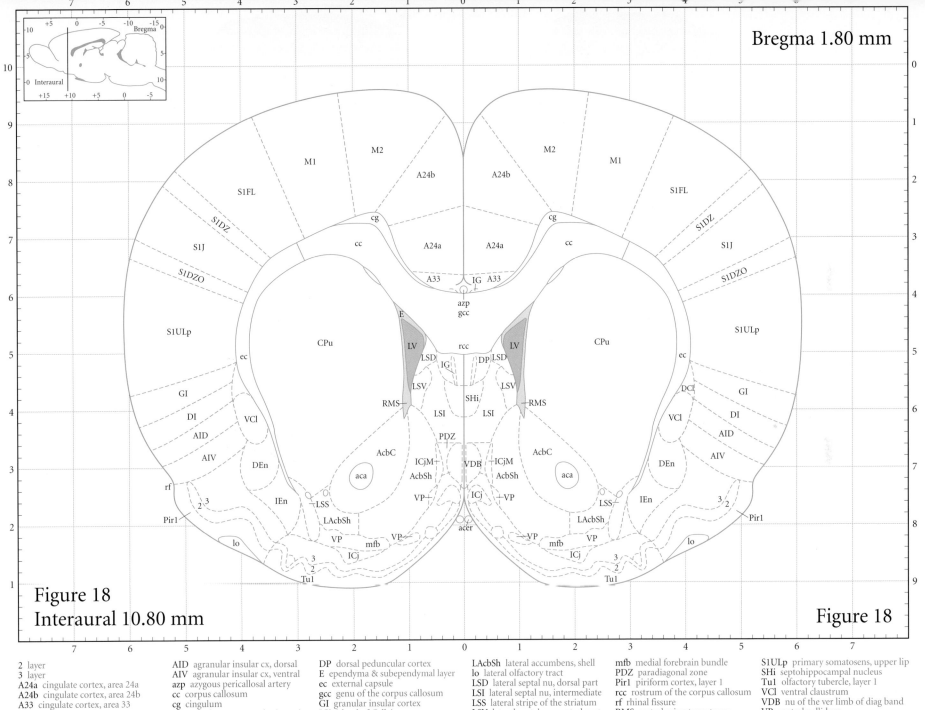

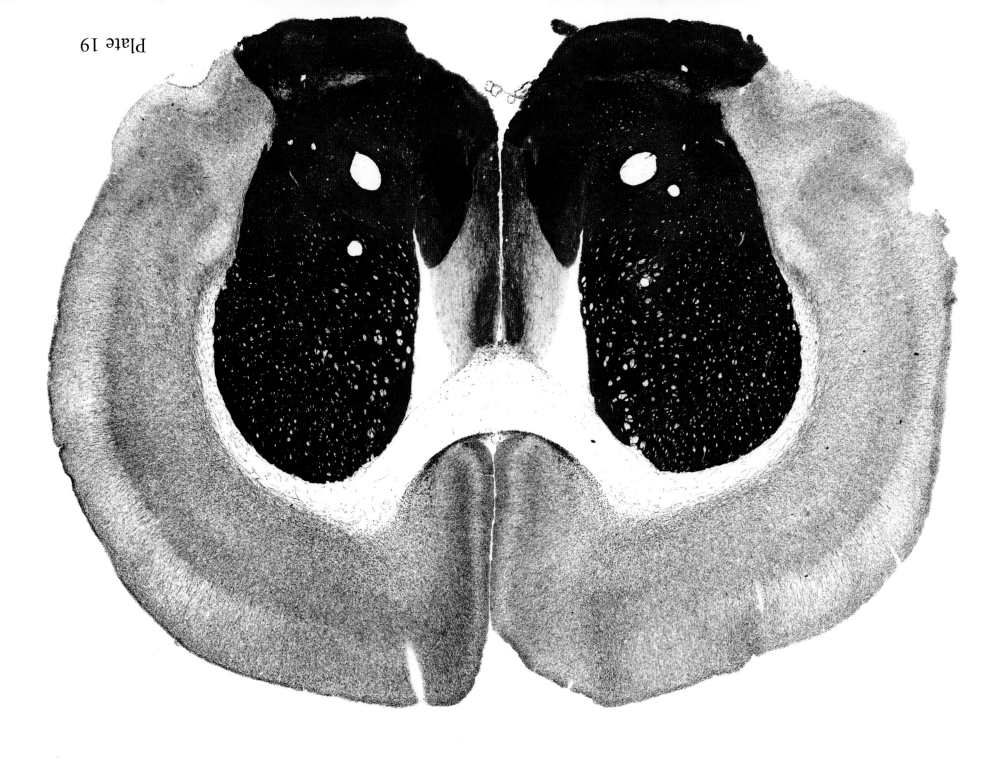

Plate 19

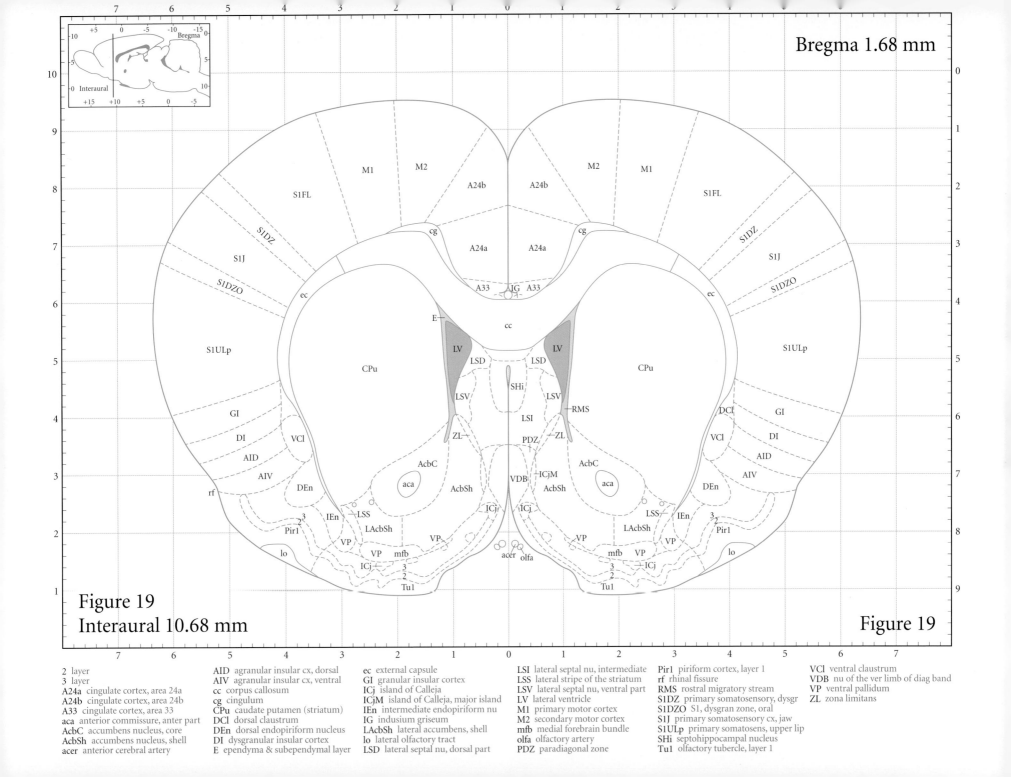

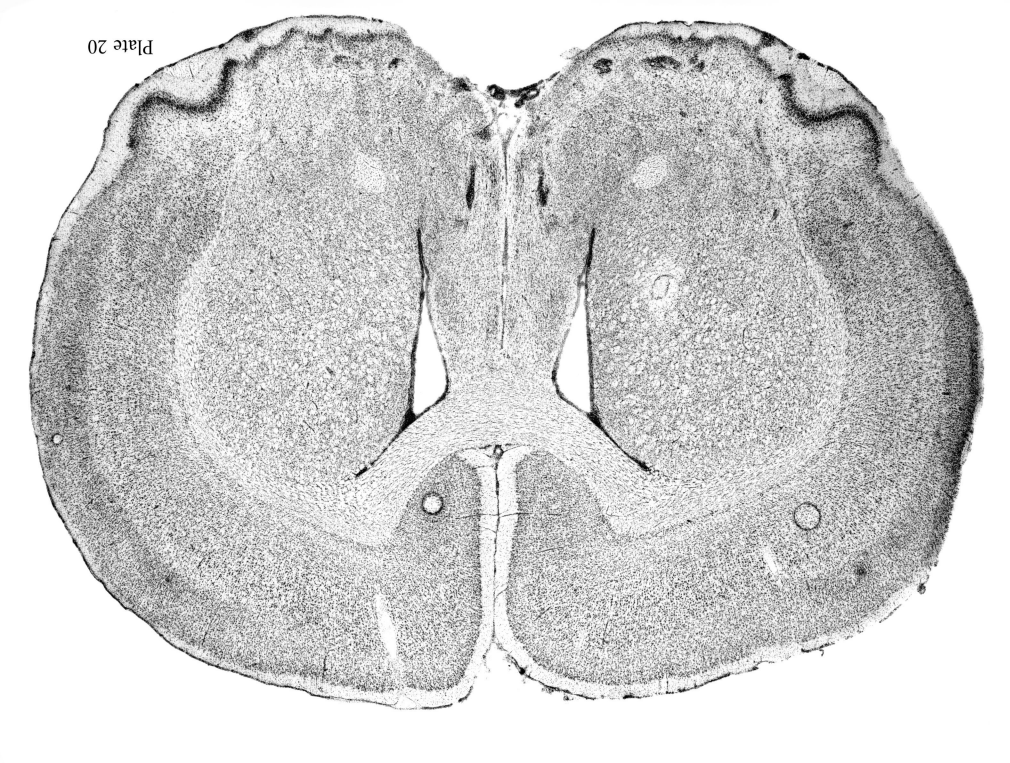

Plate 20

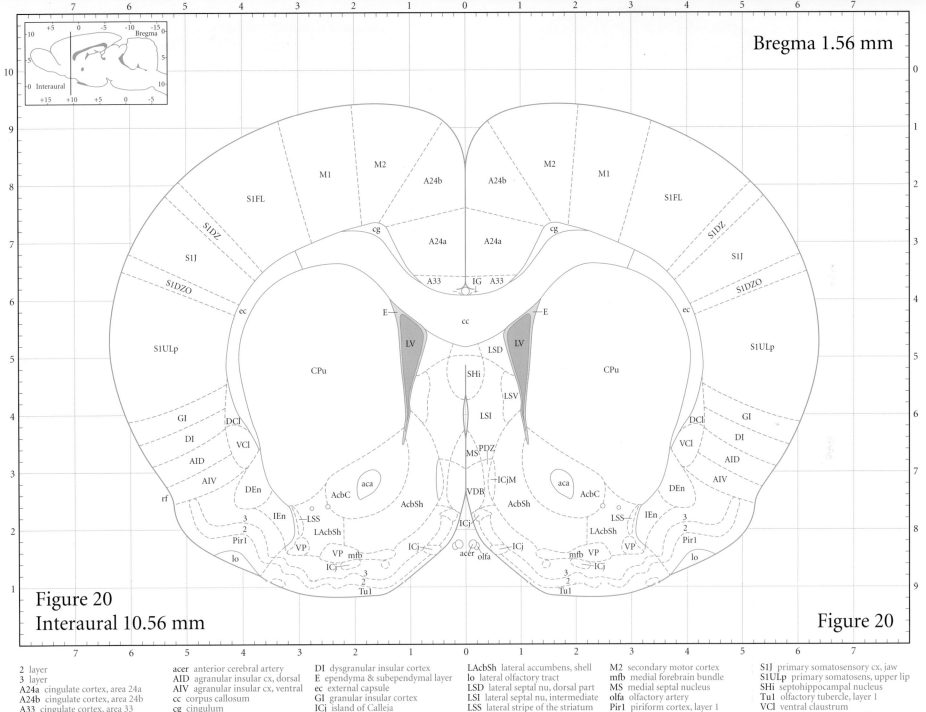

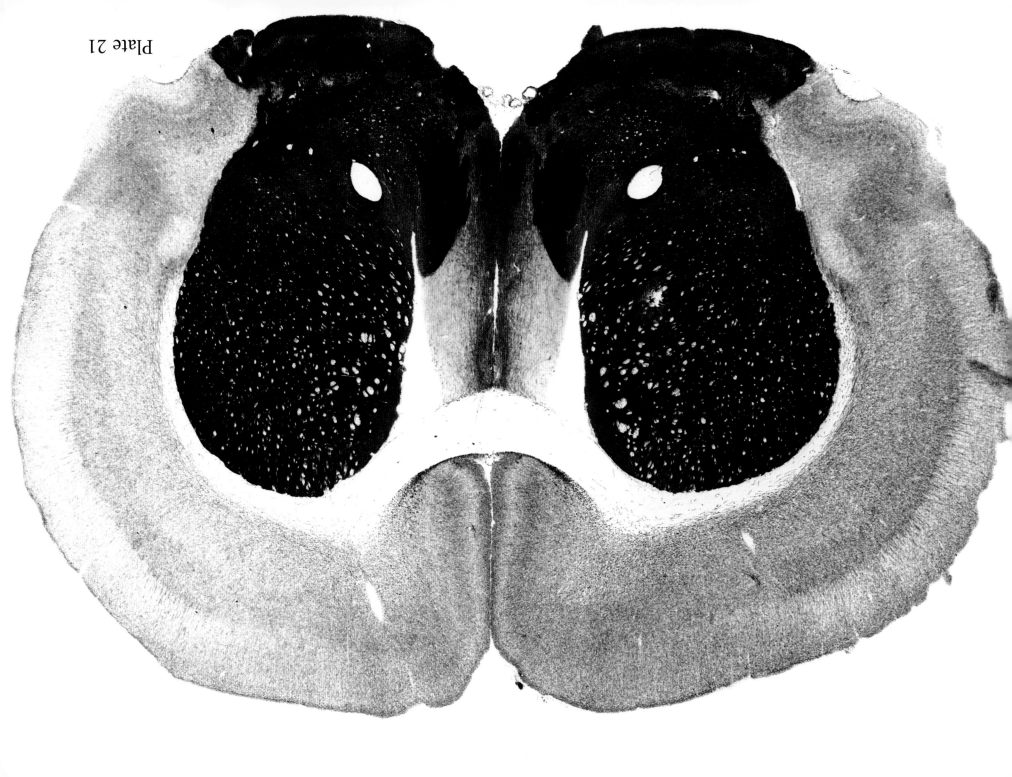

Plate 21

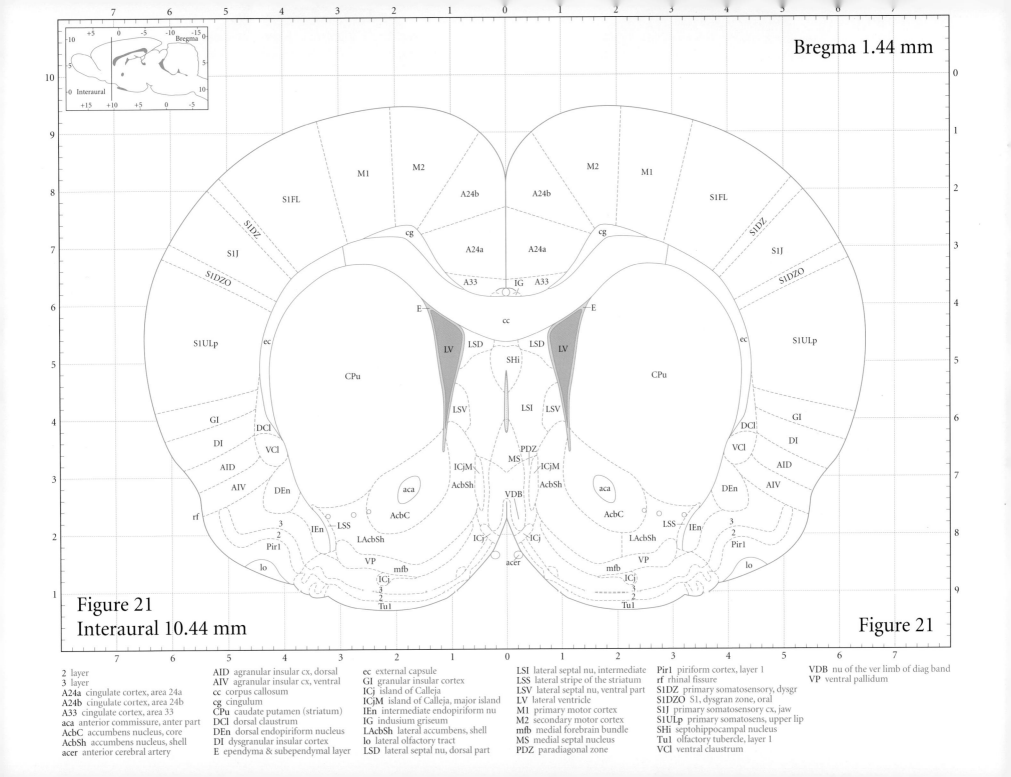

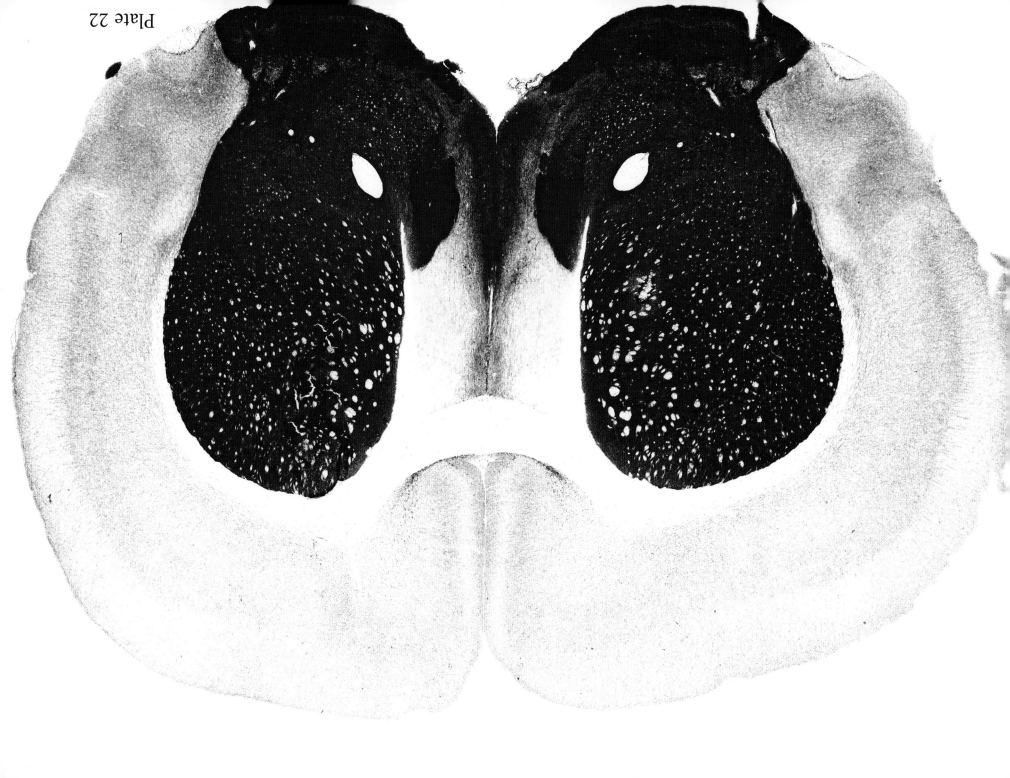

Plate 22

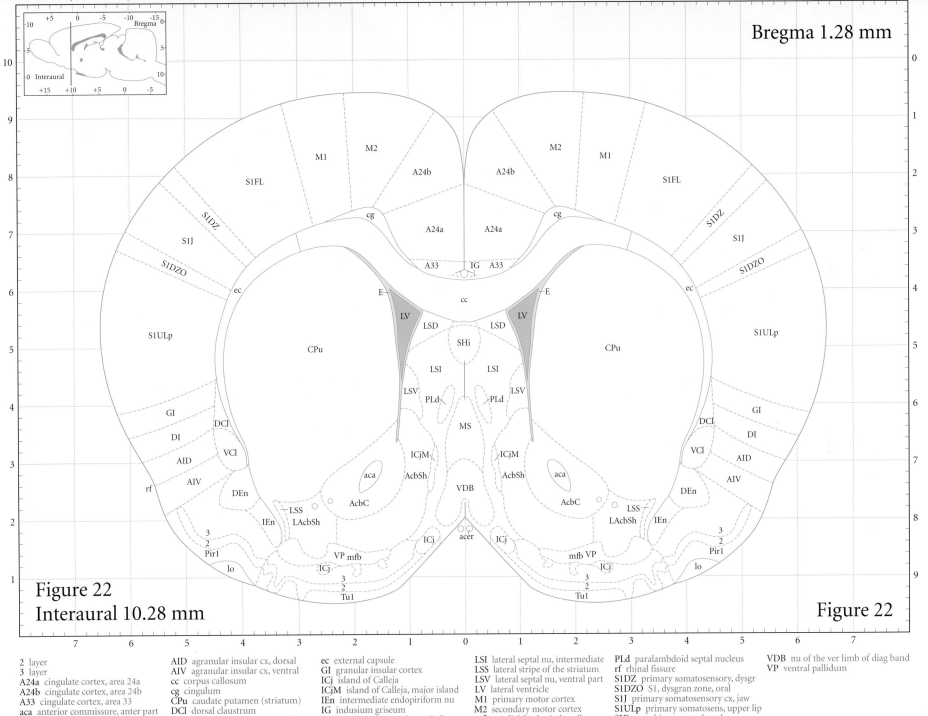

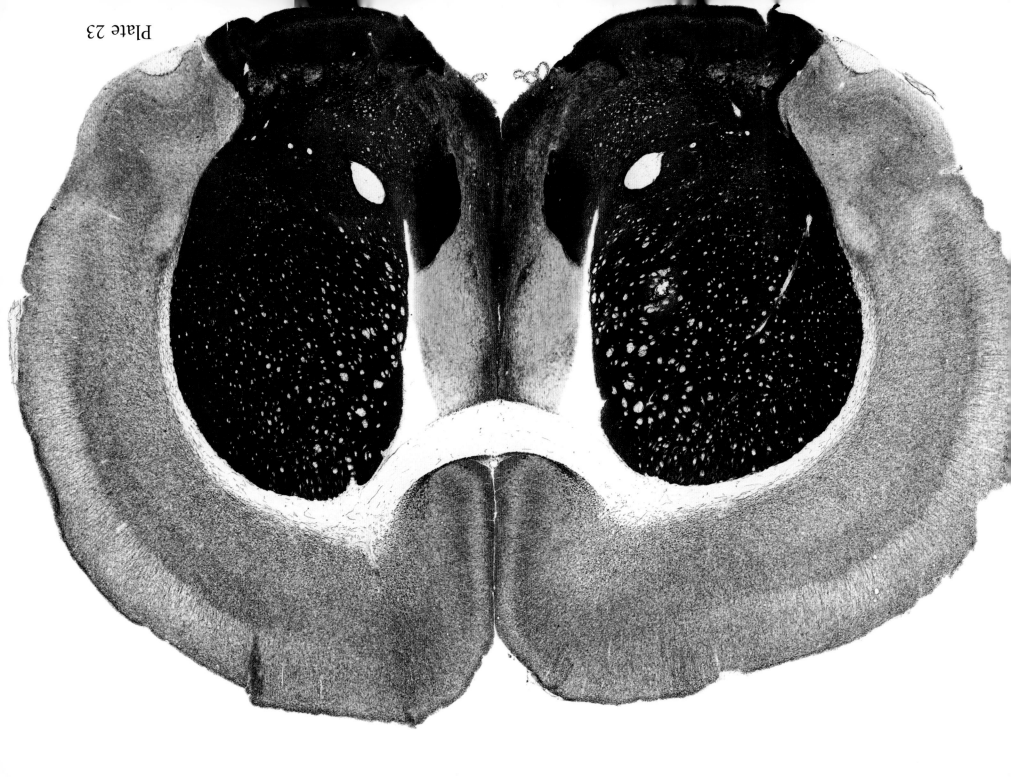

Plate 23

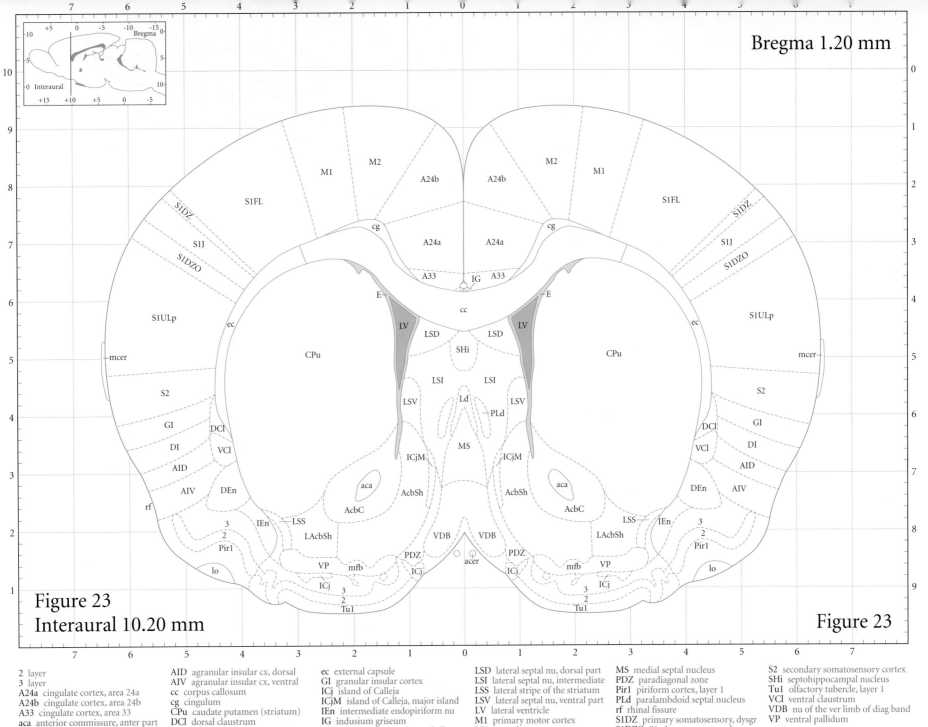

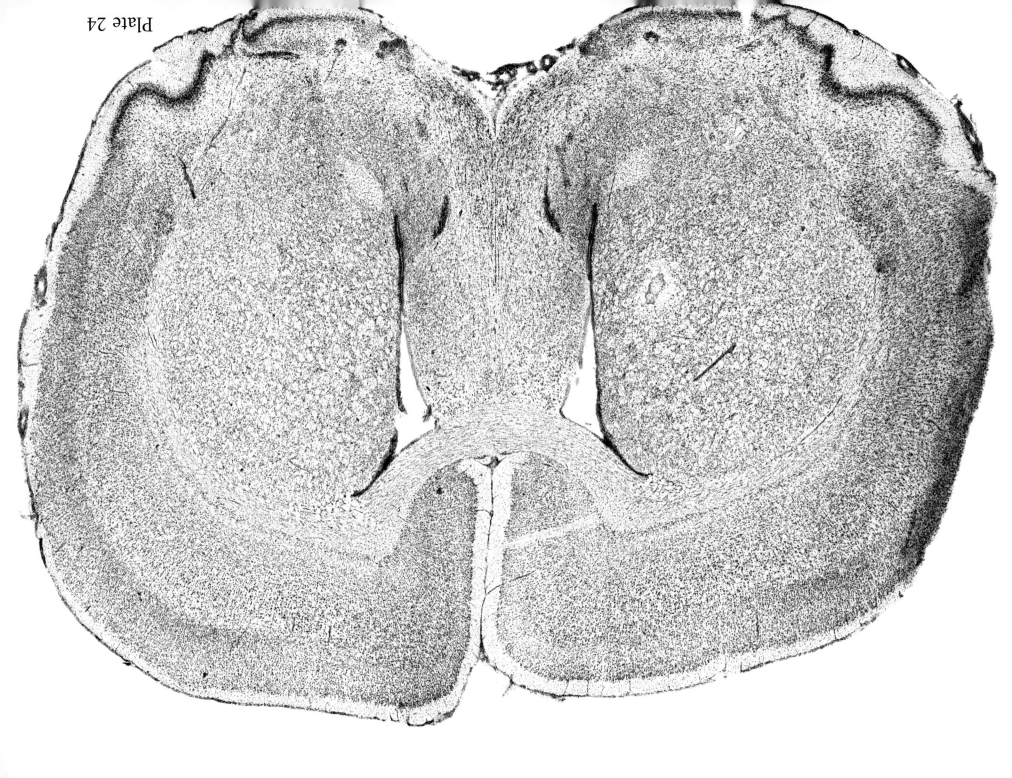

Plate 24

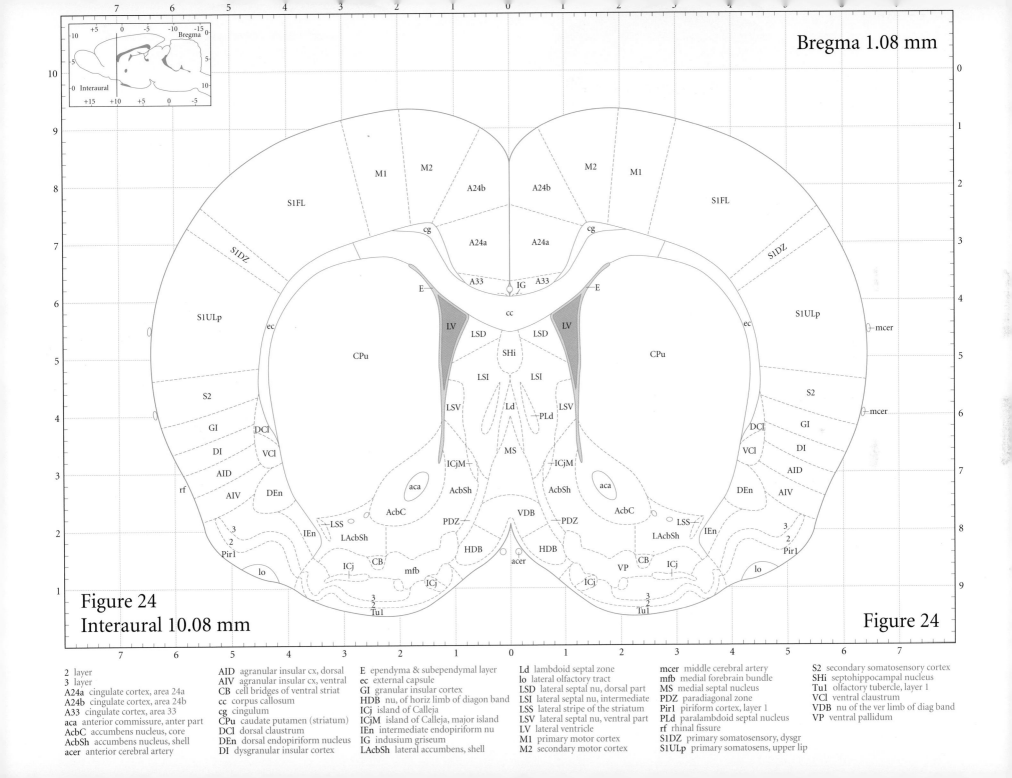

Figure 24
Interaural 10.08 mm

Bregma 1.08 mm

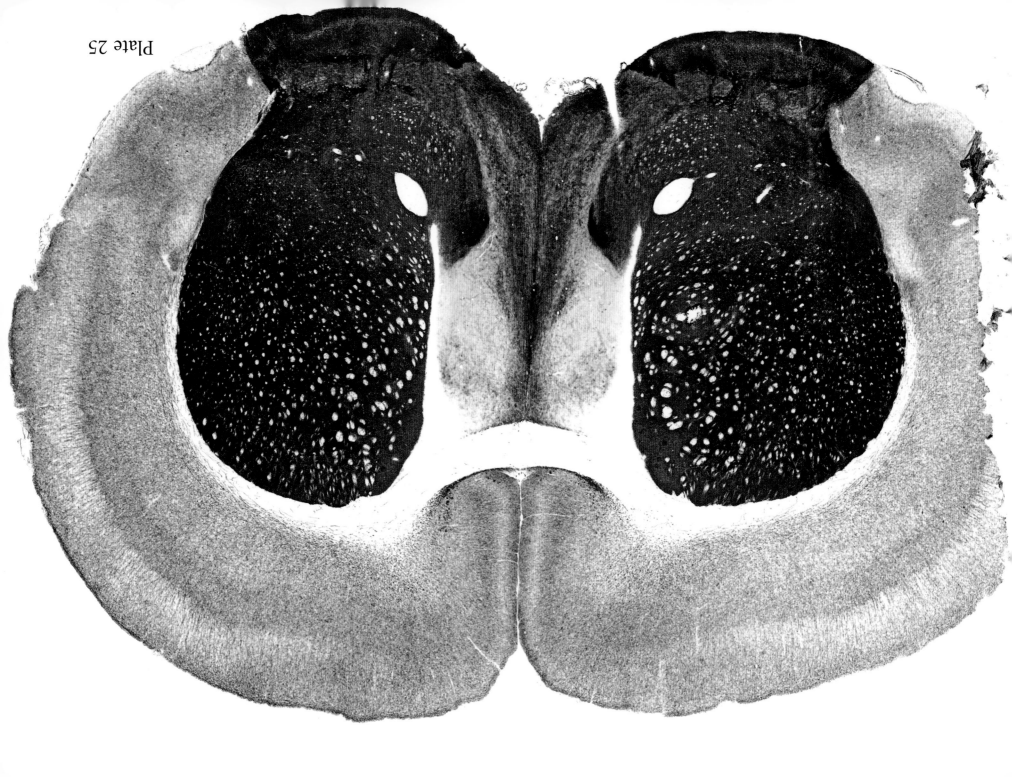

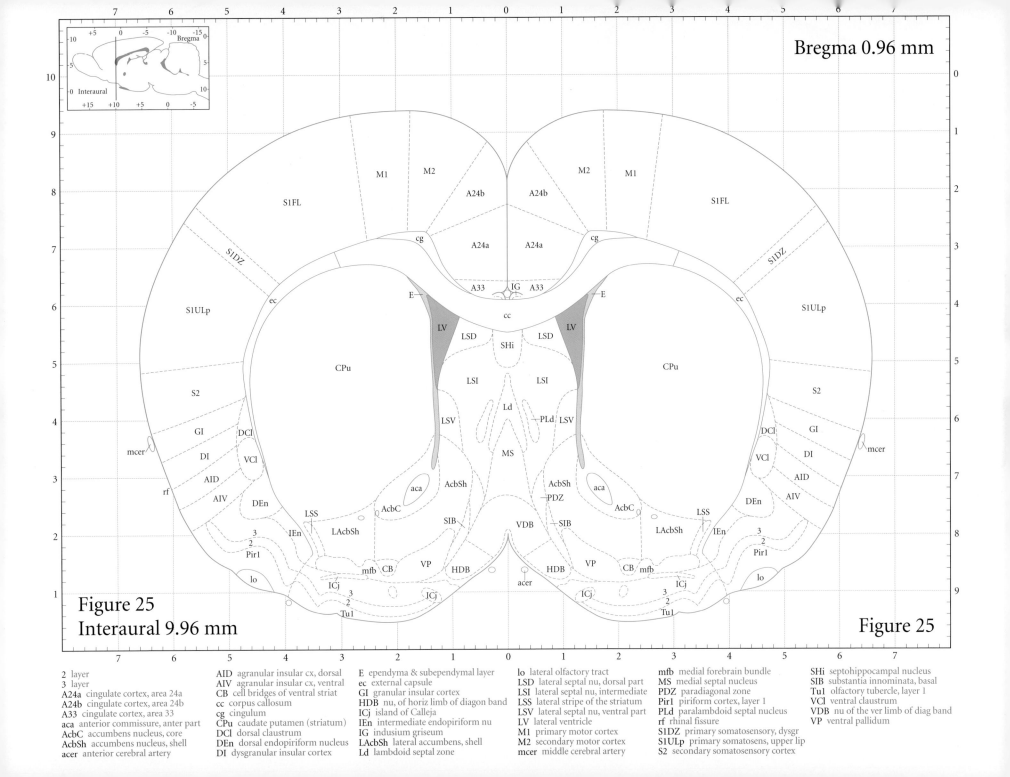

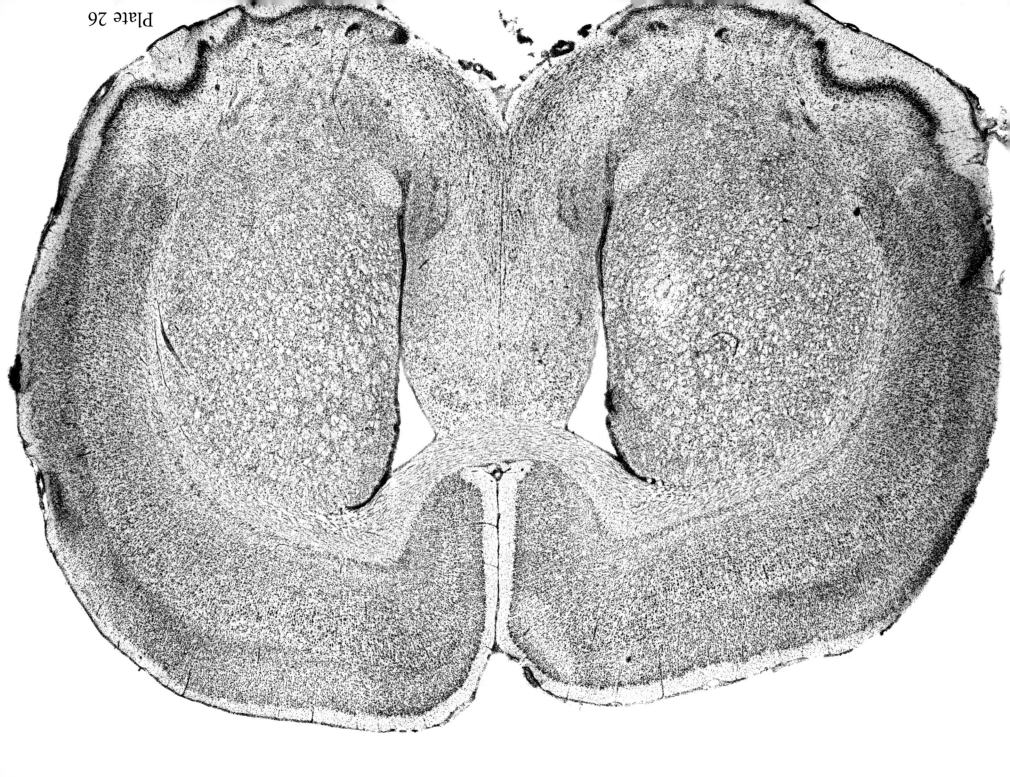

Plate 26

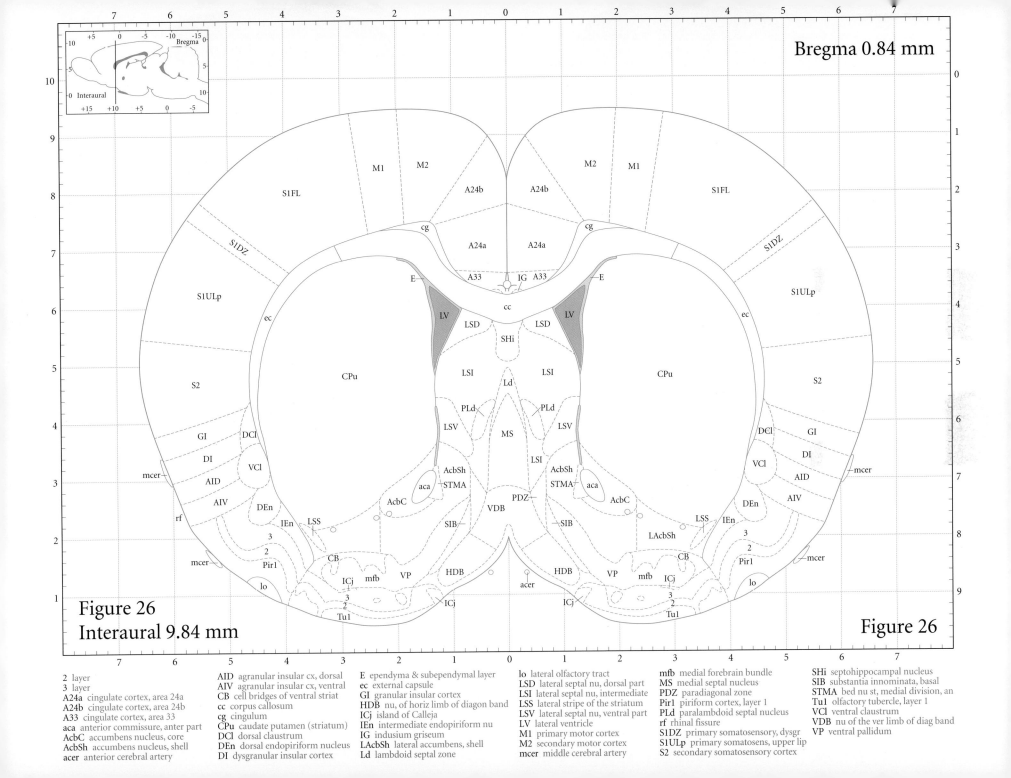

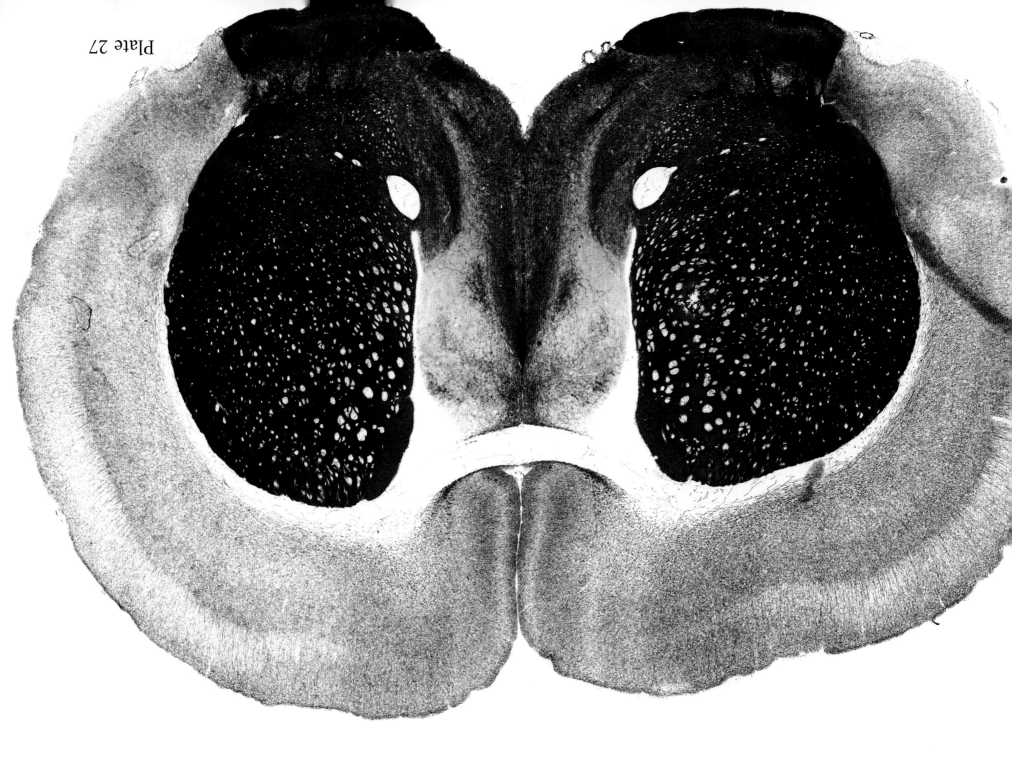

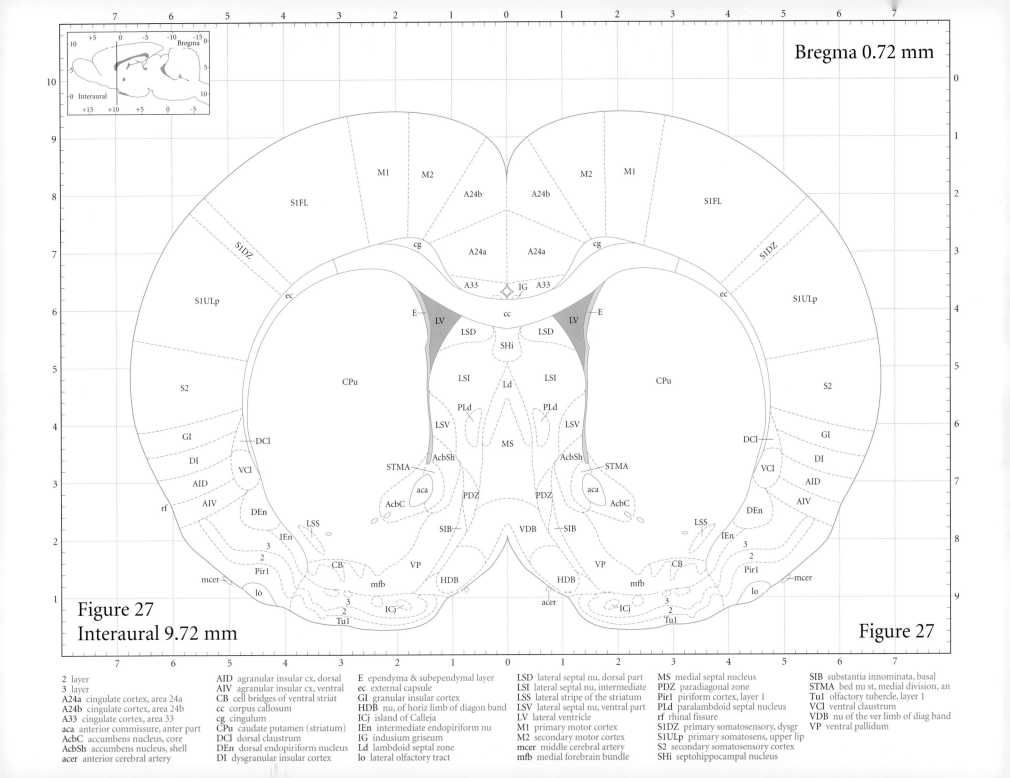

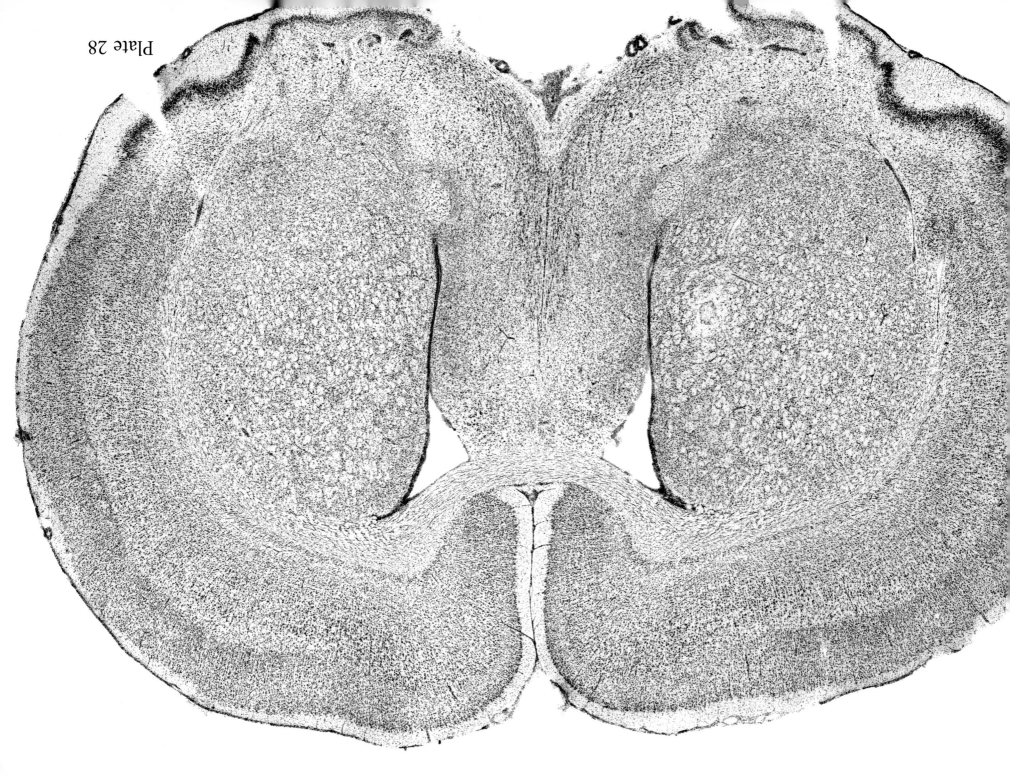

Plate 28

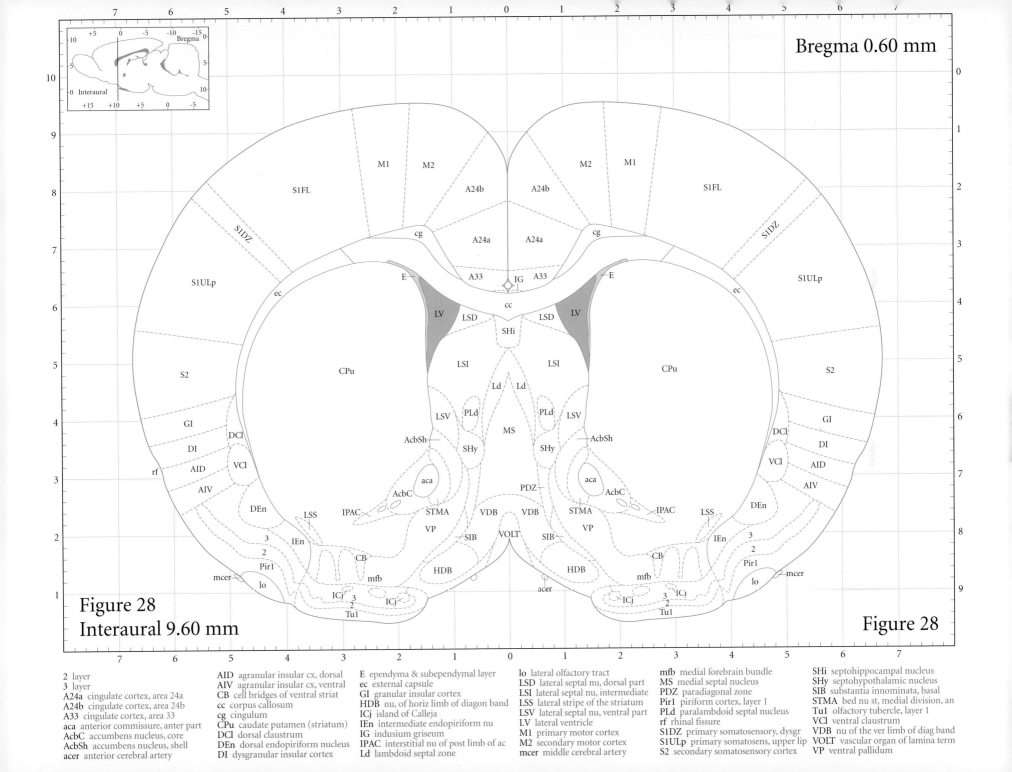

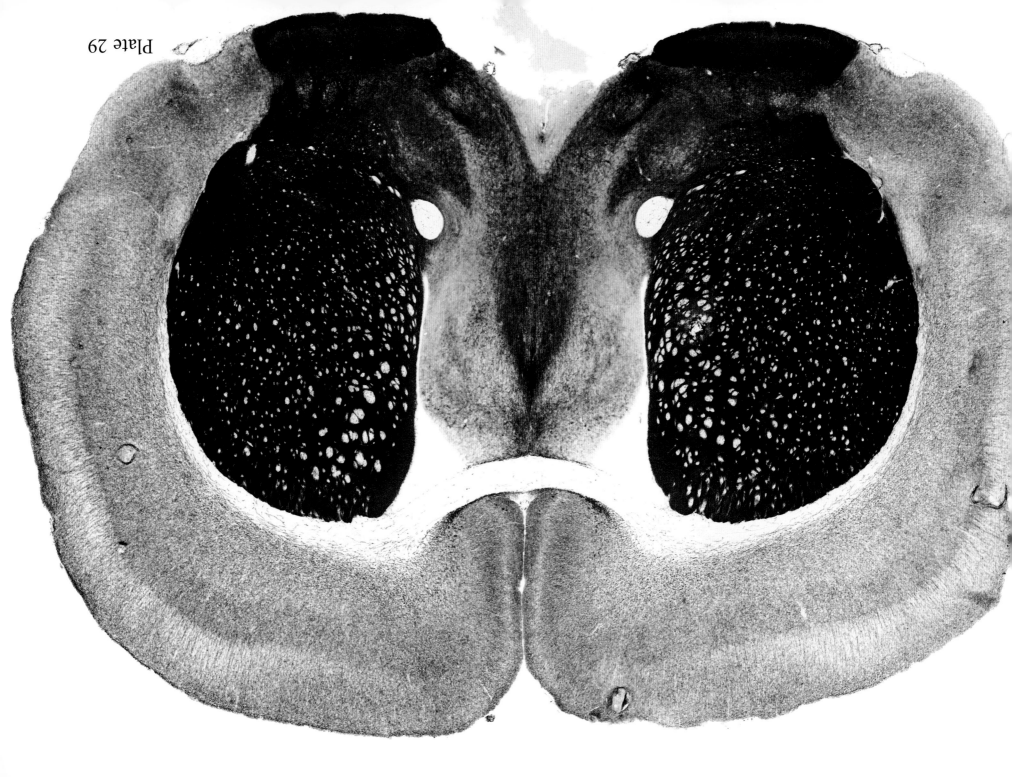

Plate 29

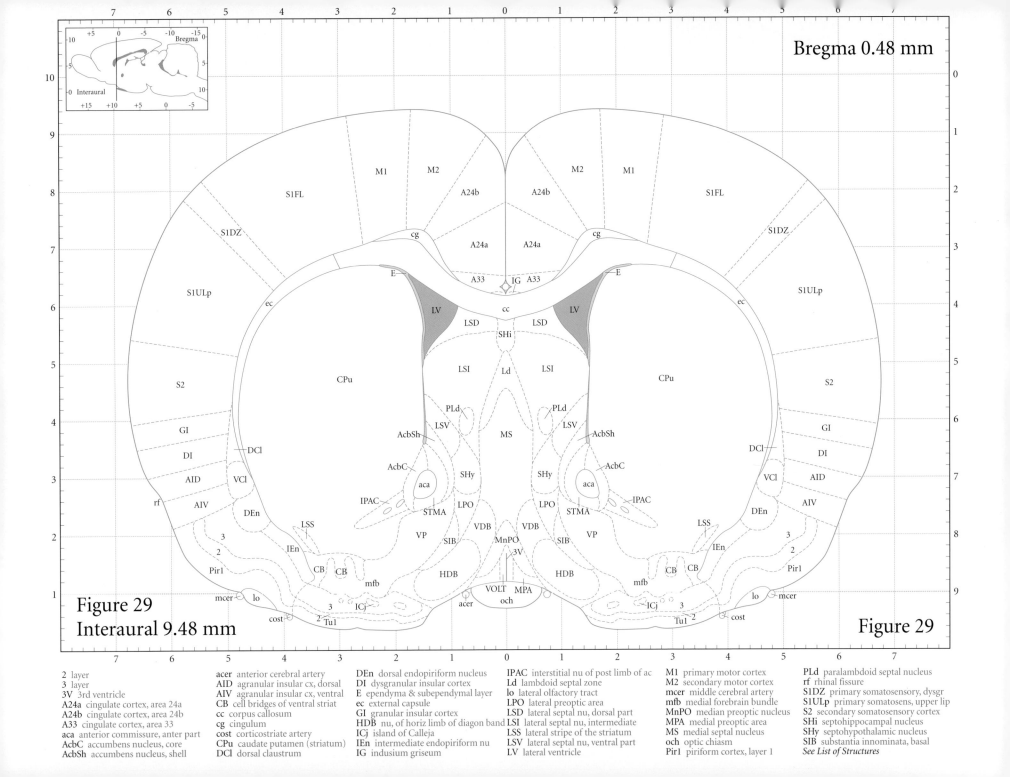

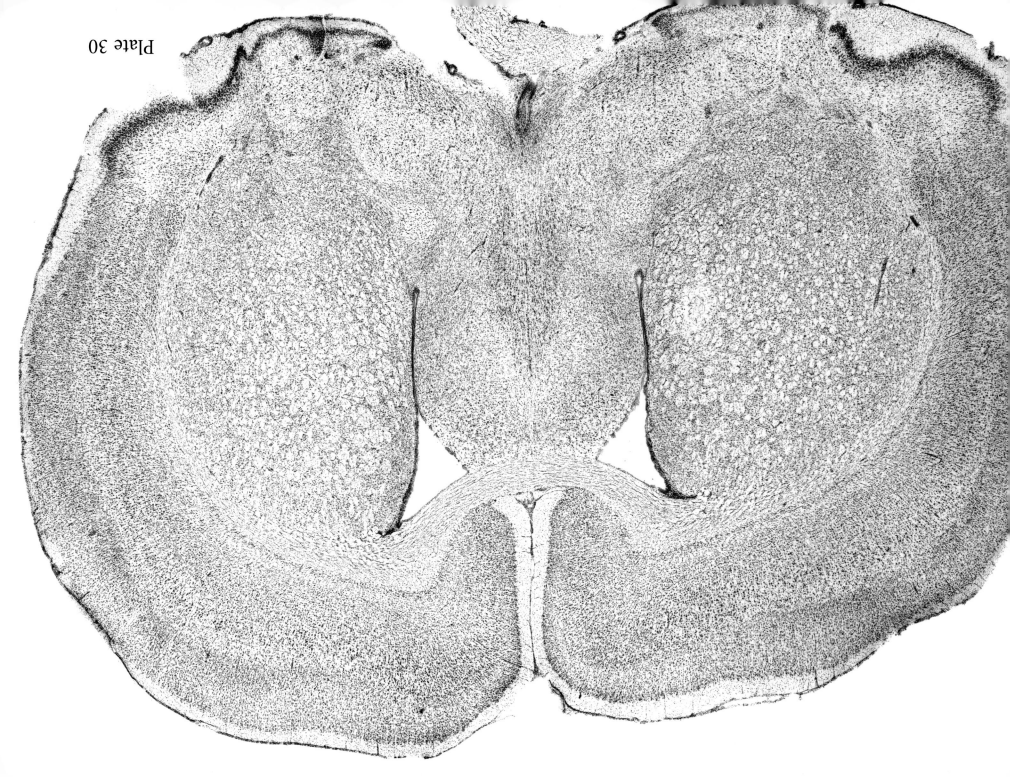

Plate 30

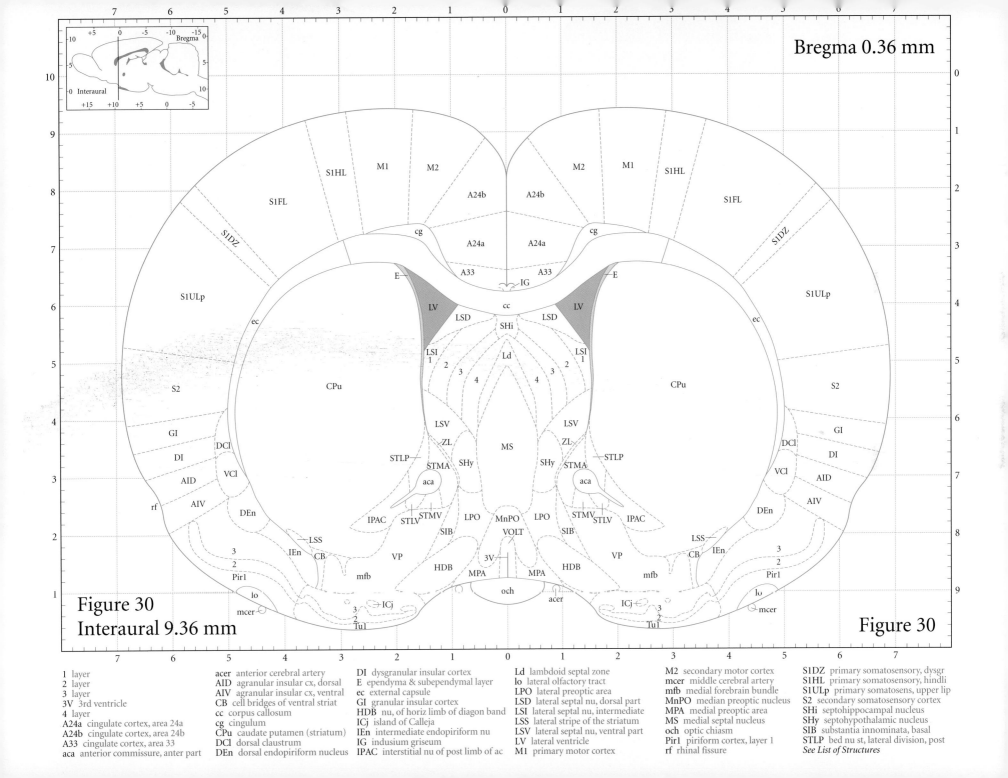

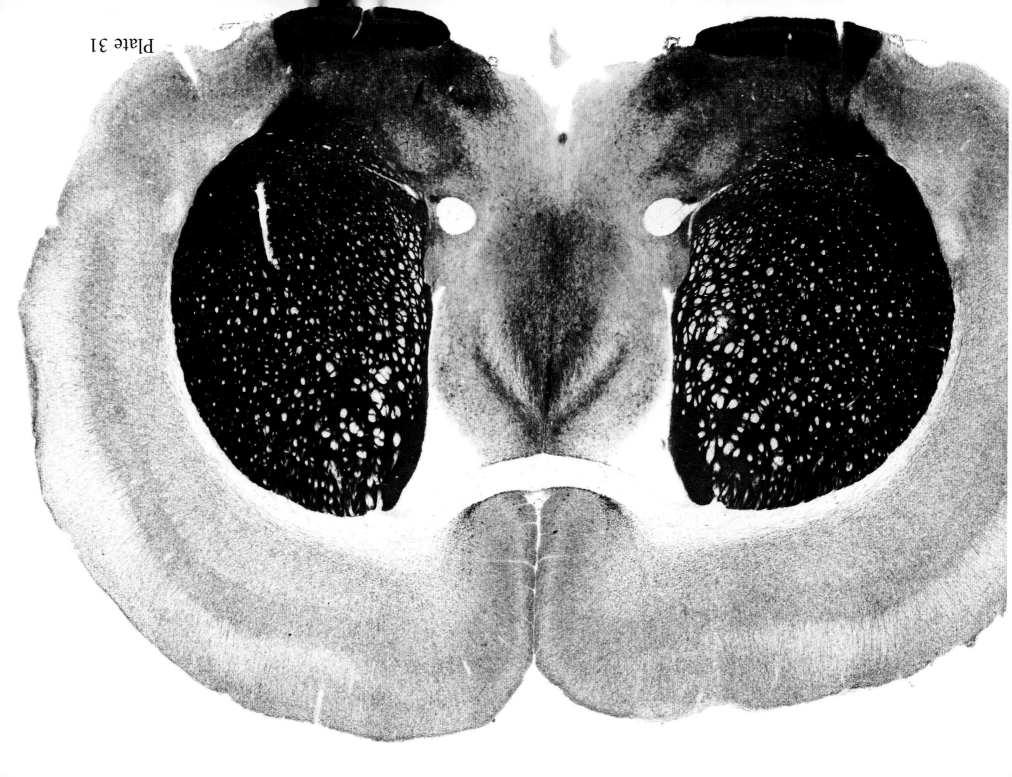
Plate 31

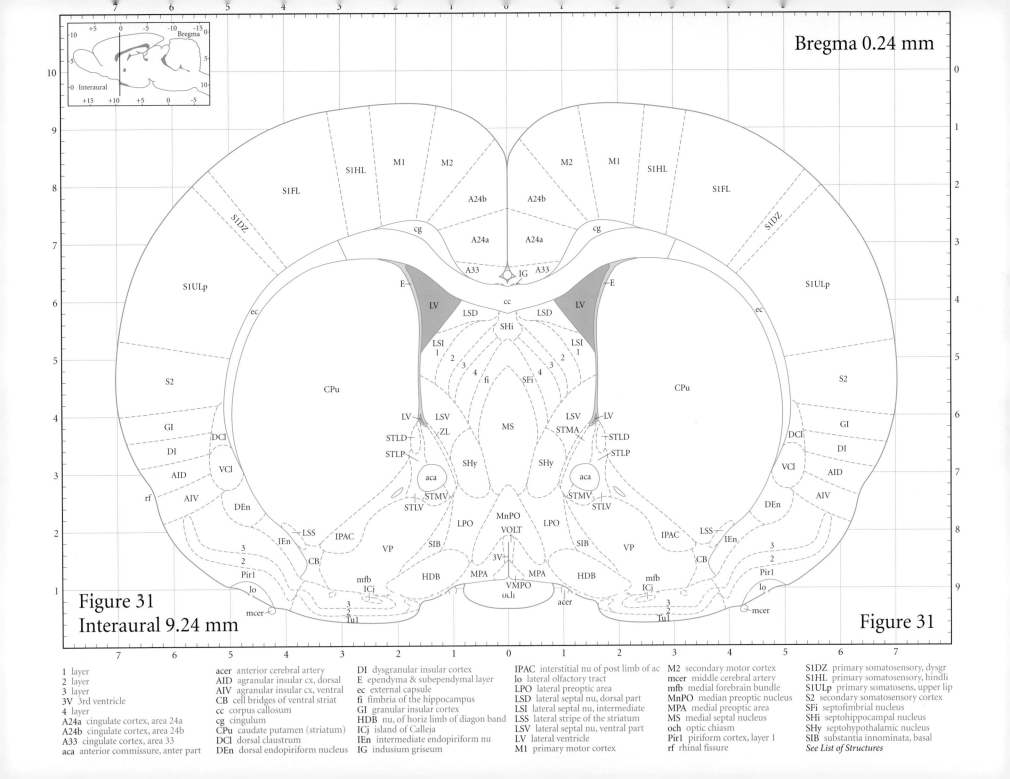

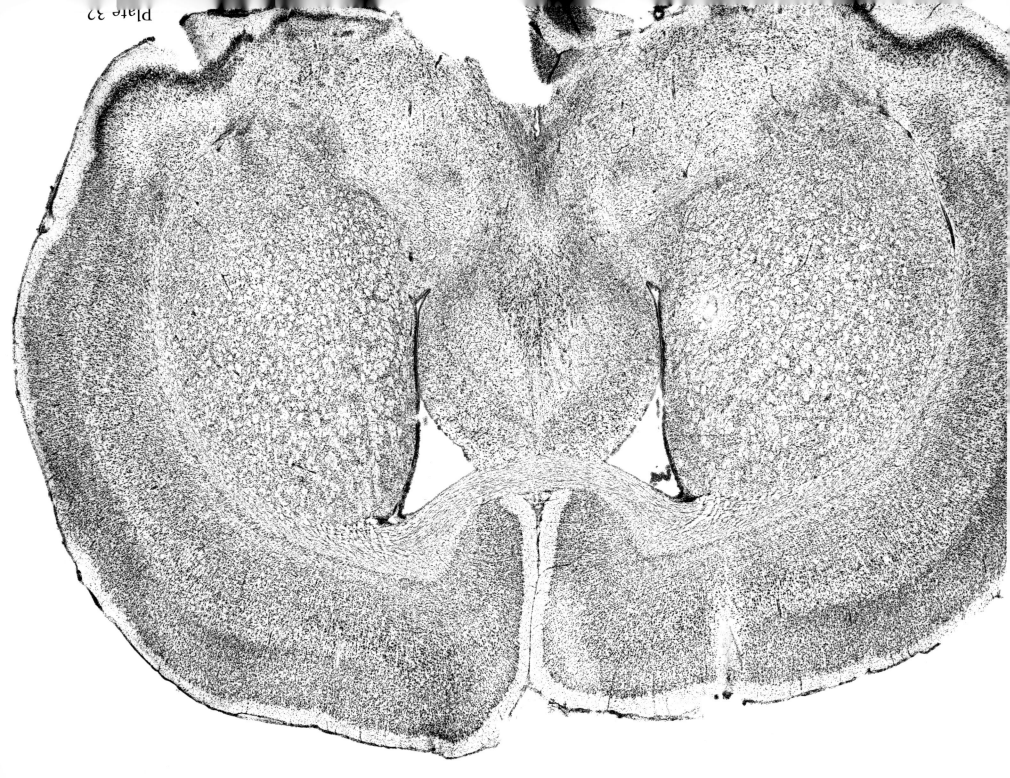

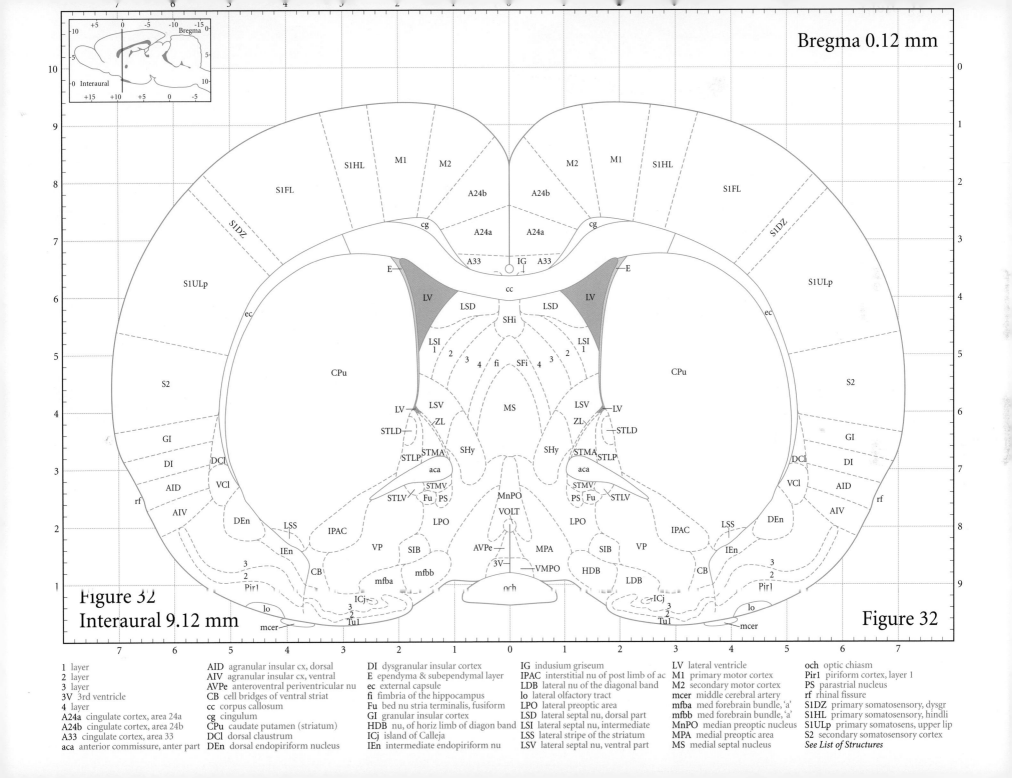

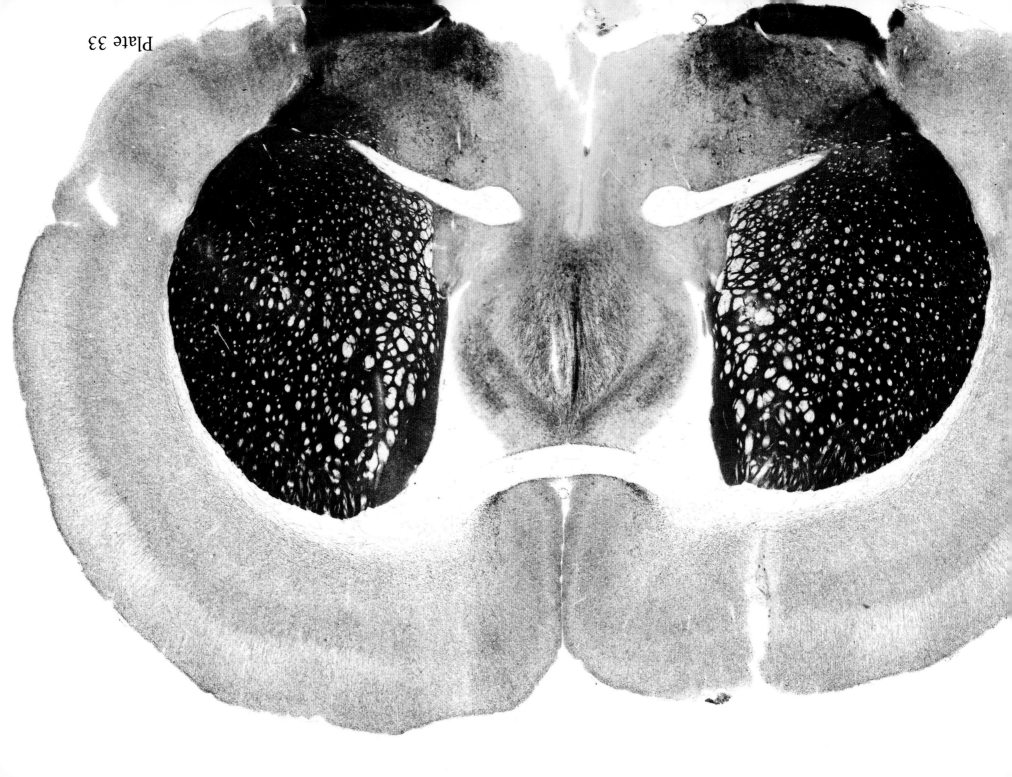

Plate 33

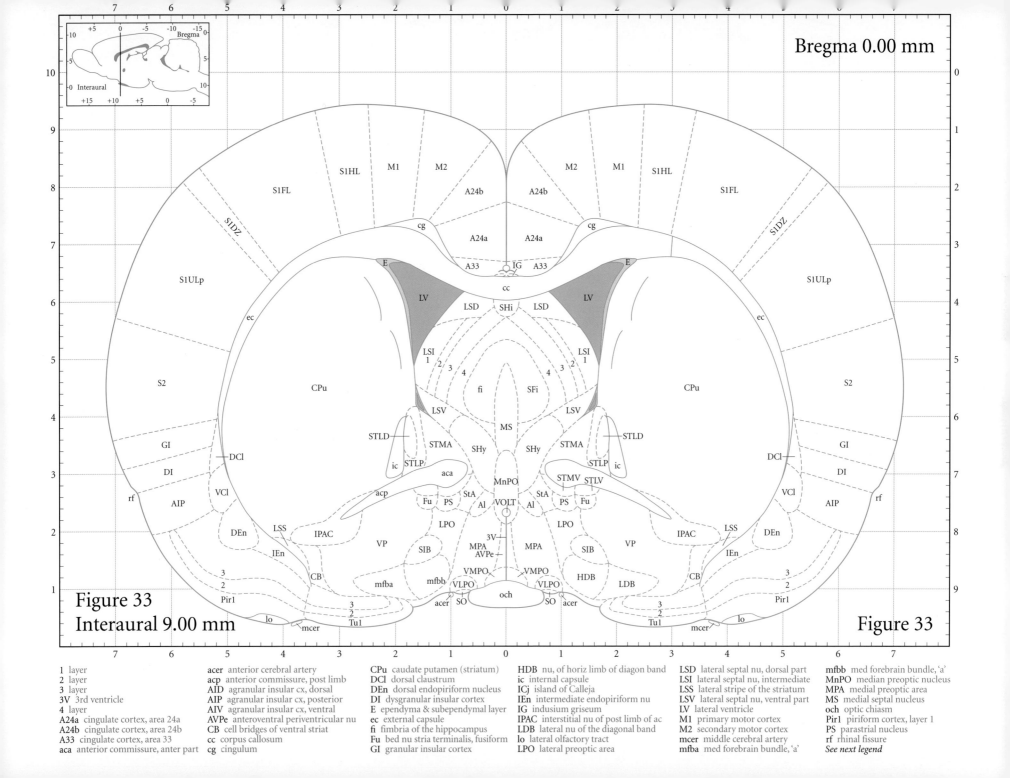

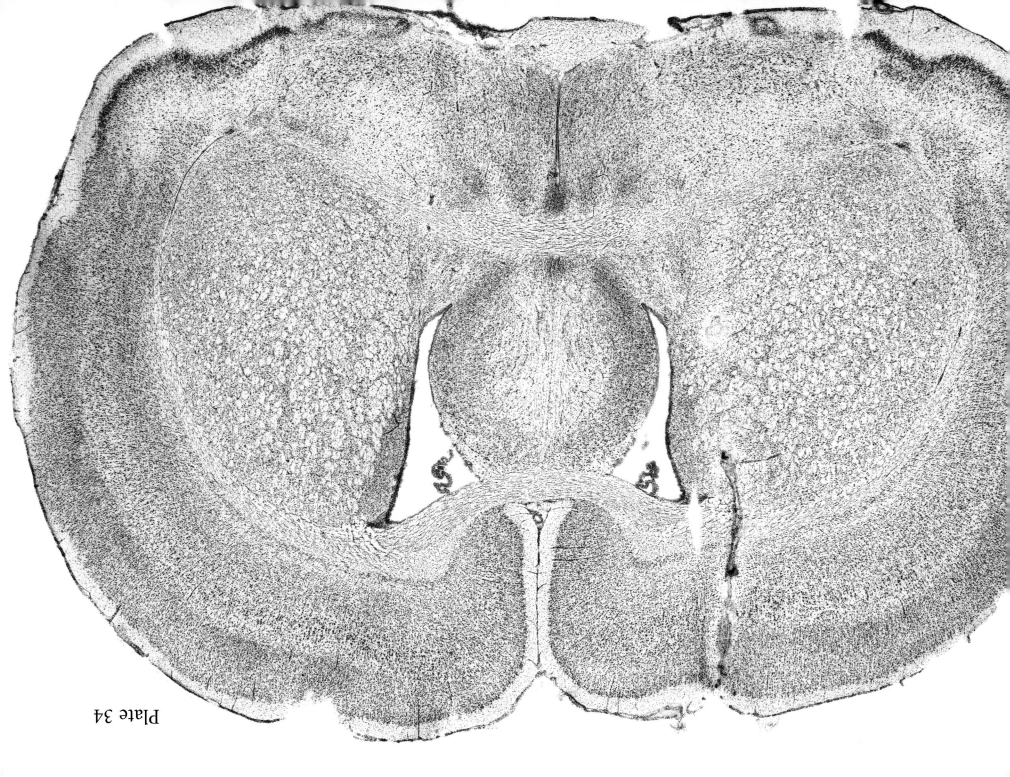

Plate 34

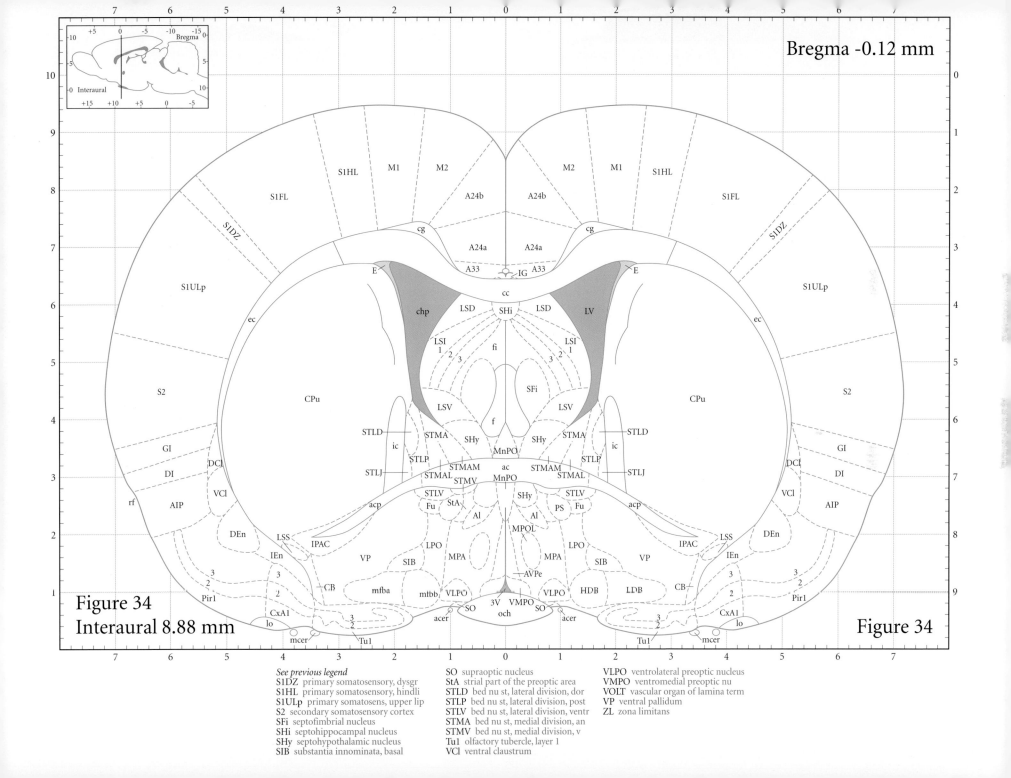

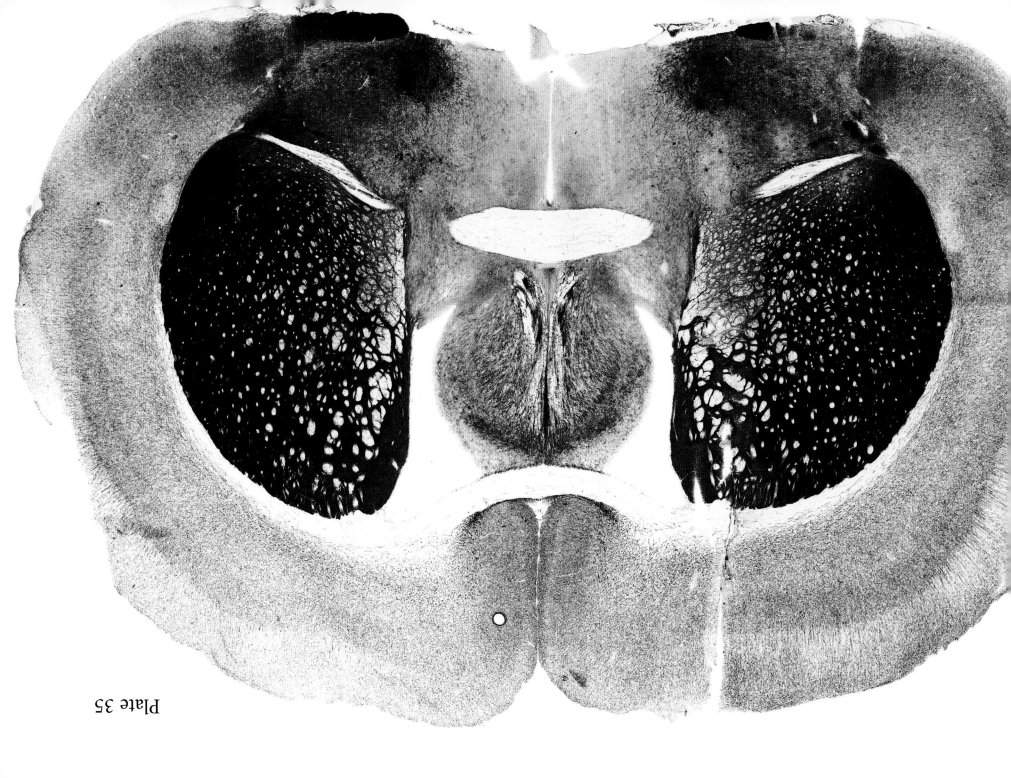

Plate 35

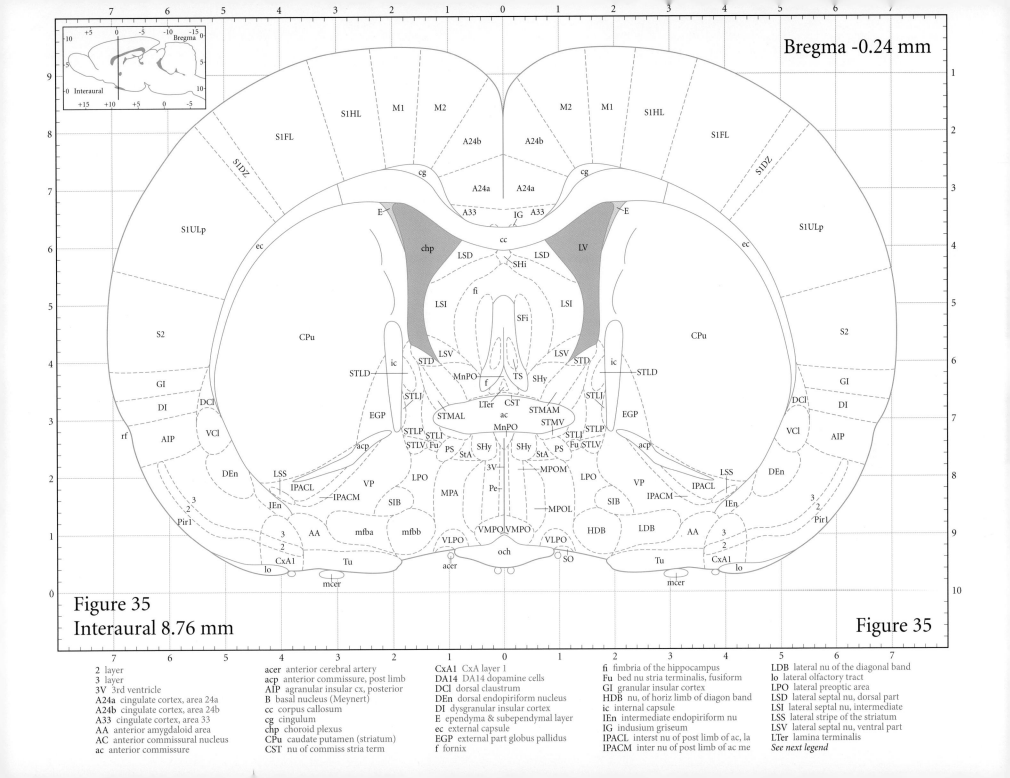

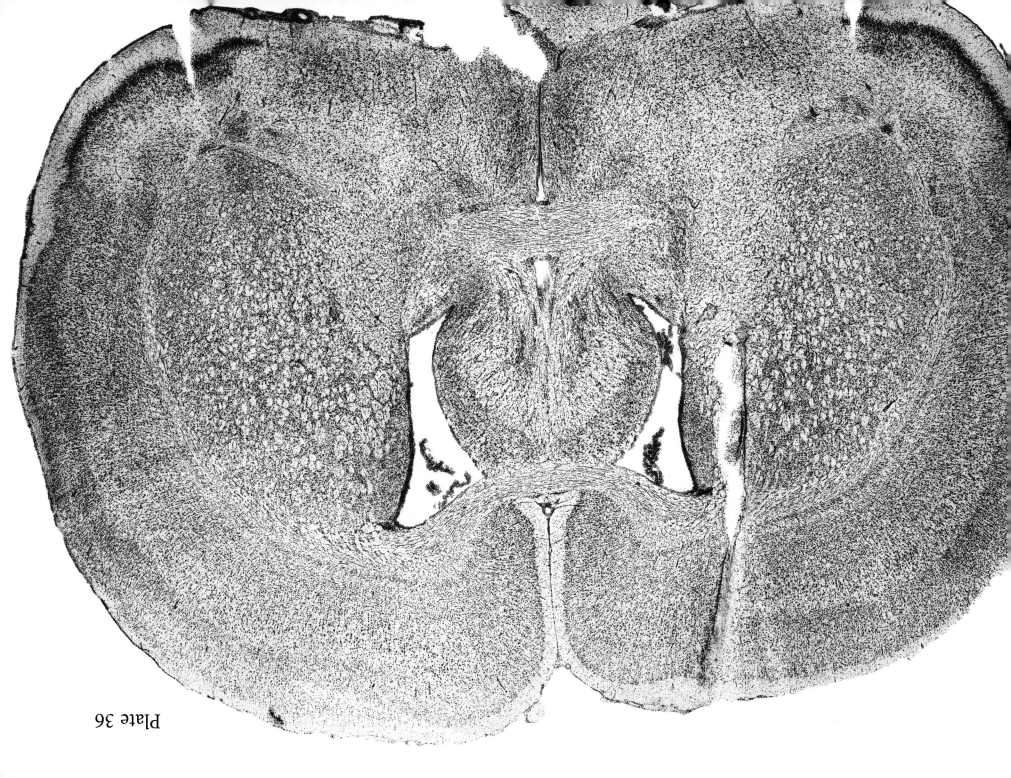

Plate 36

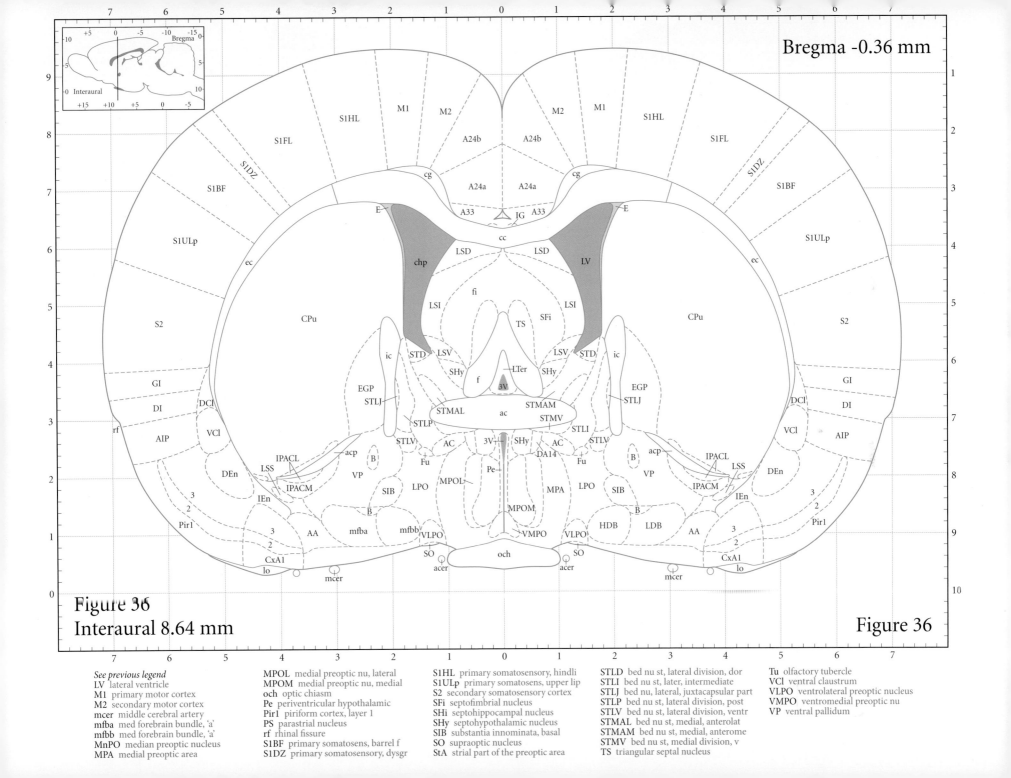

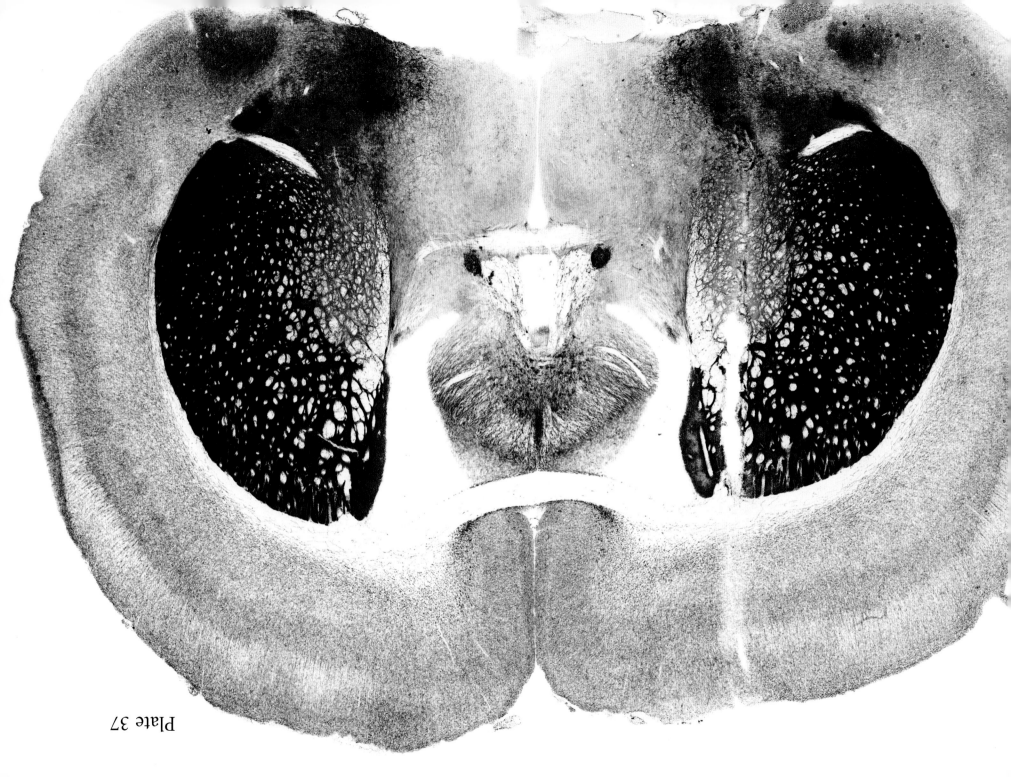

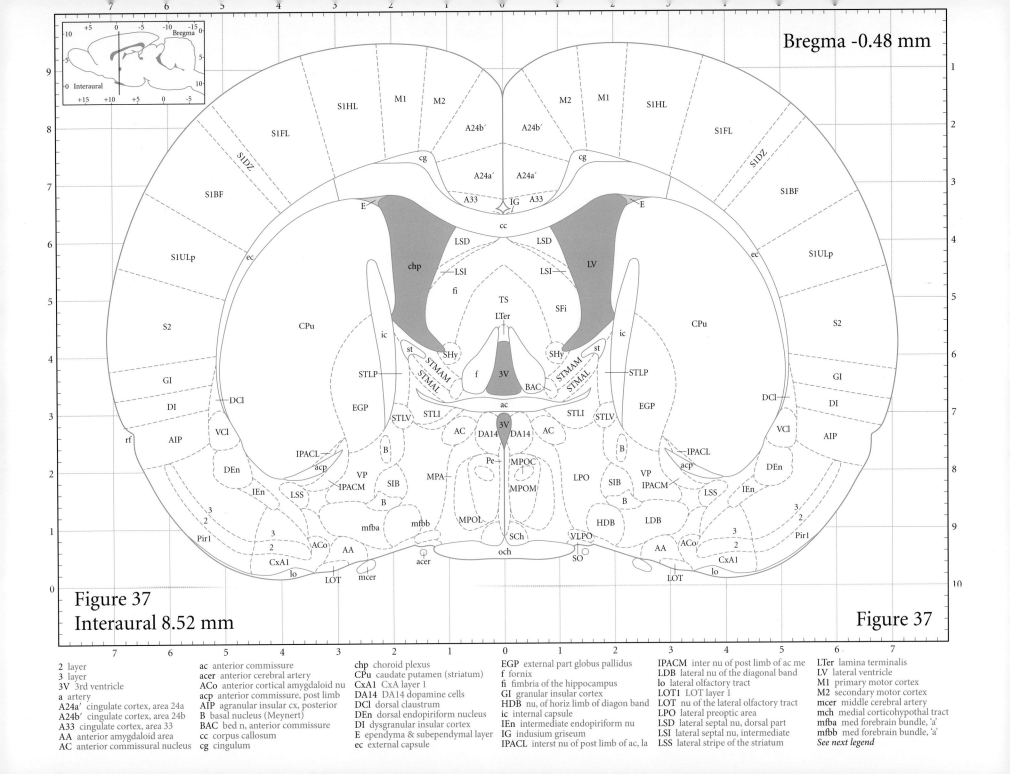

Figure 37
Interaural 8.52 mm / Bregma -0.48 mm

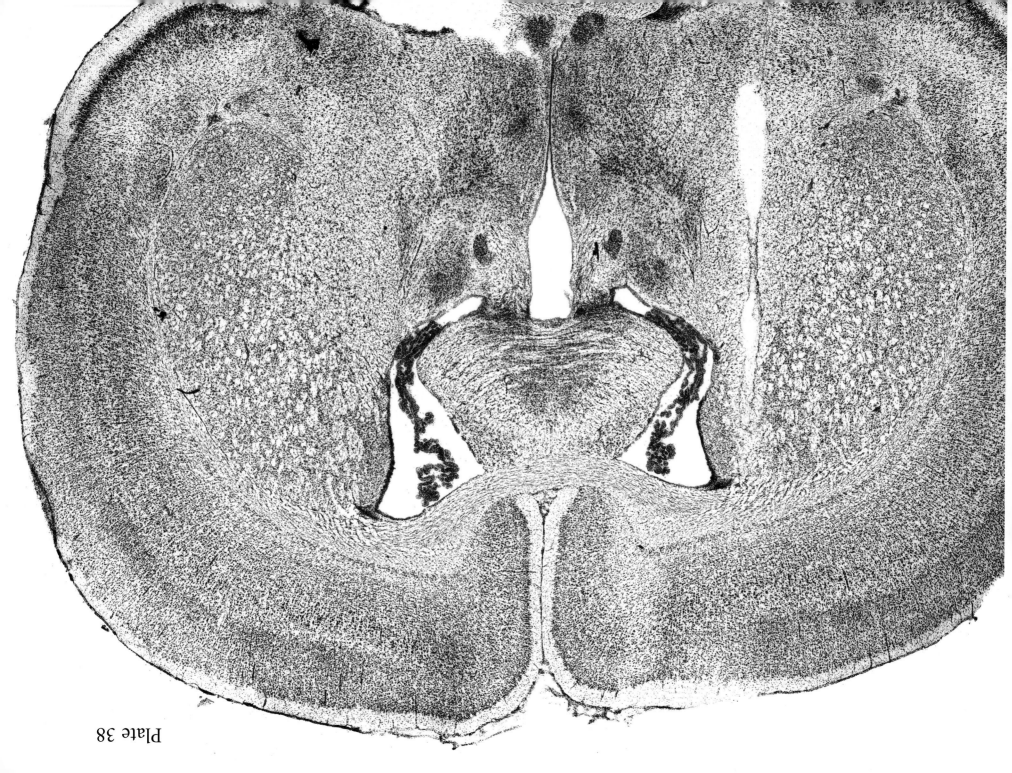

Plate 38

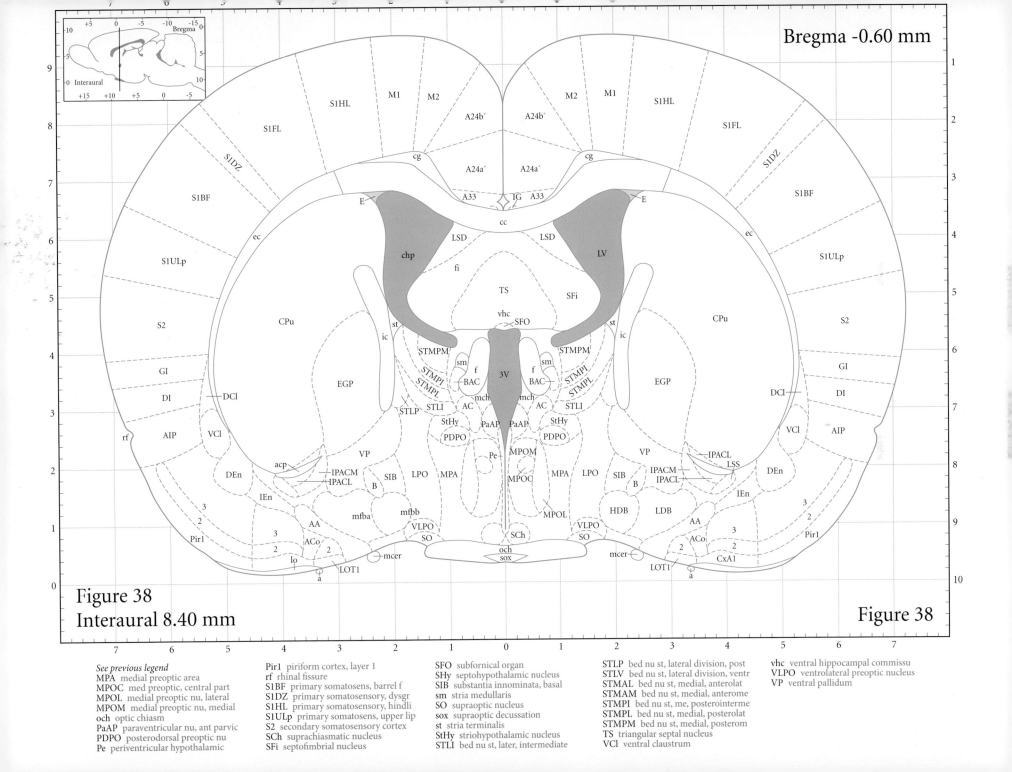

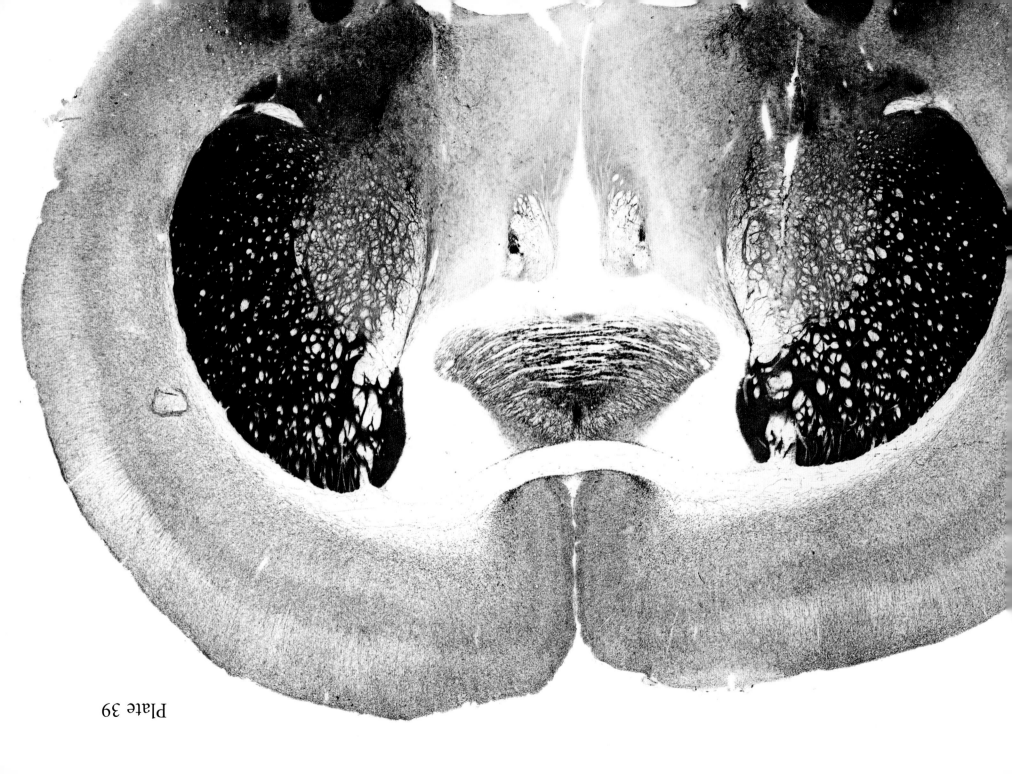

Plate 39

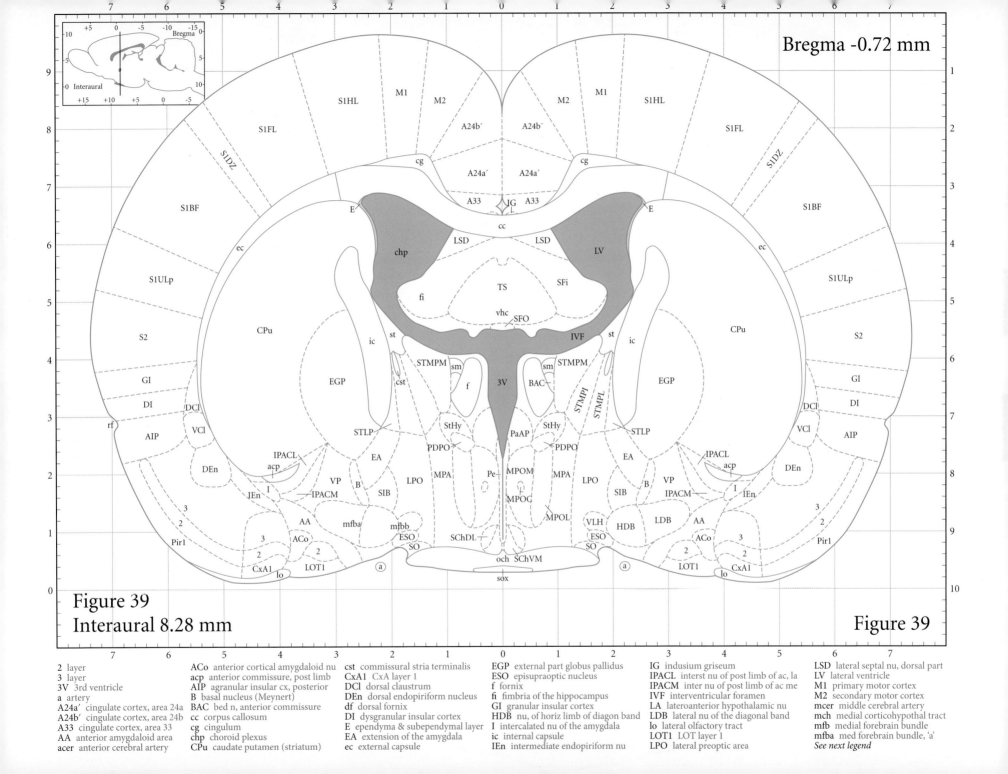

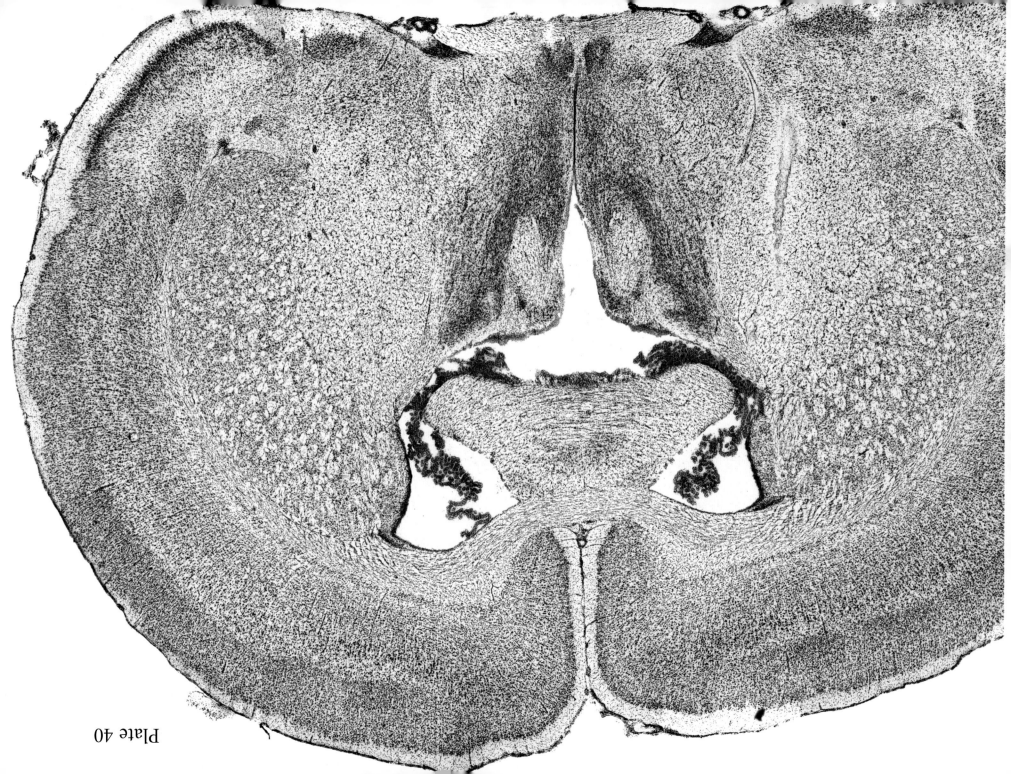

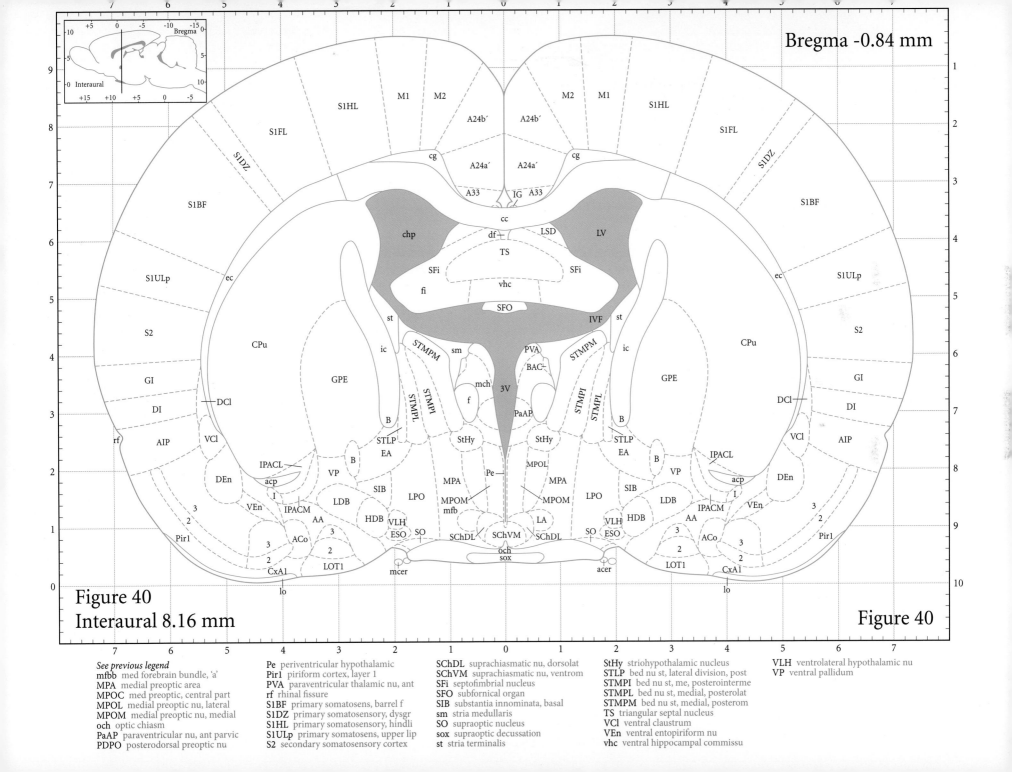

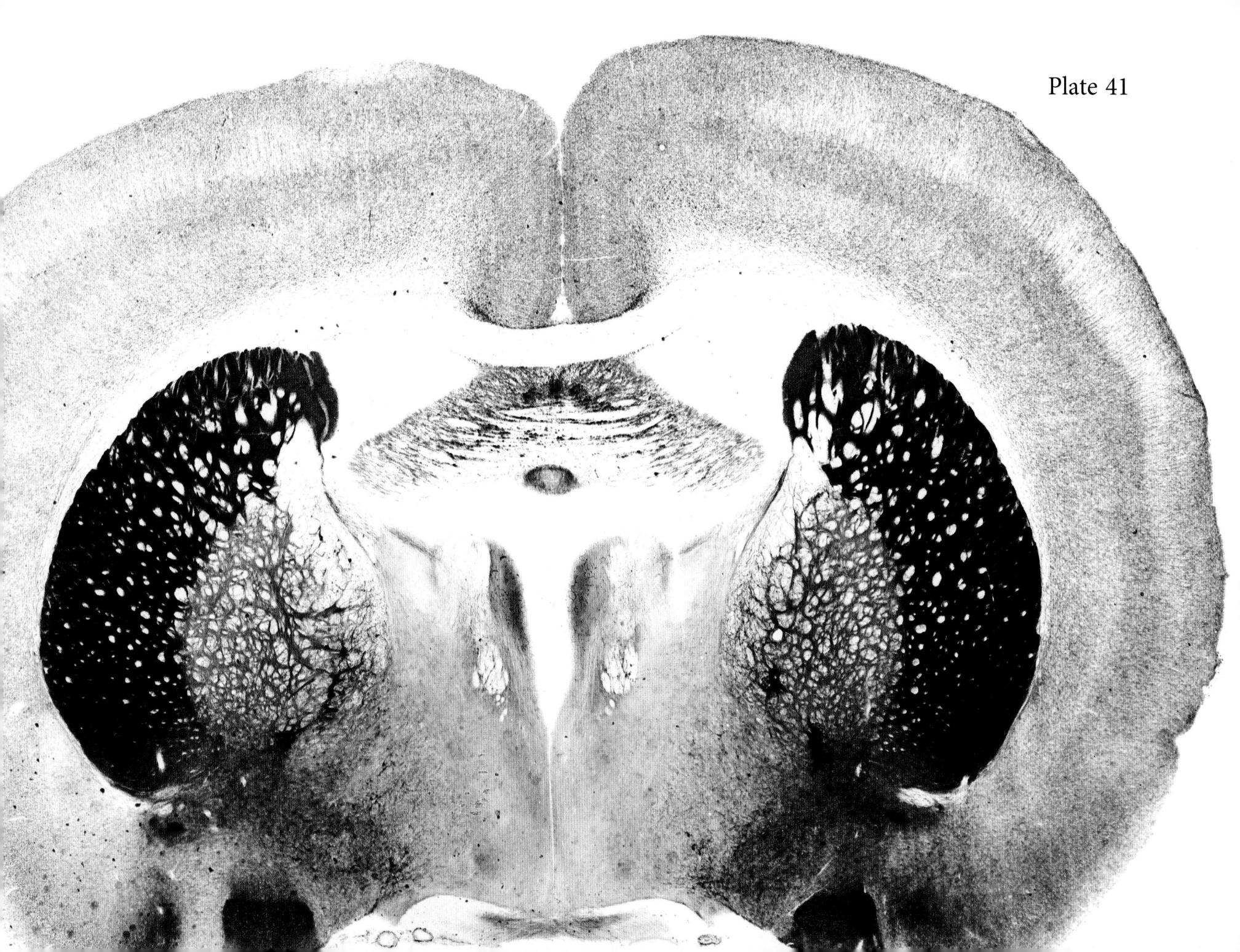

Plate 41

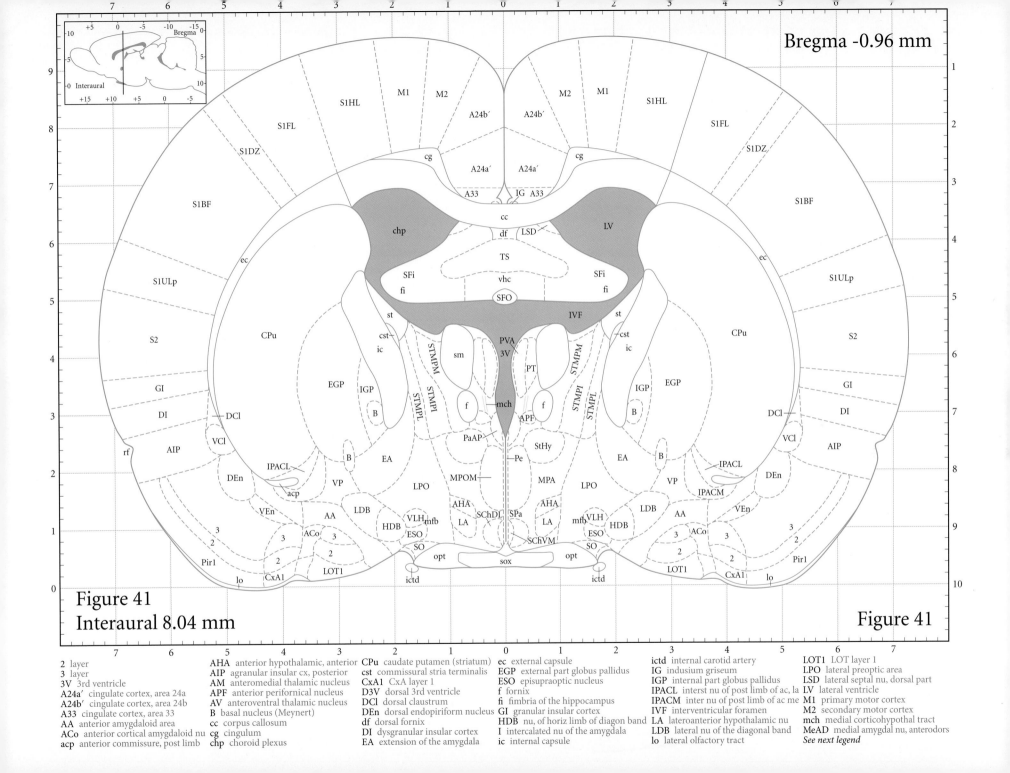

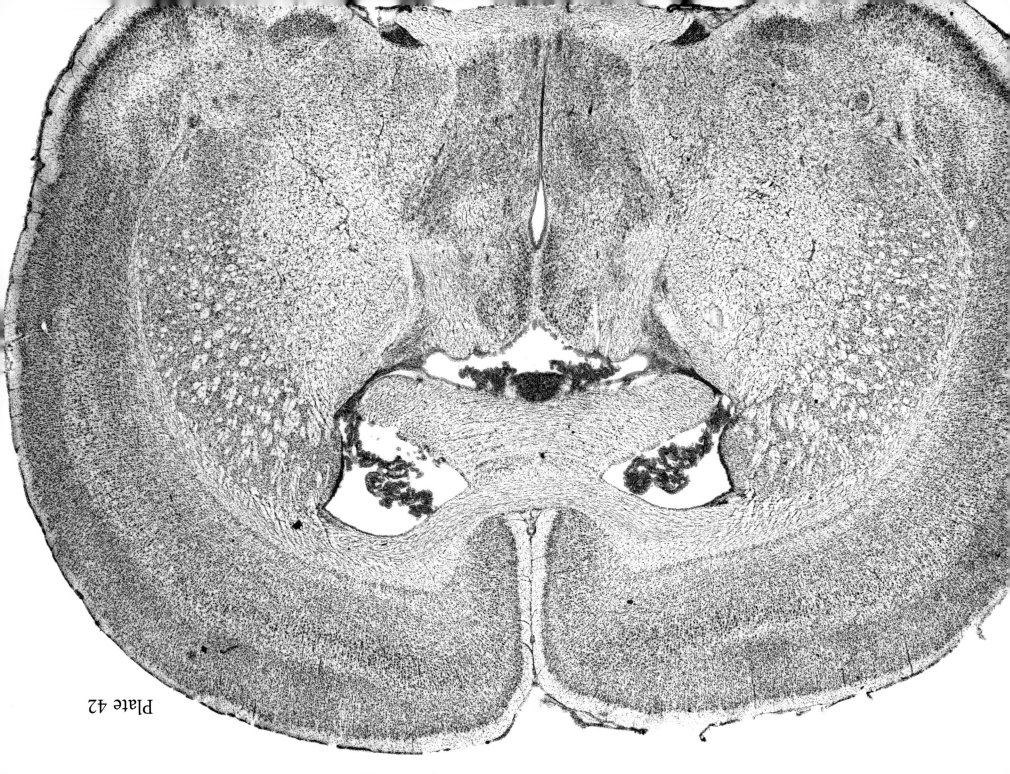

Plate 42

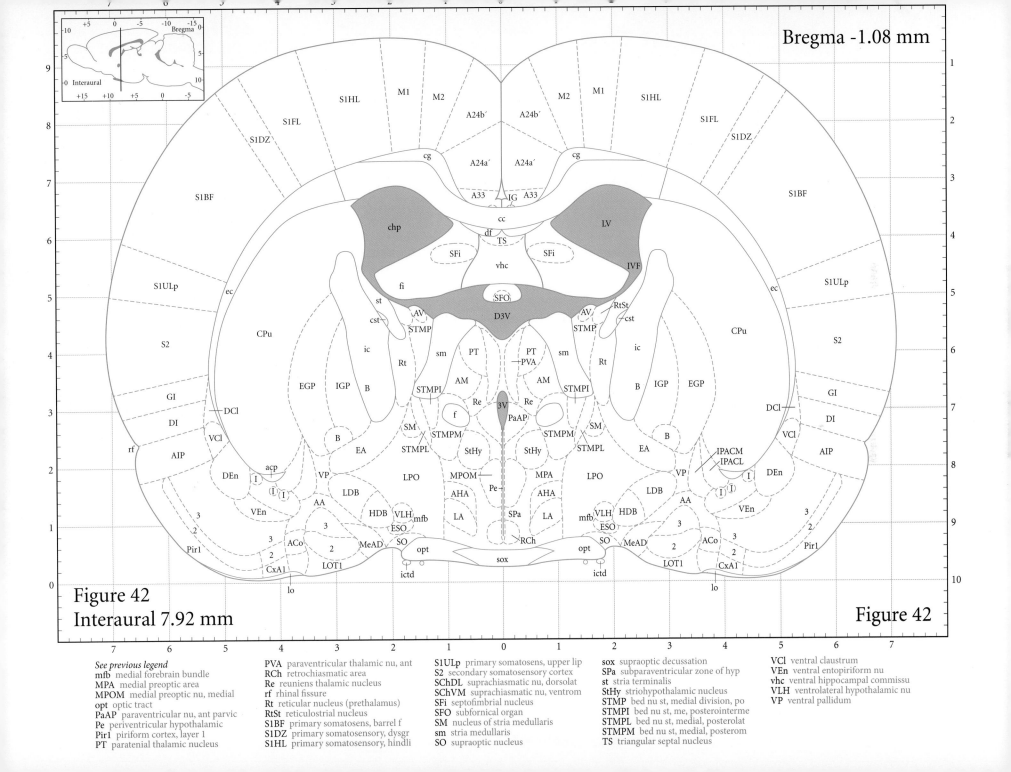

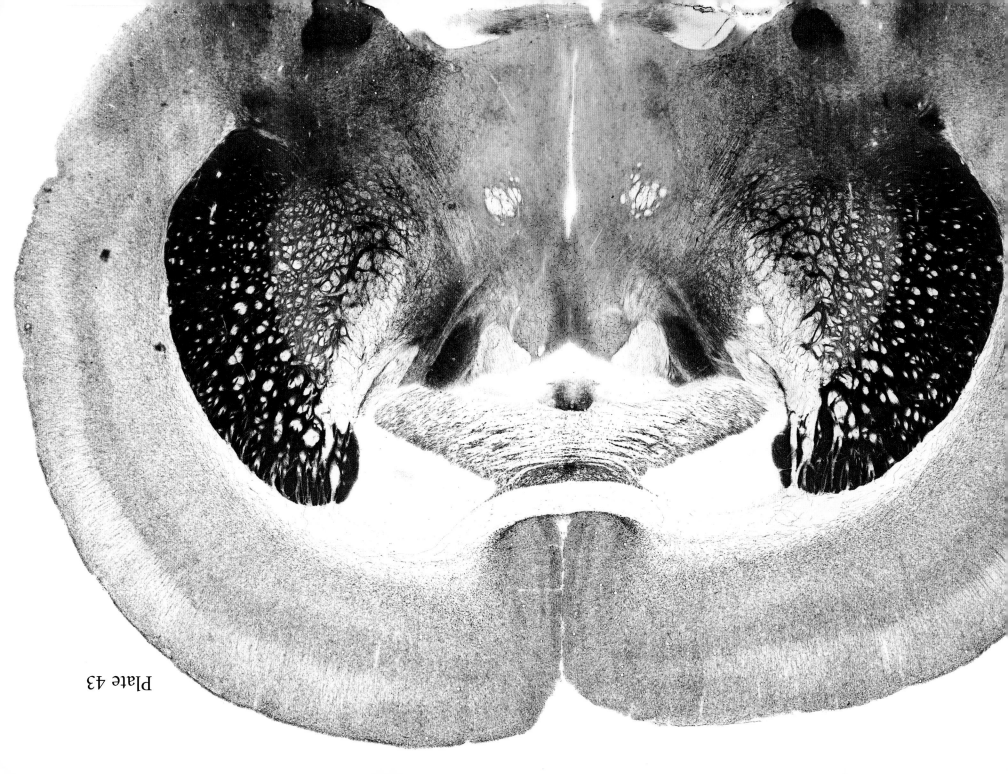

Plate 43

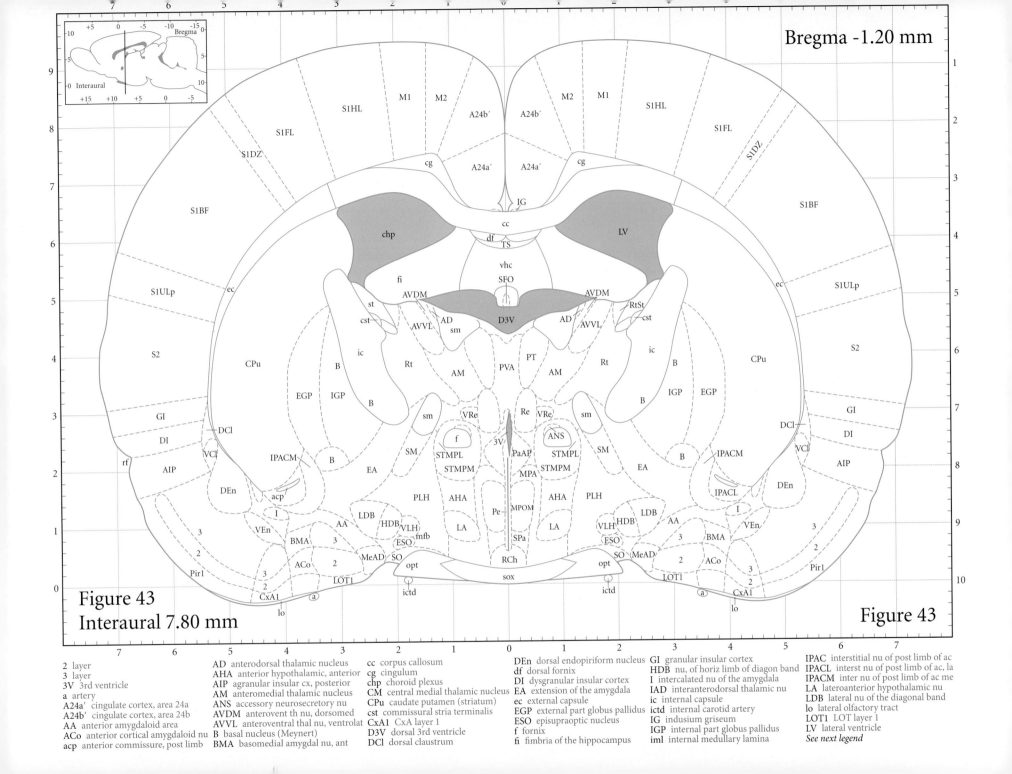

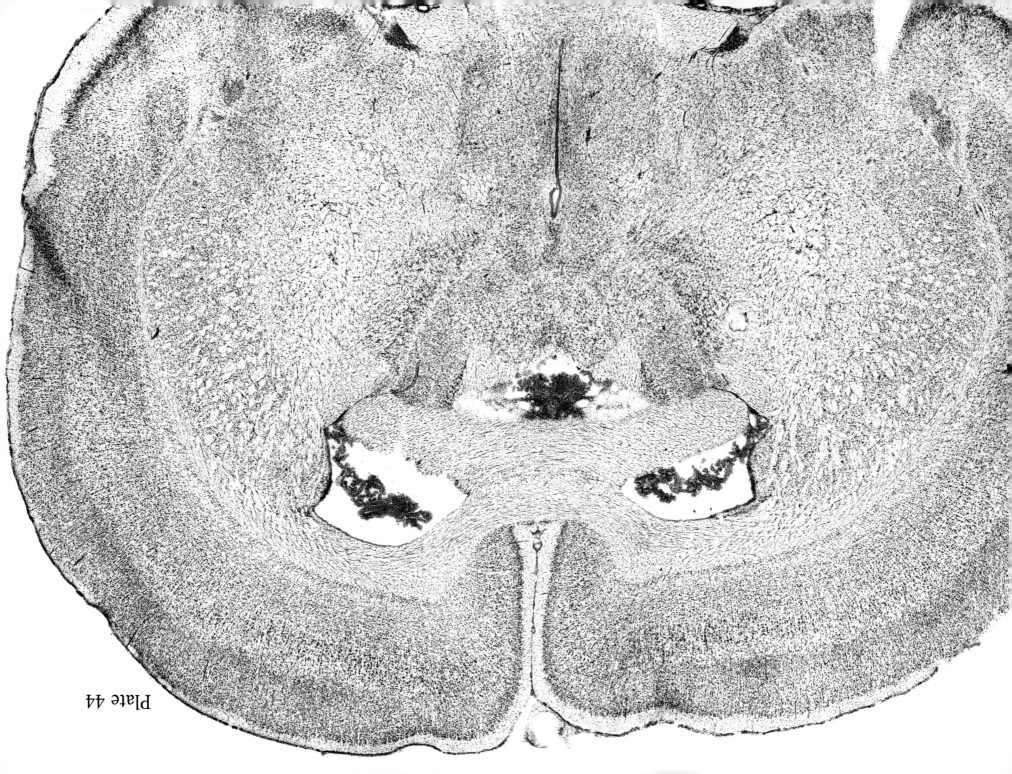

Plate 44

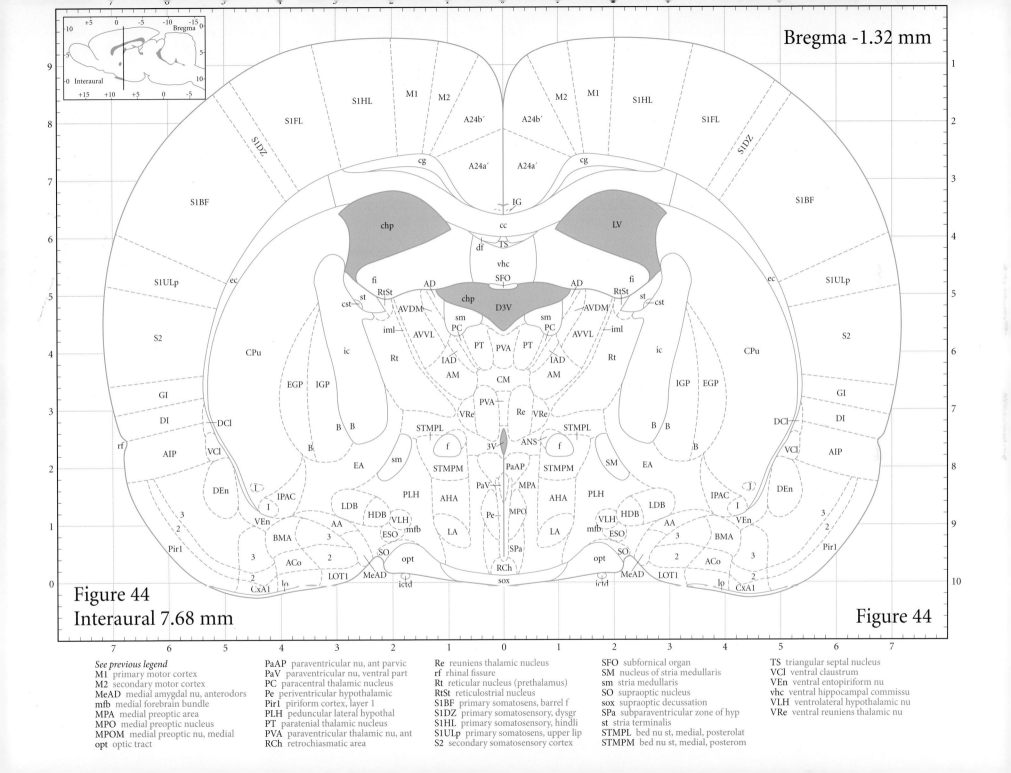

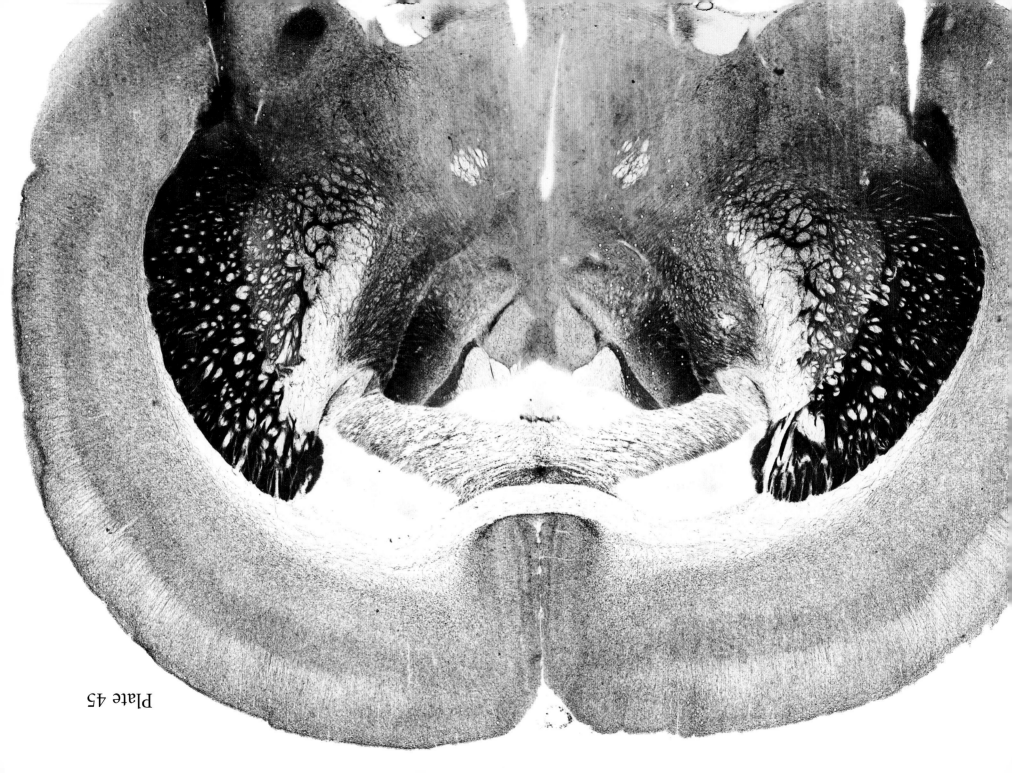

Plate 45

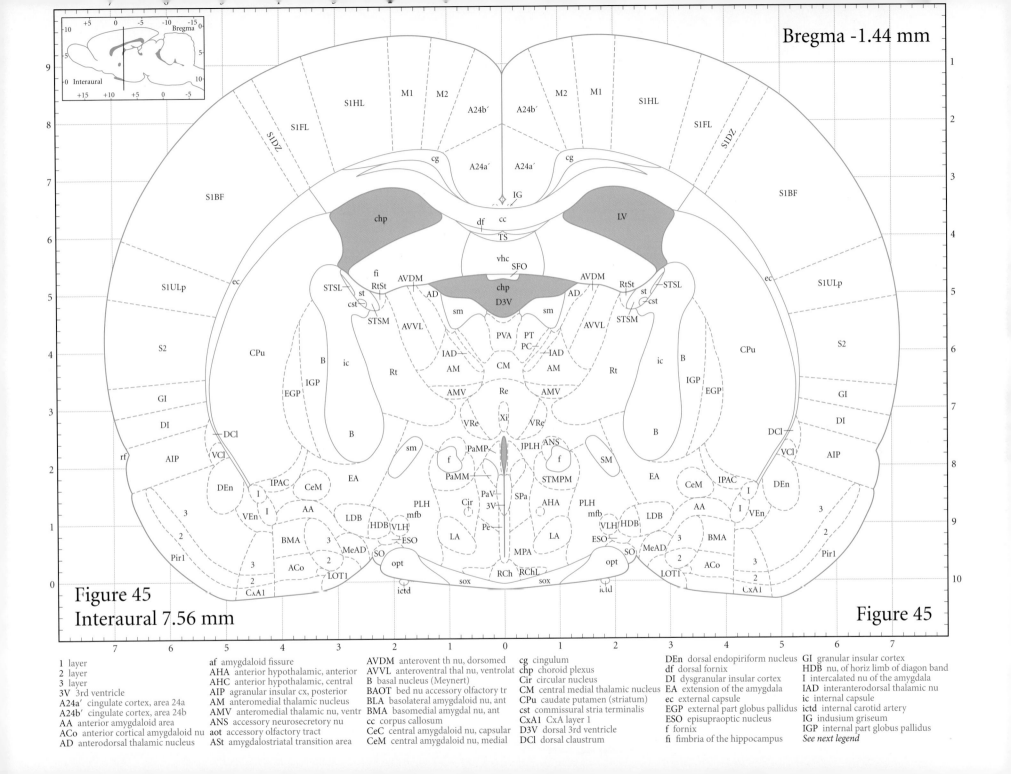

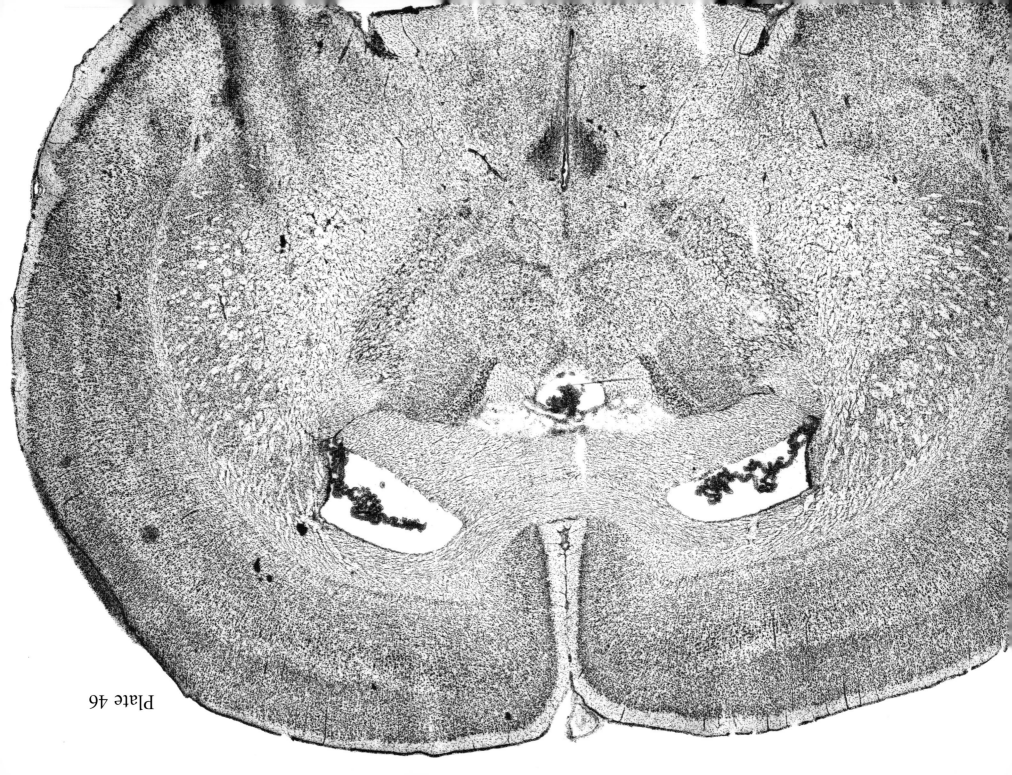

Plate 46

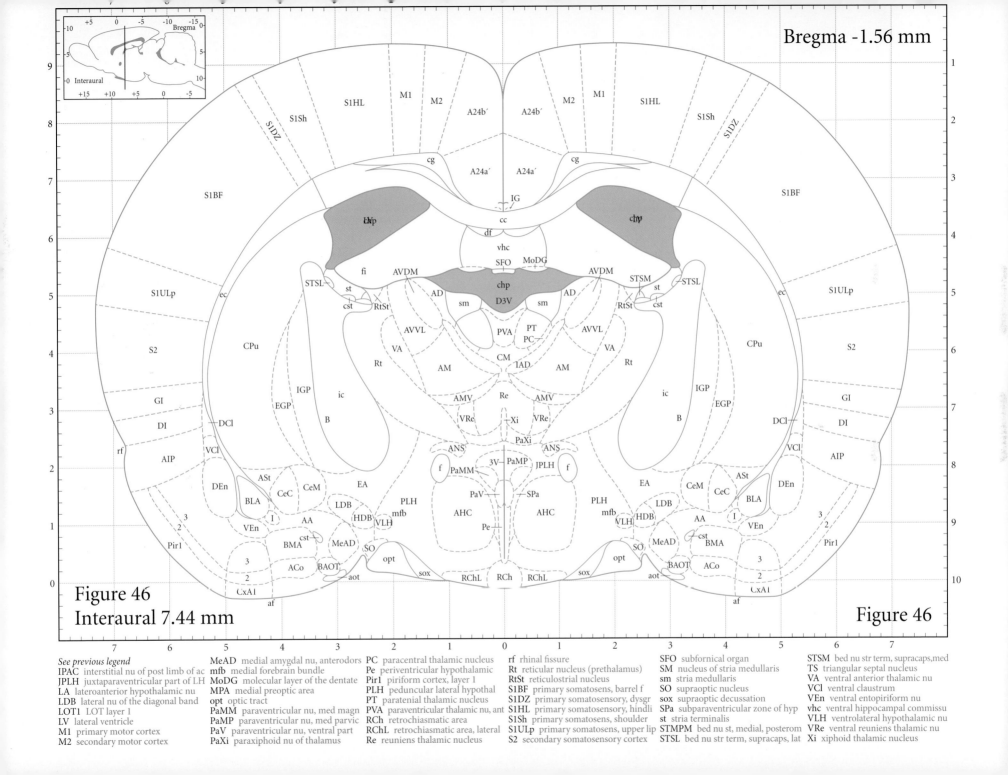

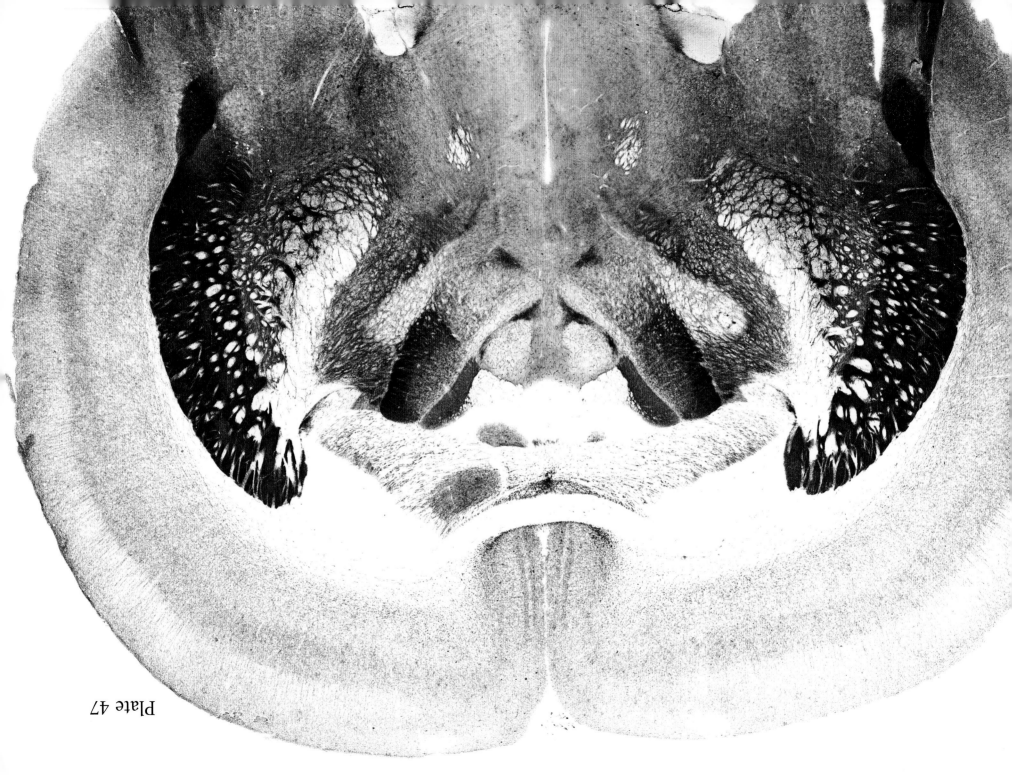

Plate 47

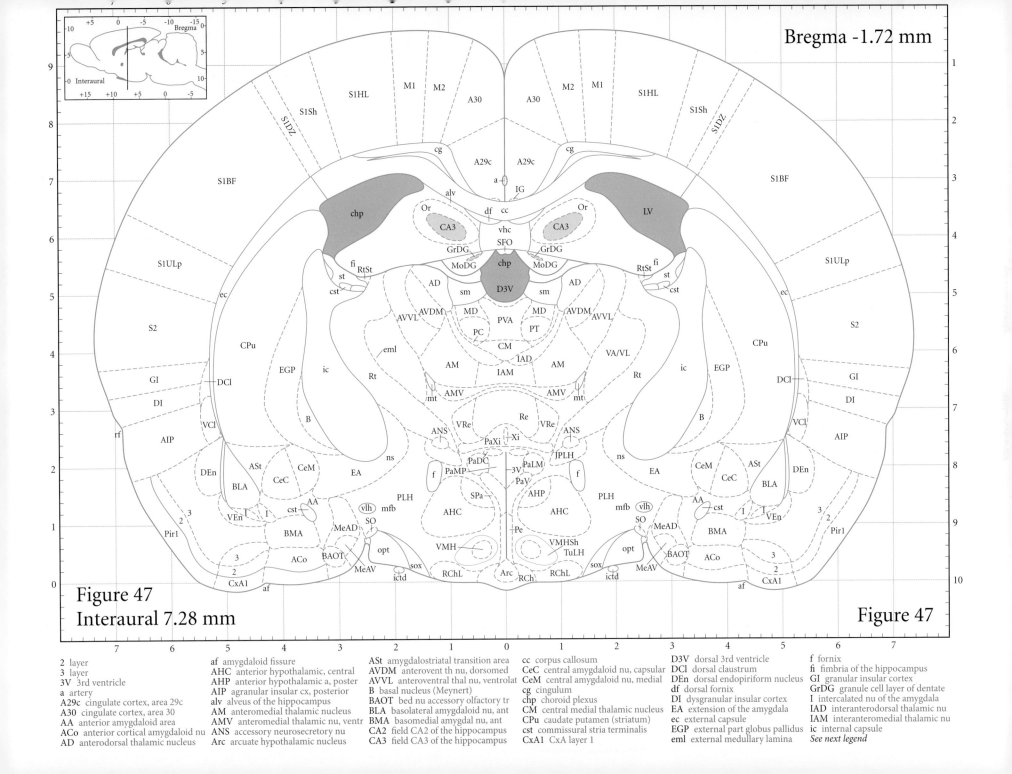

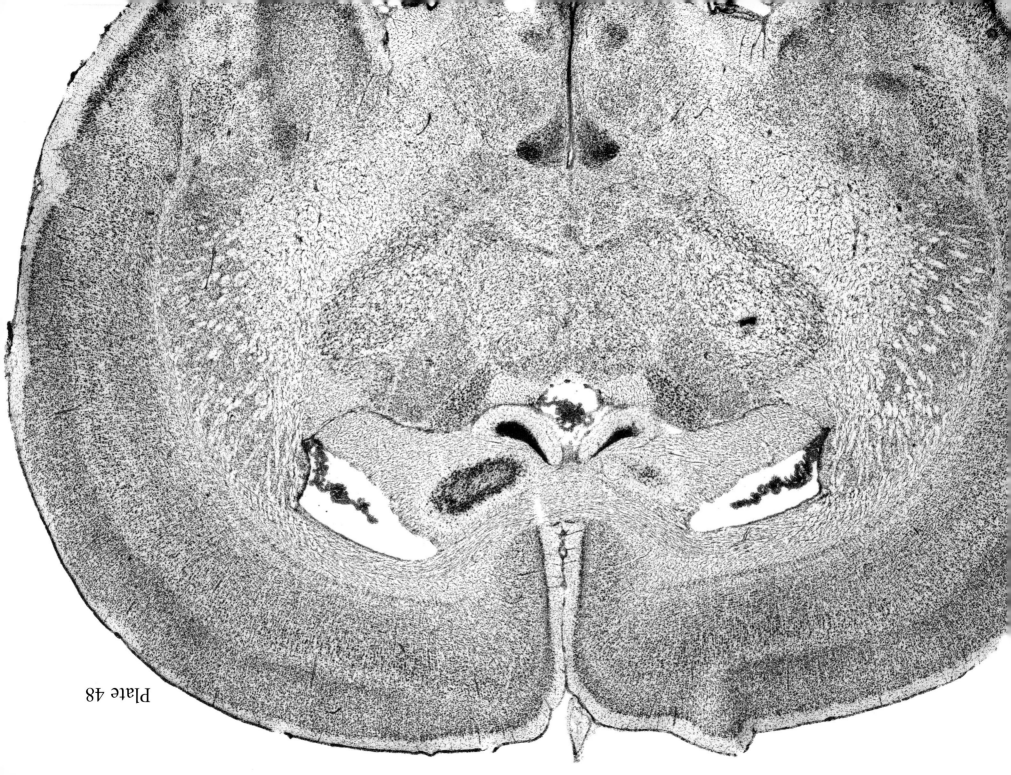

Plate 48

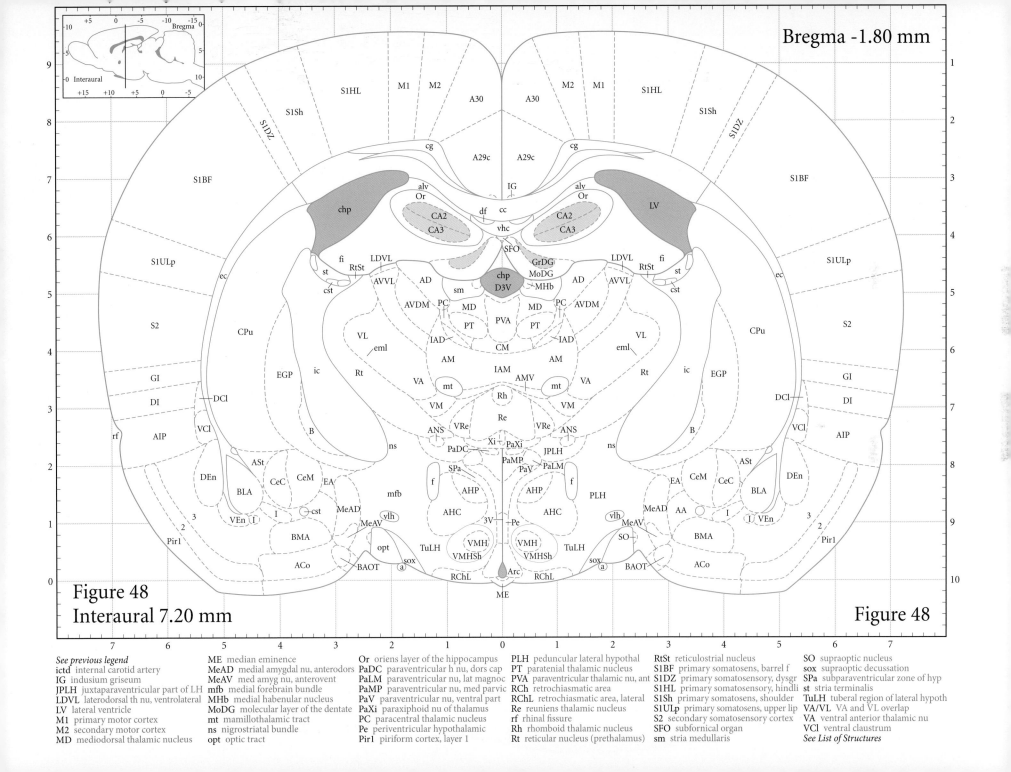

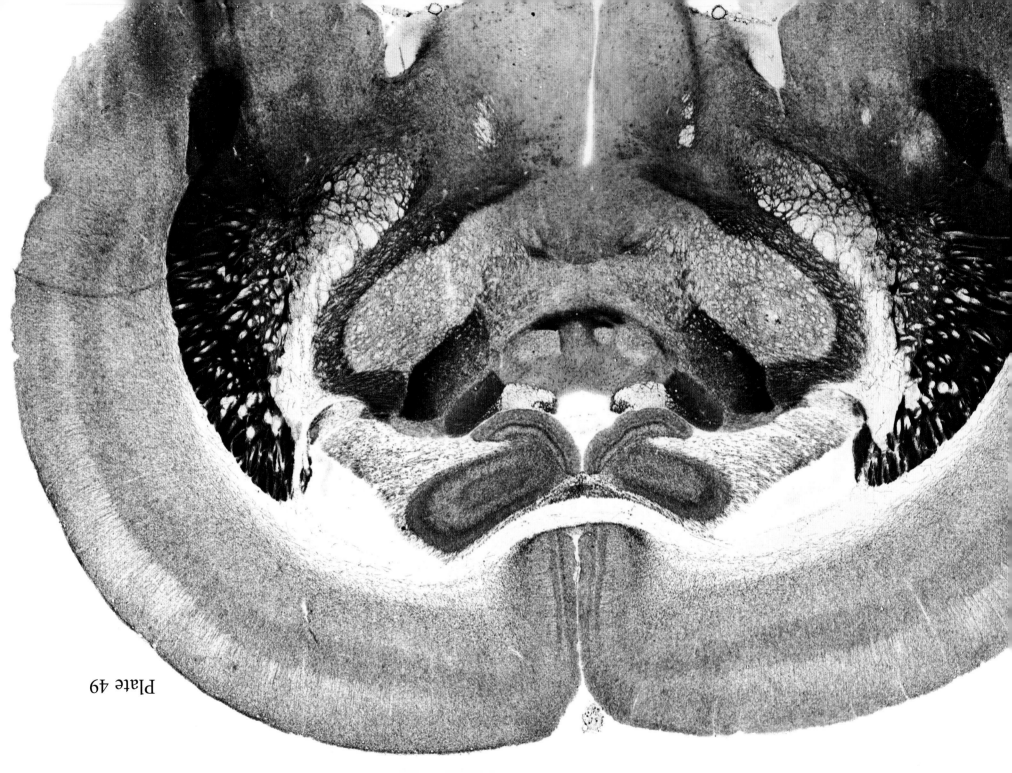

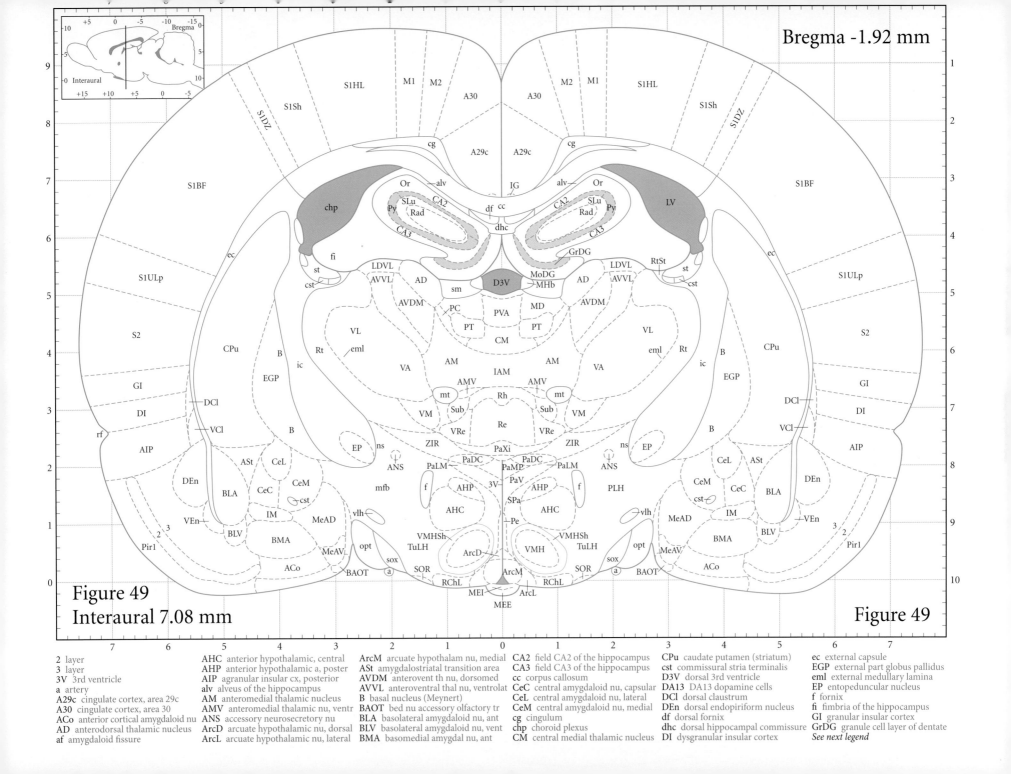

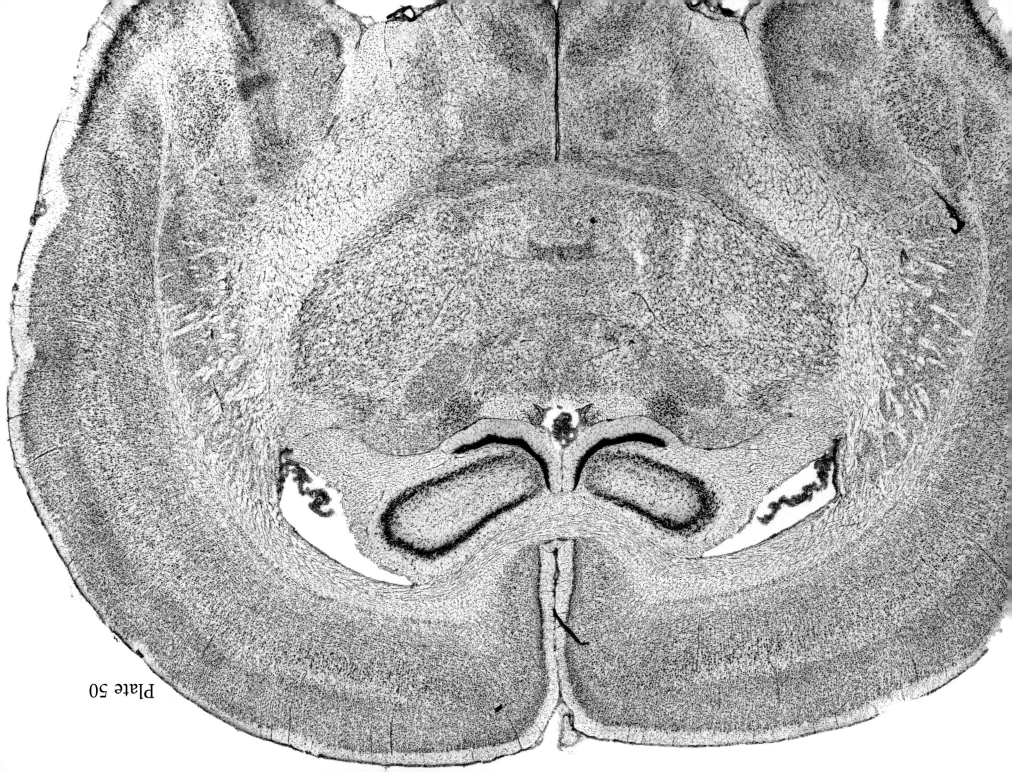

Plate 50

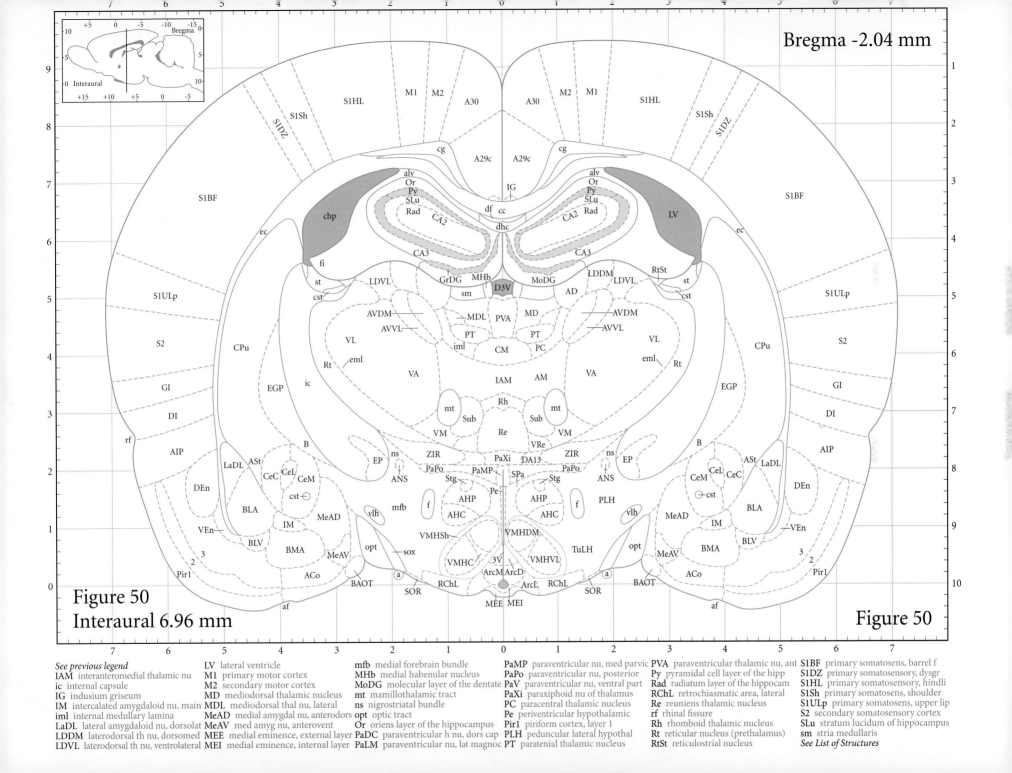

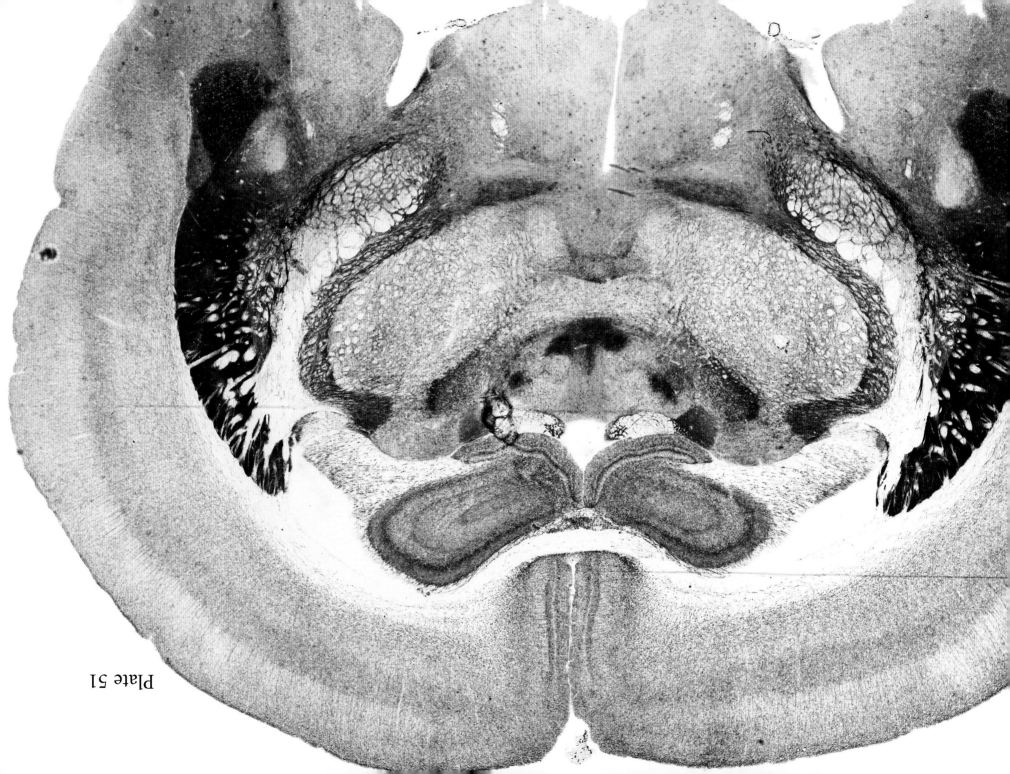

Plate 51

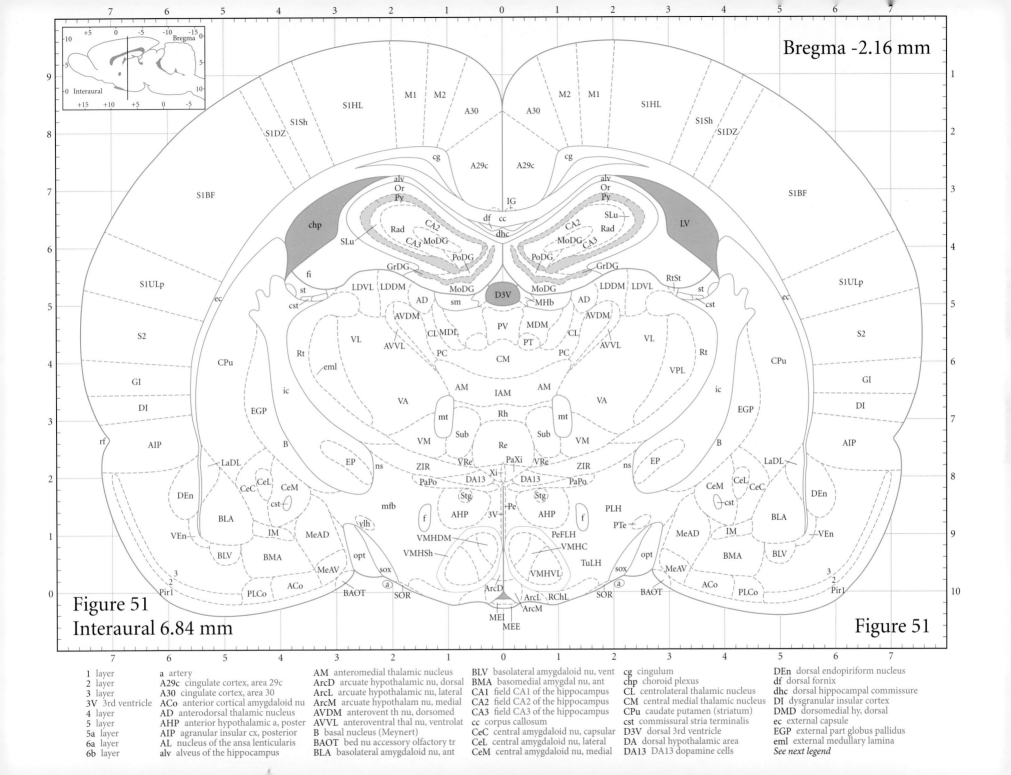

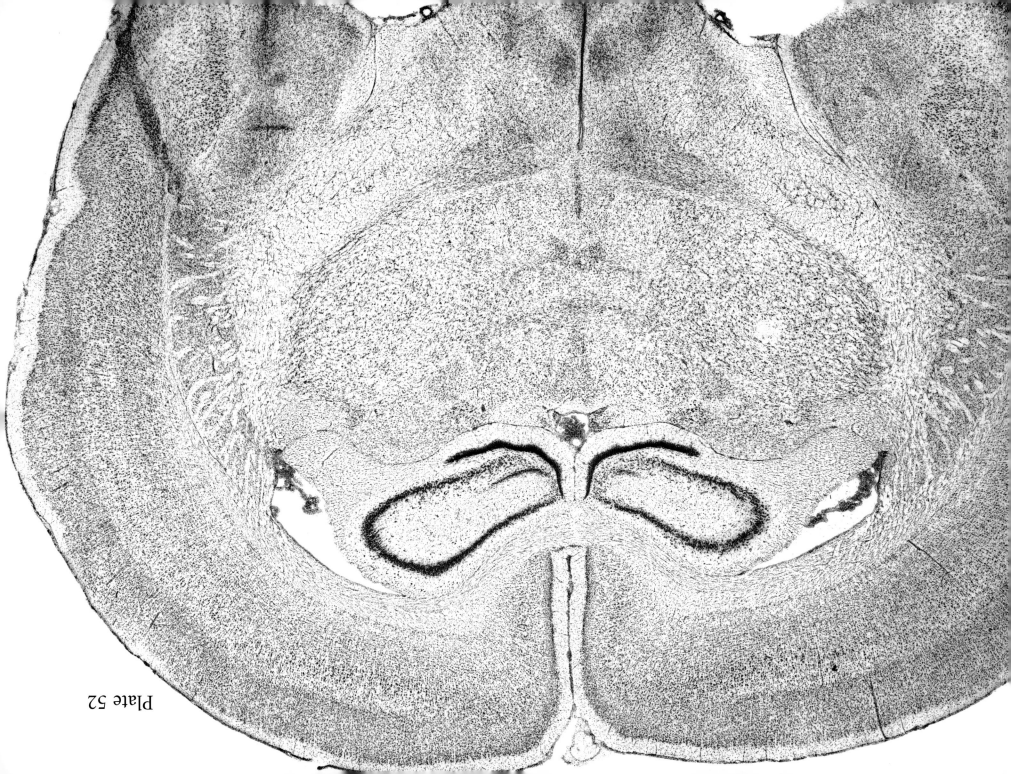

Plate 52

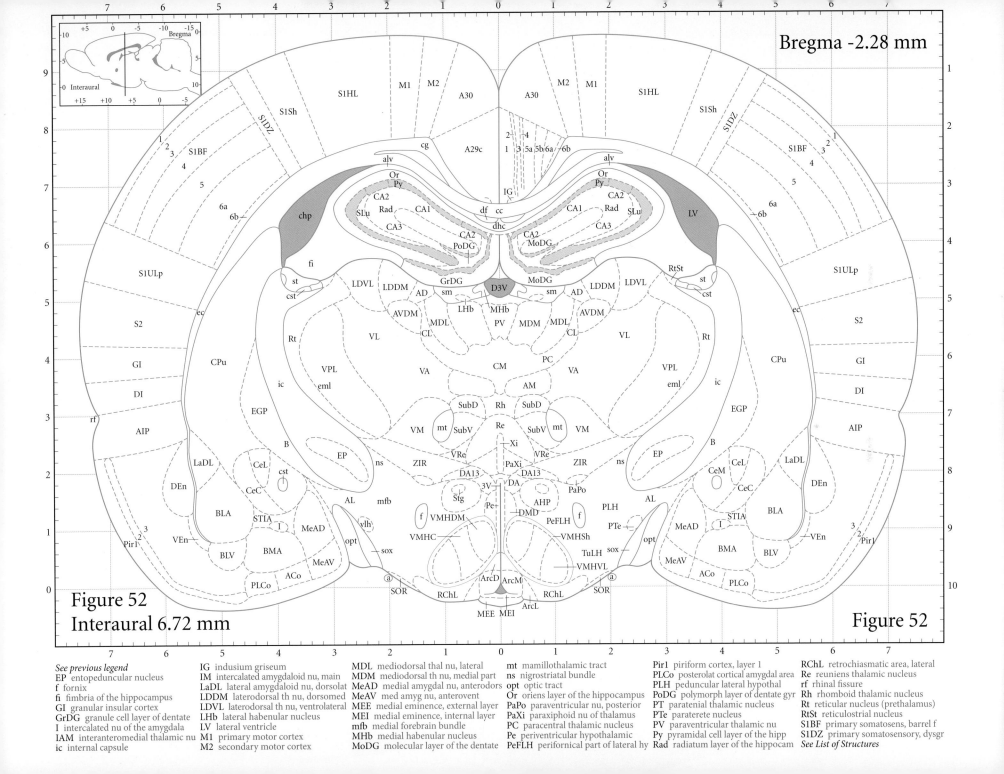

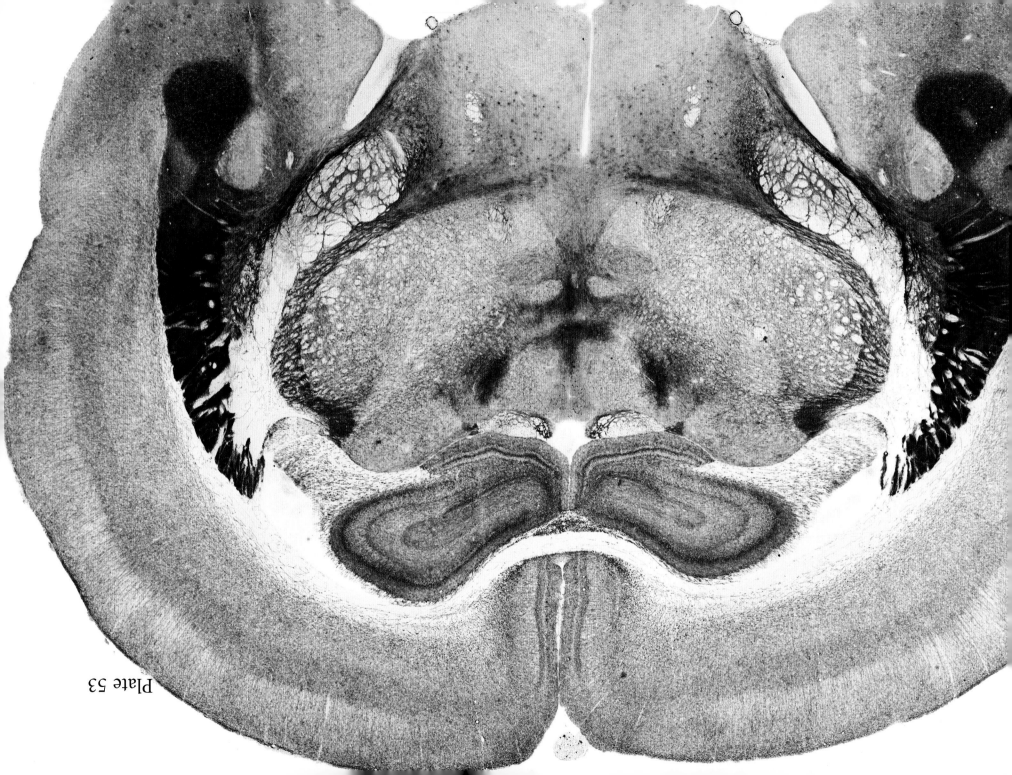

Plate 53

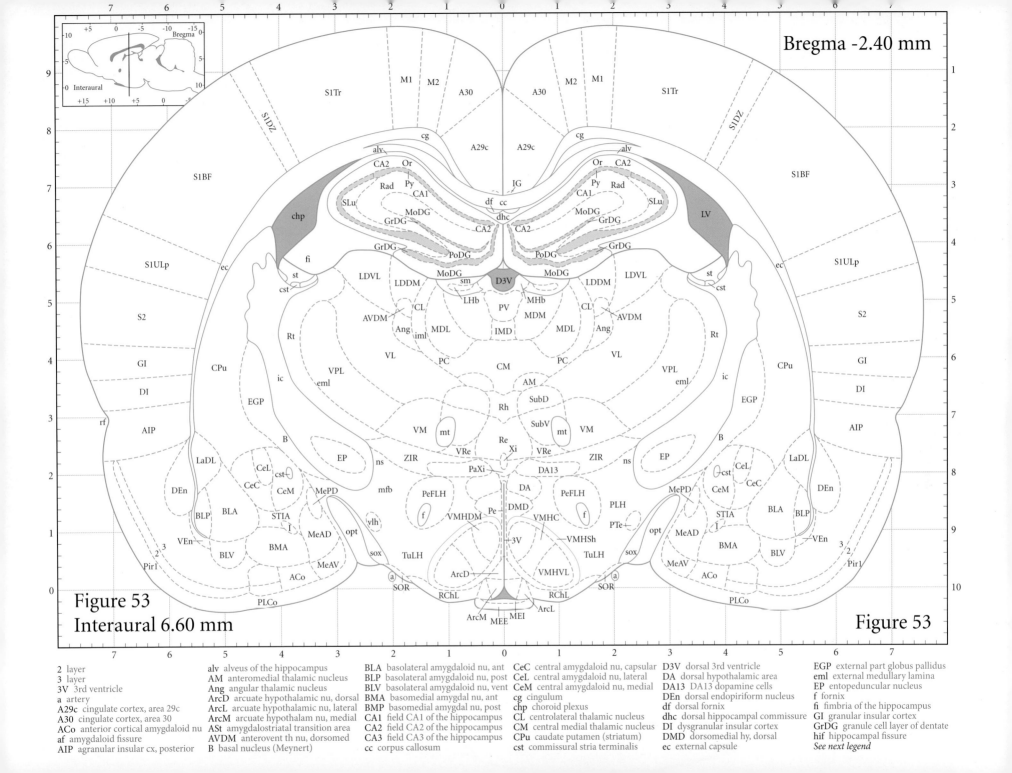

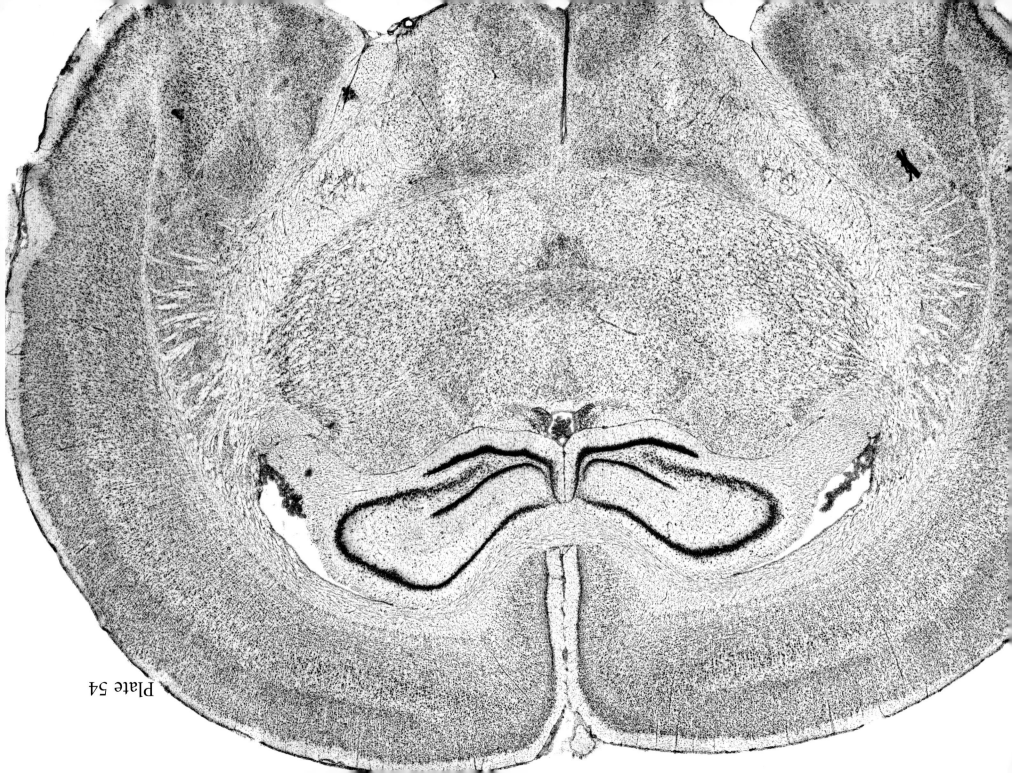

Plate 54

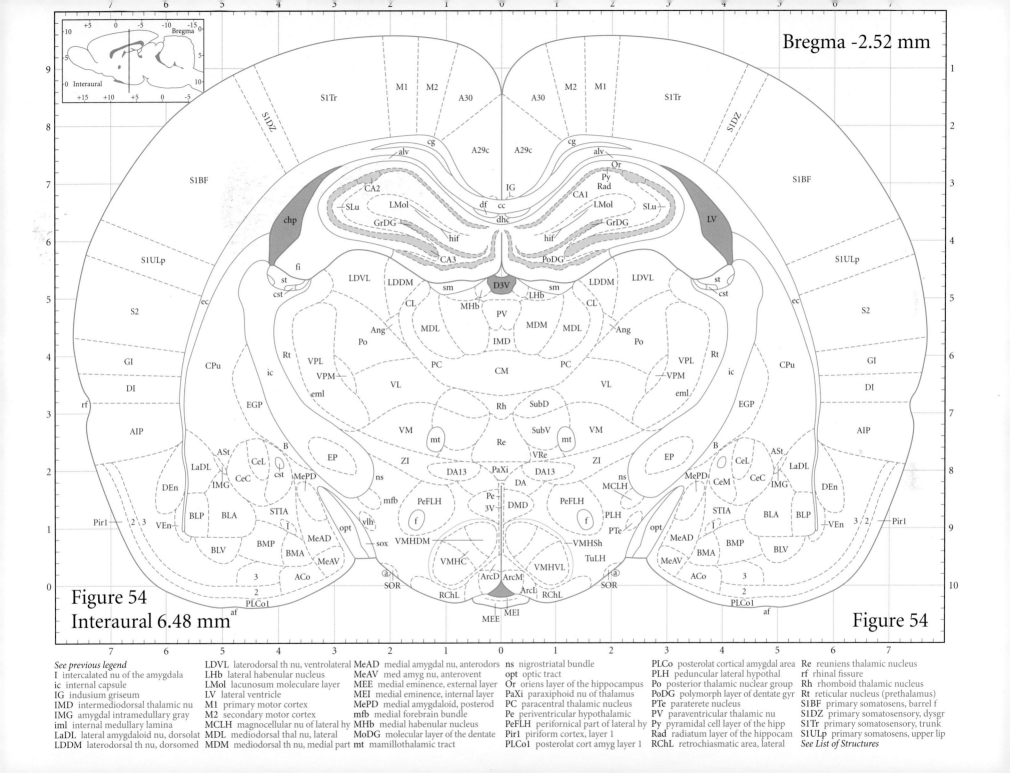

Figure 54
Interaural 6.48 mm
Bregma -2.52 mm

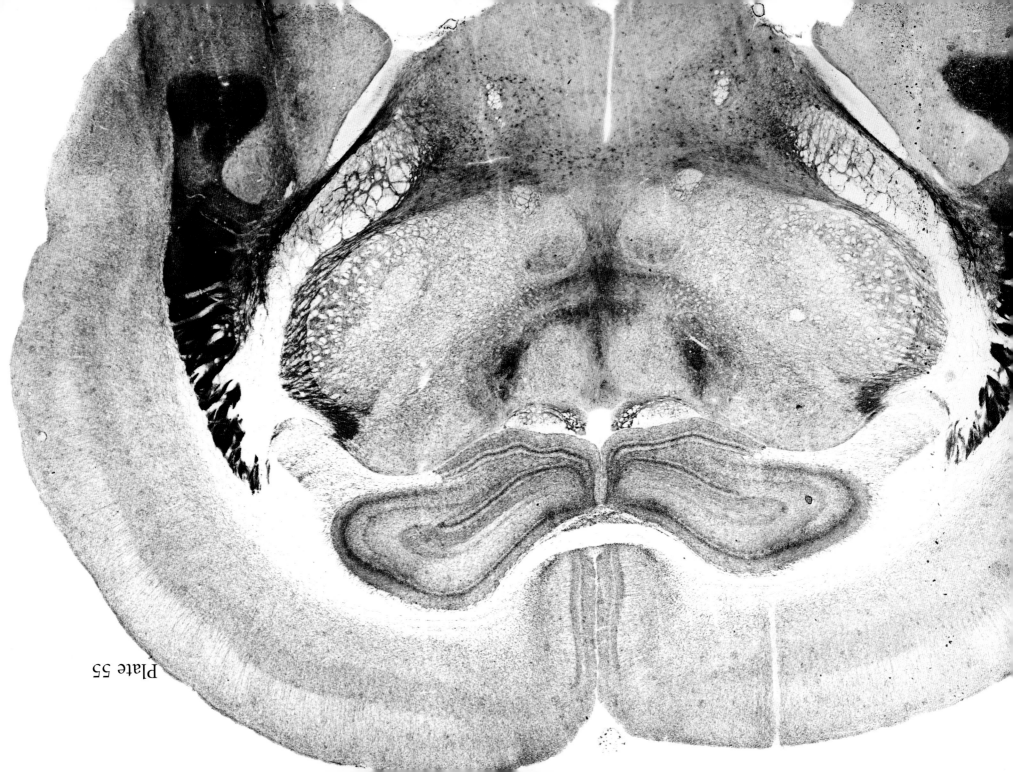

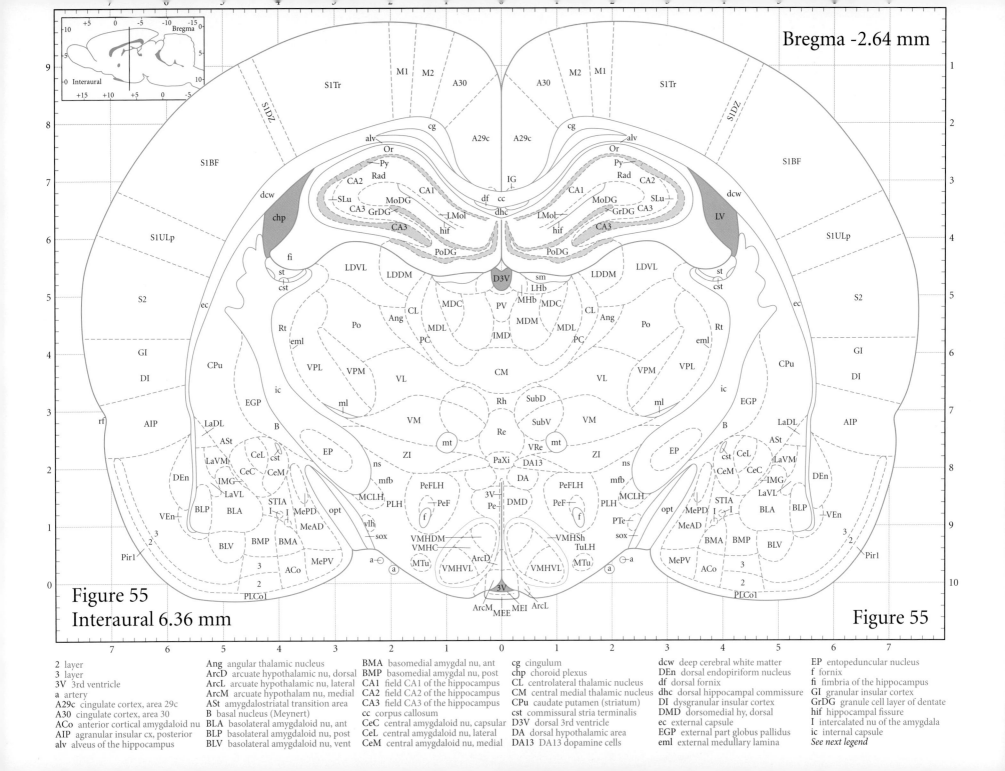

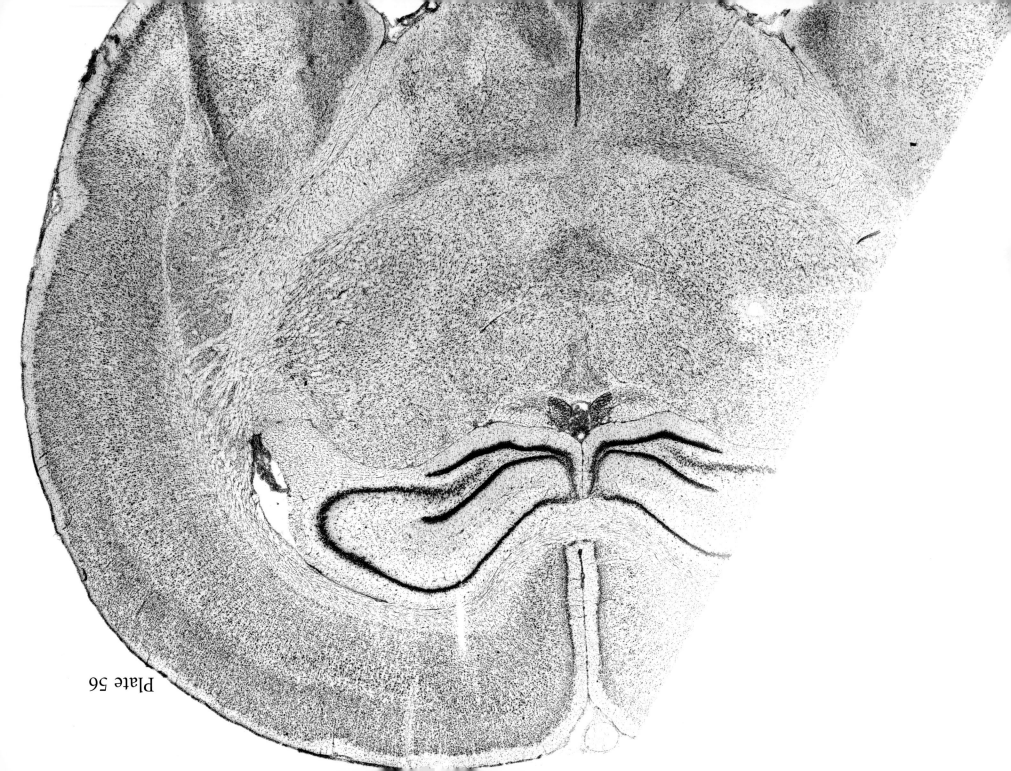

Plate 56

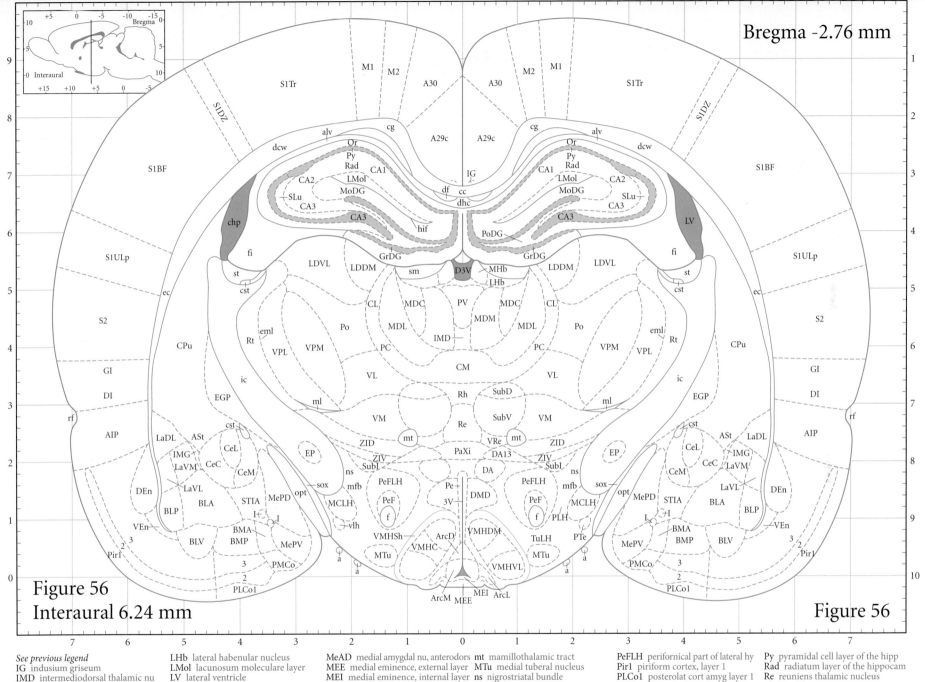

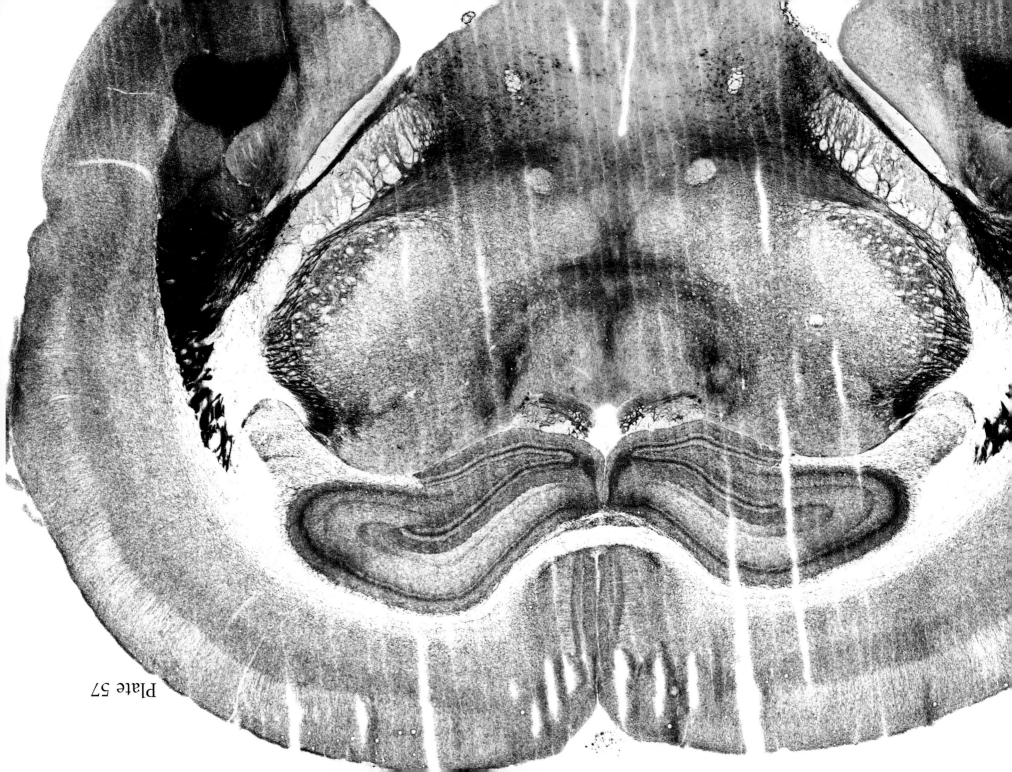

Plate 57

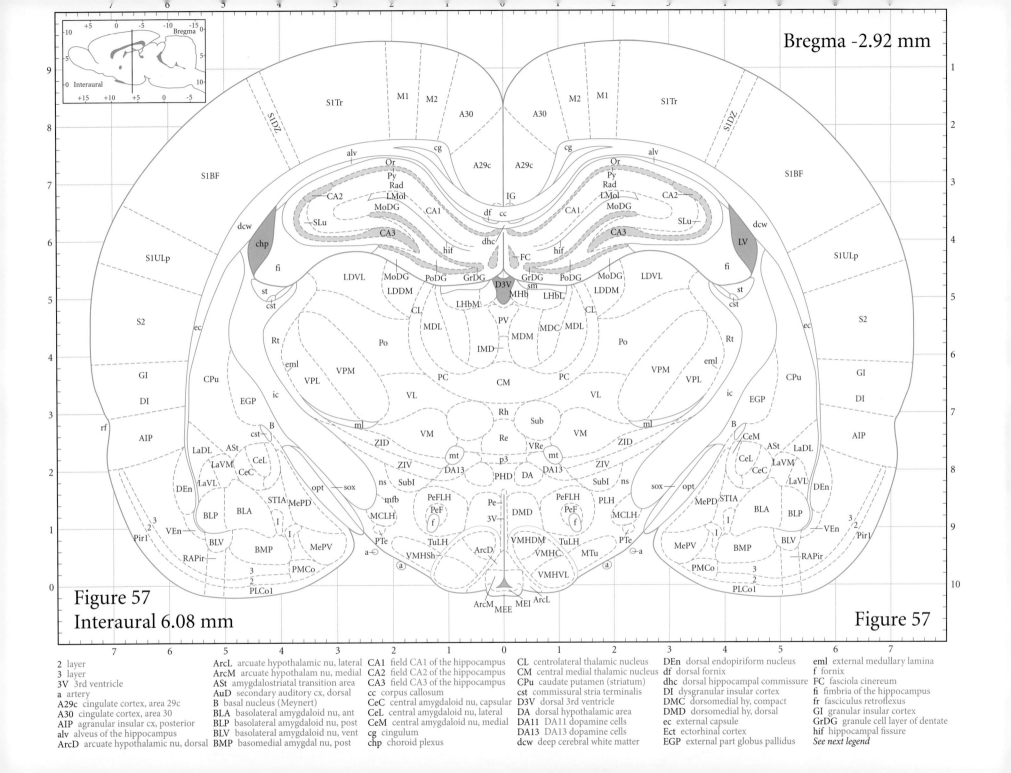

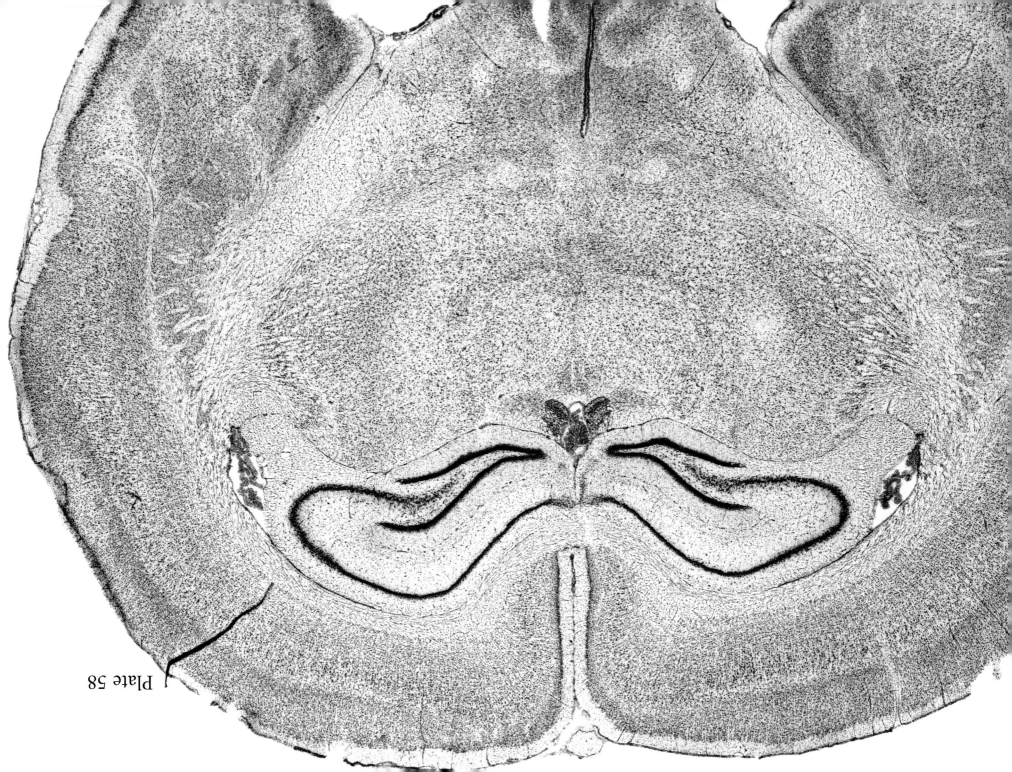

Plate 58

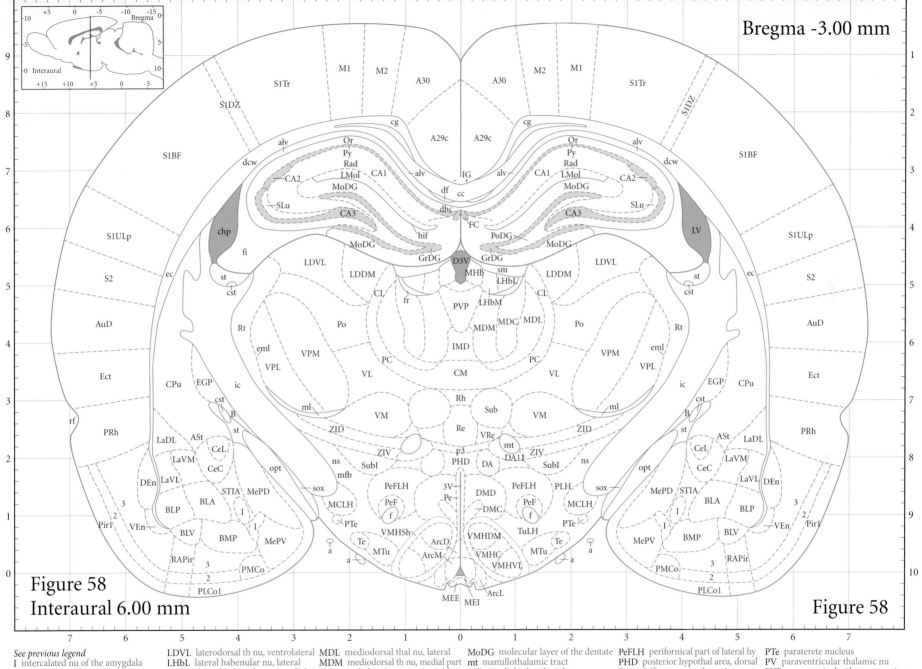

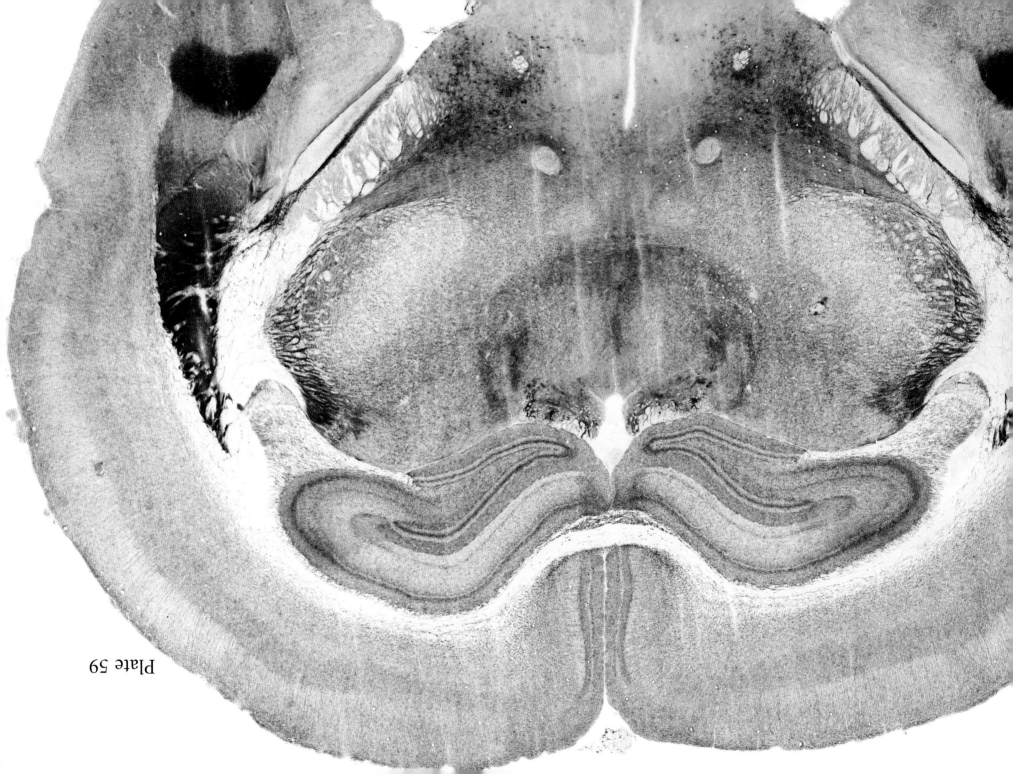

Plate 59

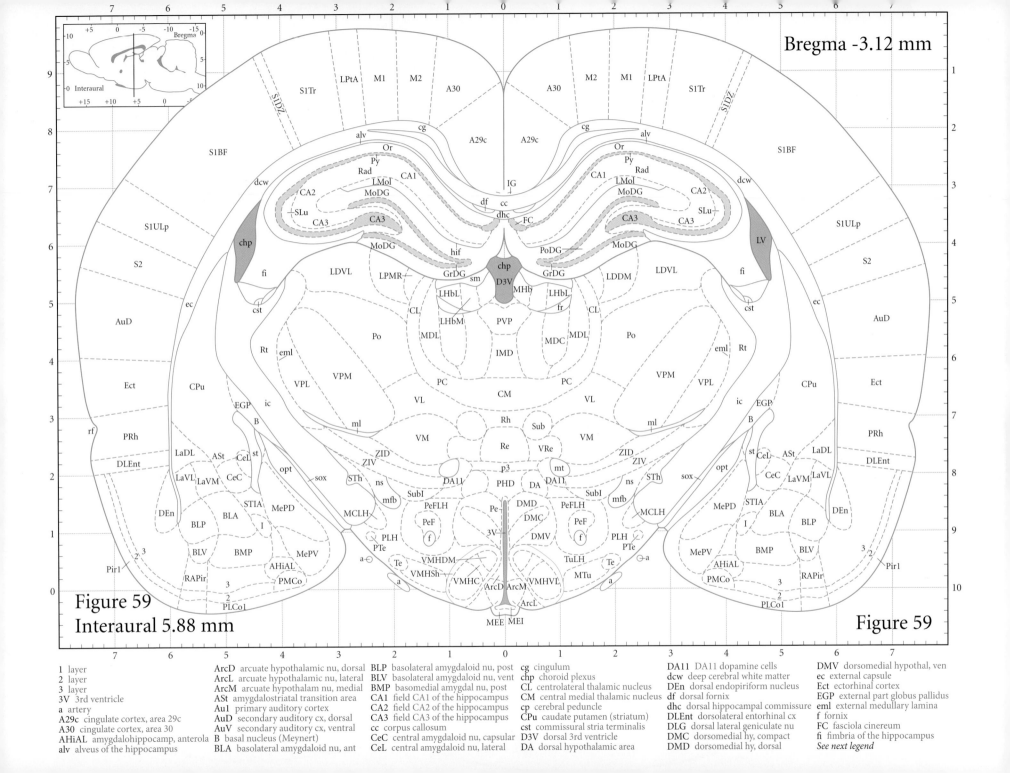

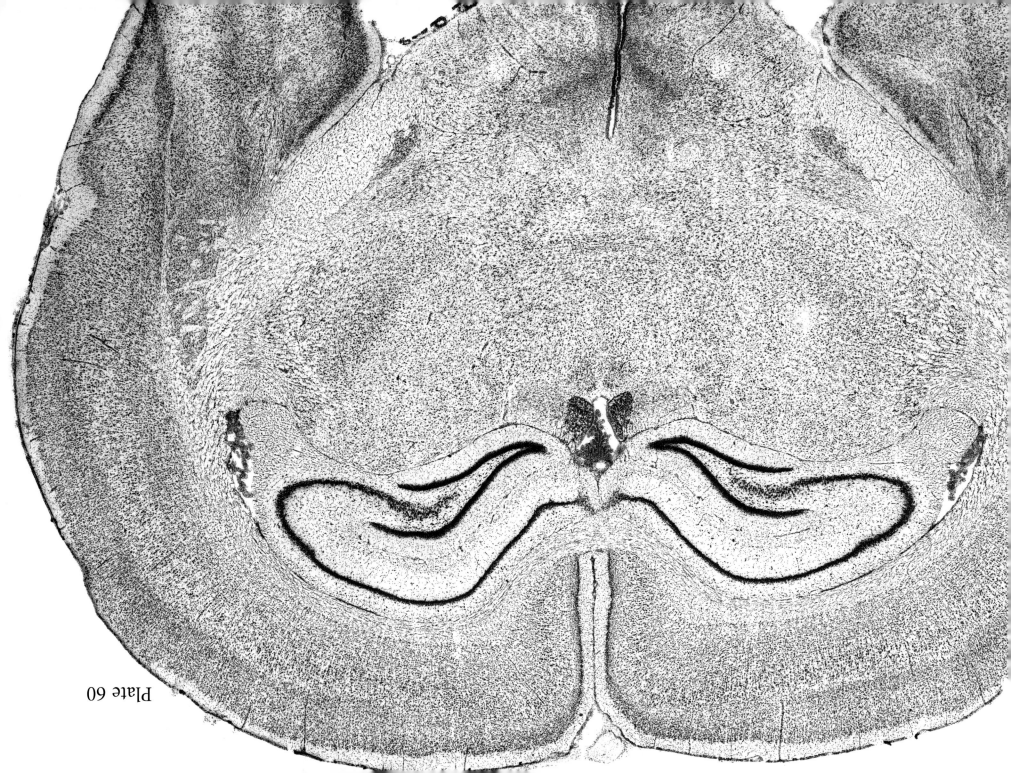

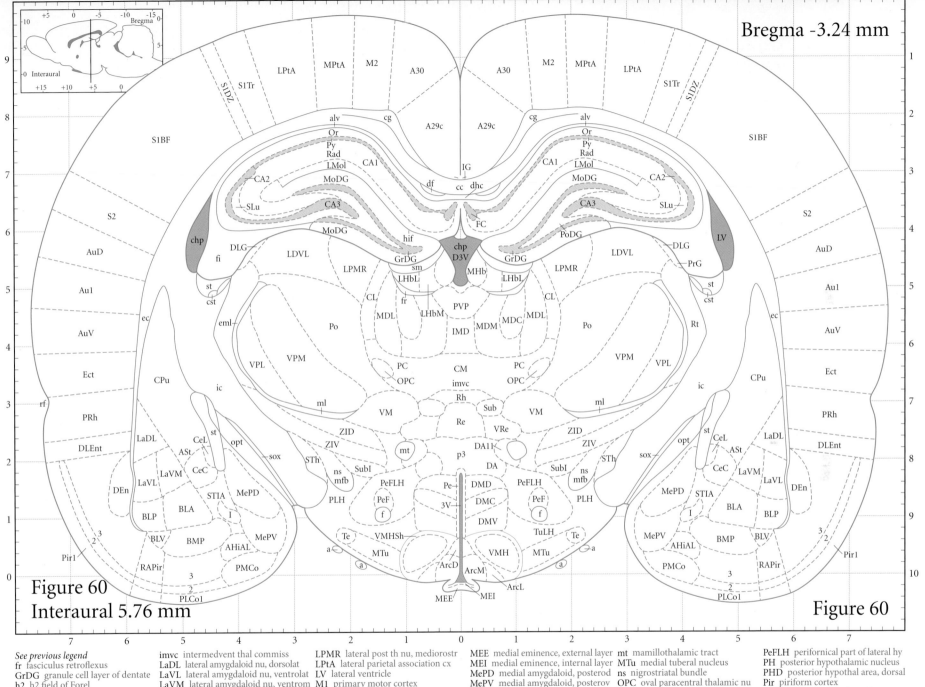

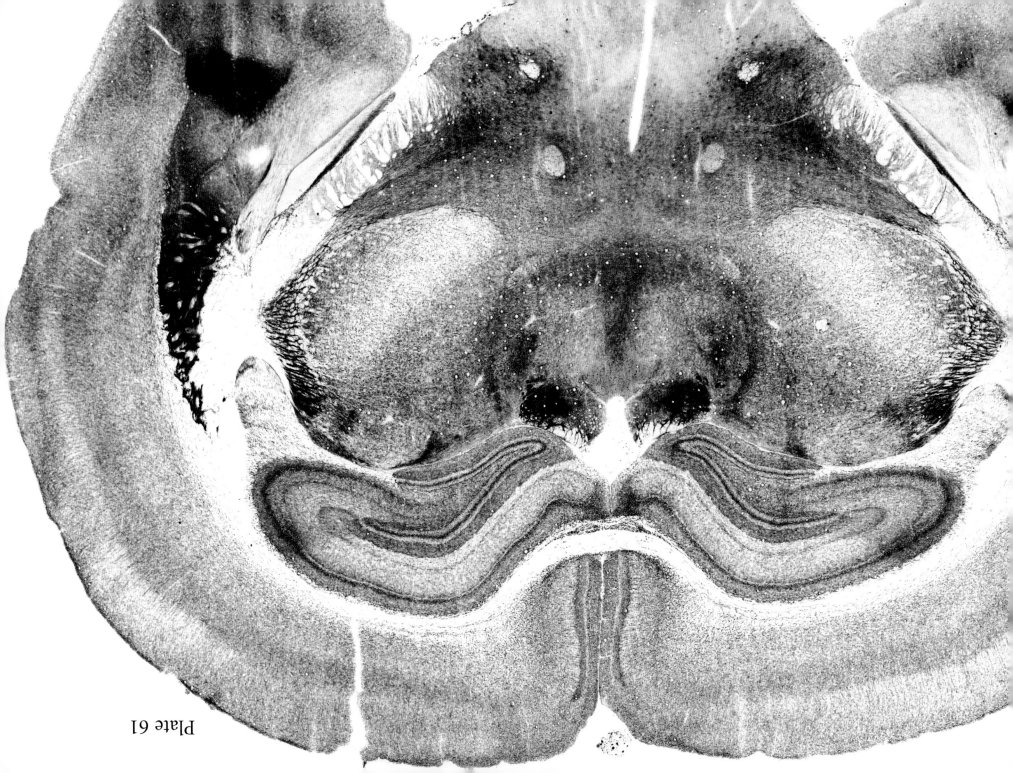

Plate 61

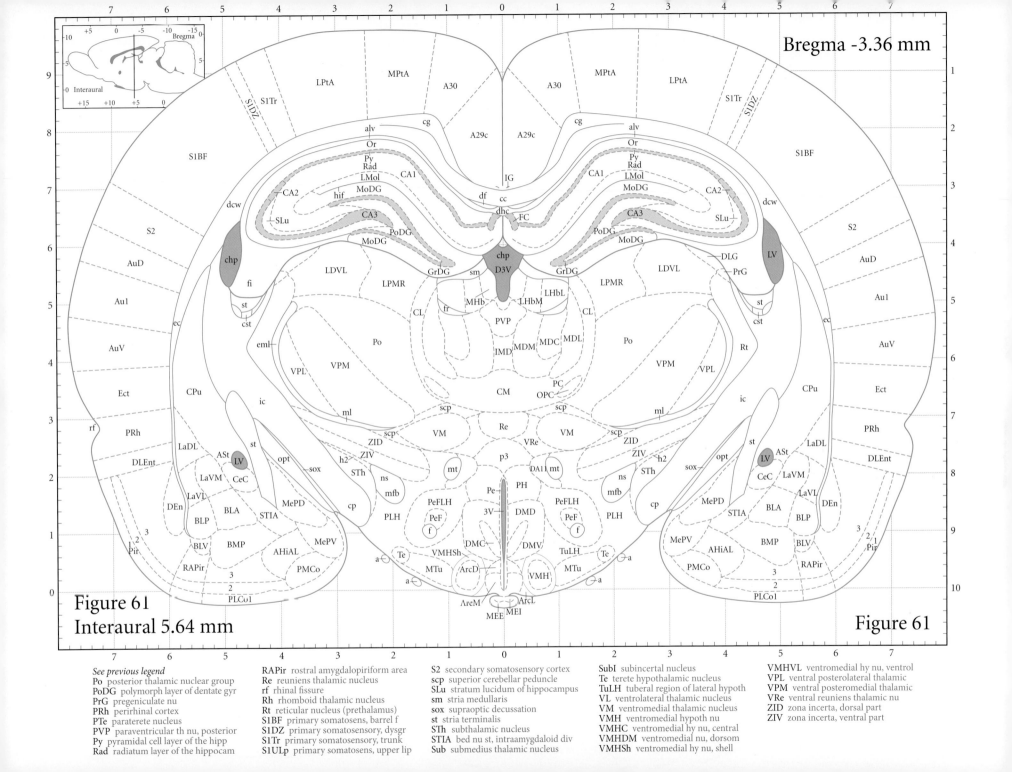

Figure 61
Interaural 5.64 mm
Bregma -3.36 mm

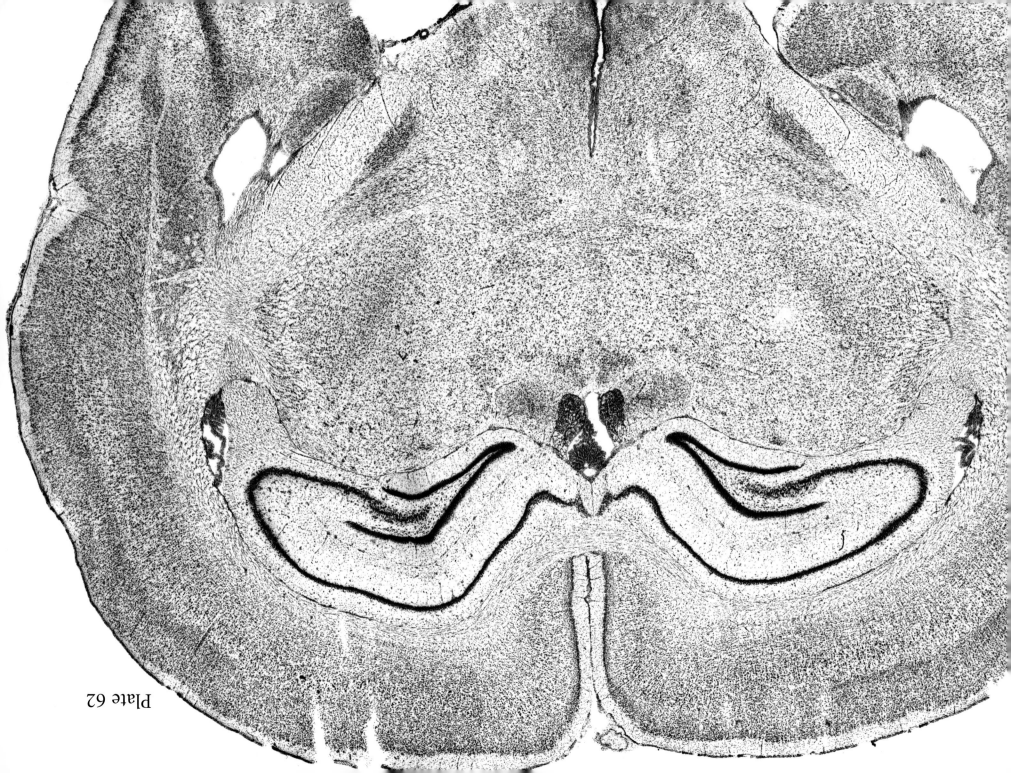

Plate 62

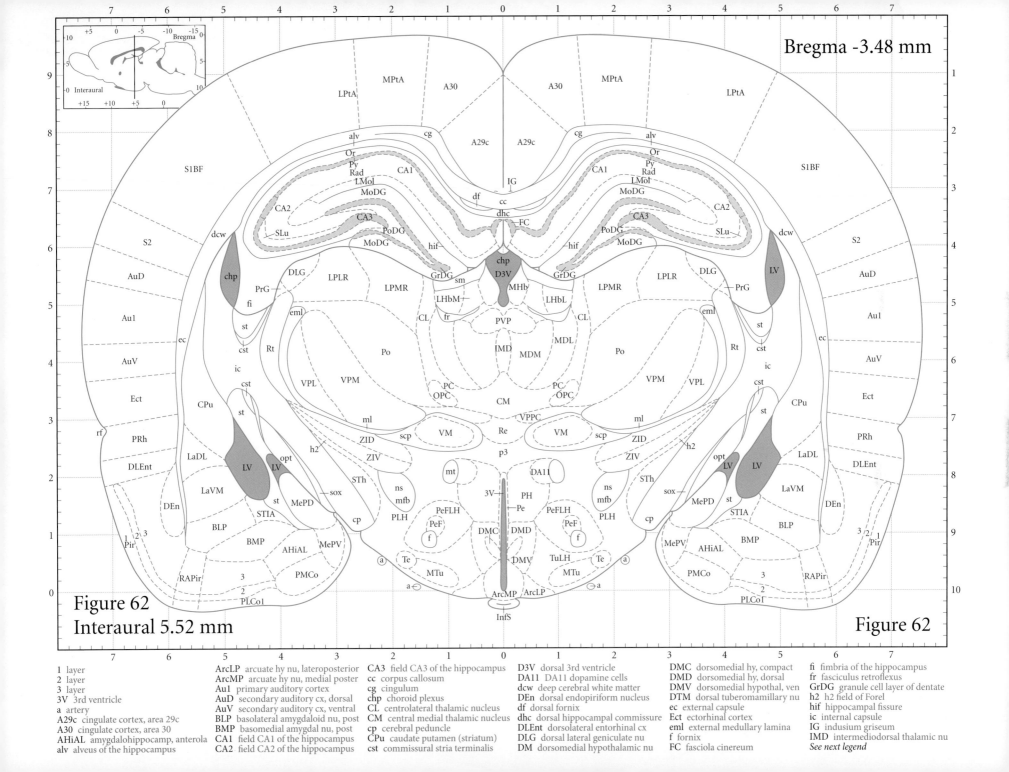

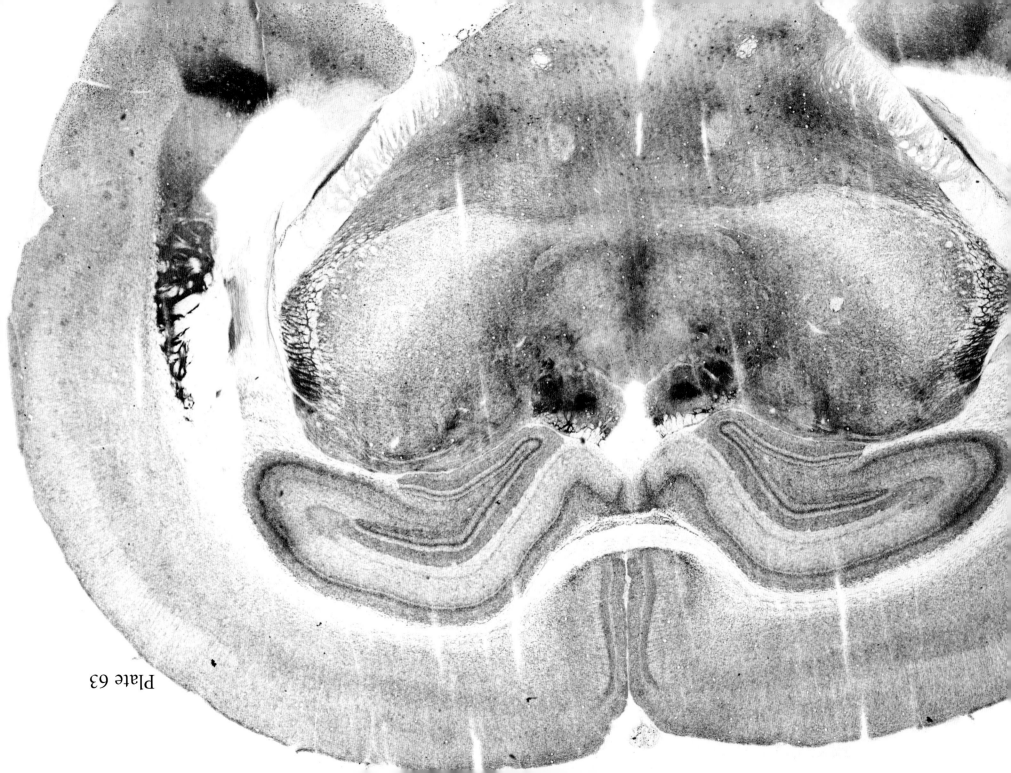

Plate 63

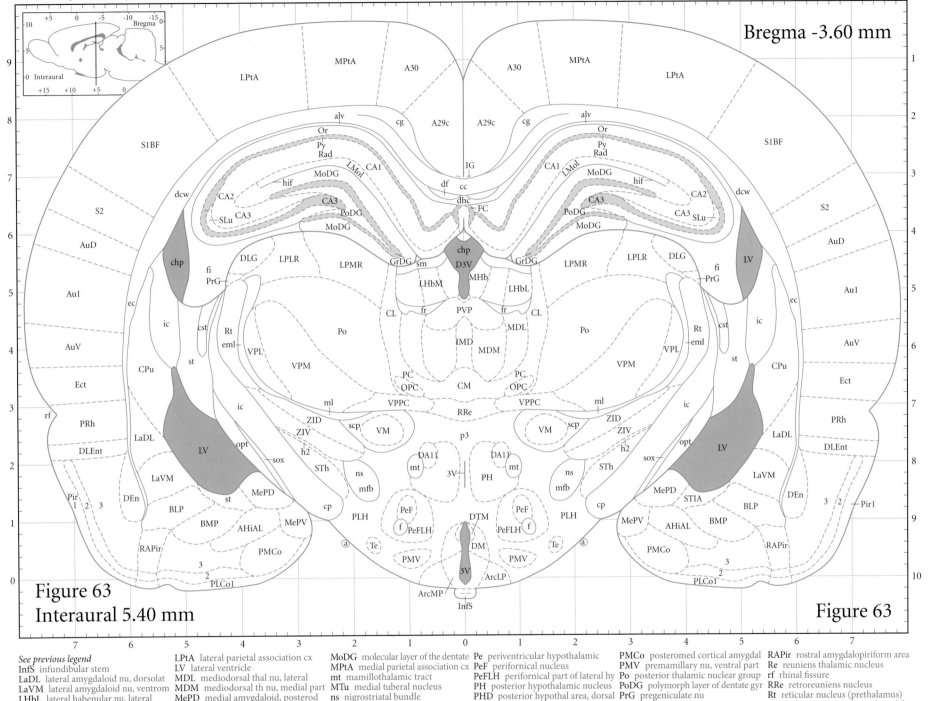

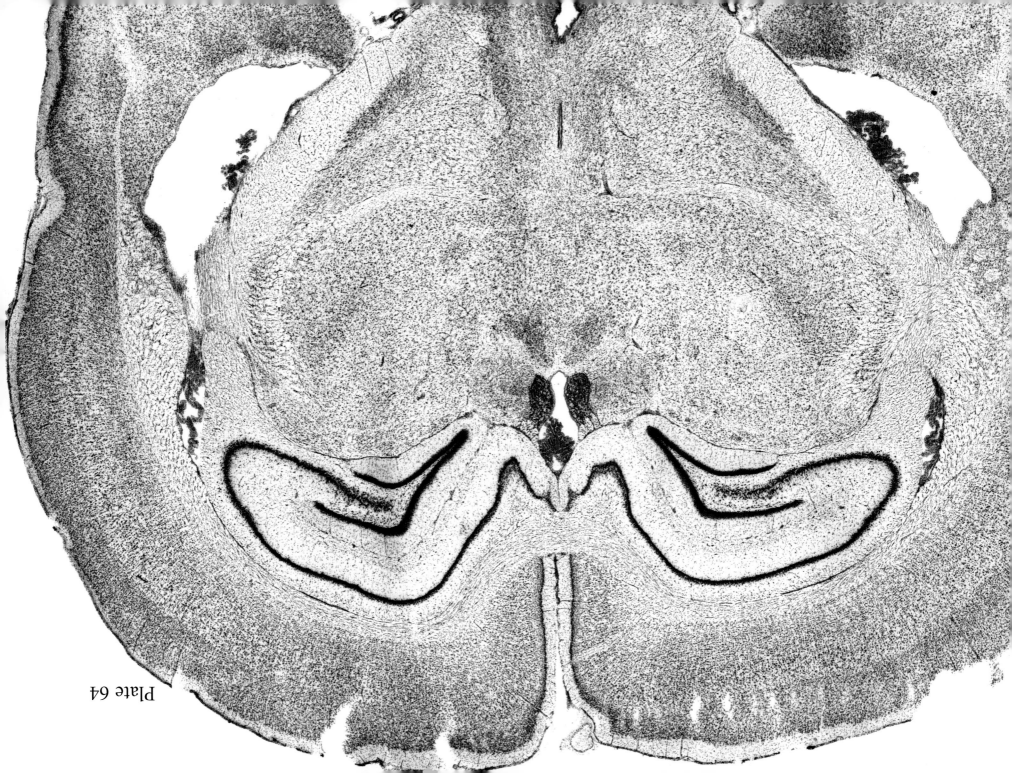

Plate 64

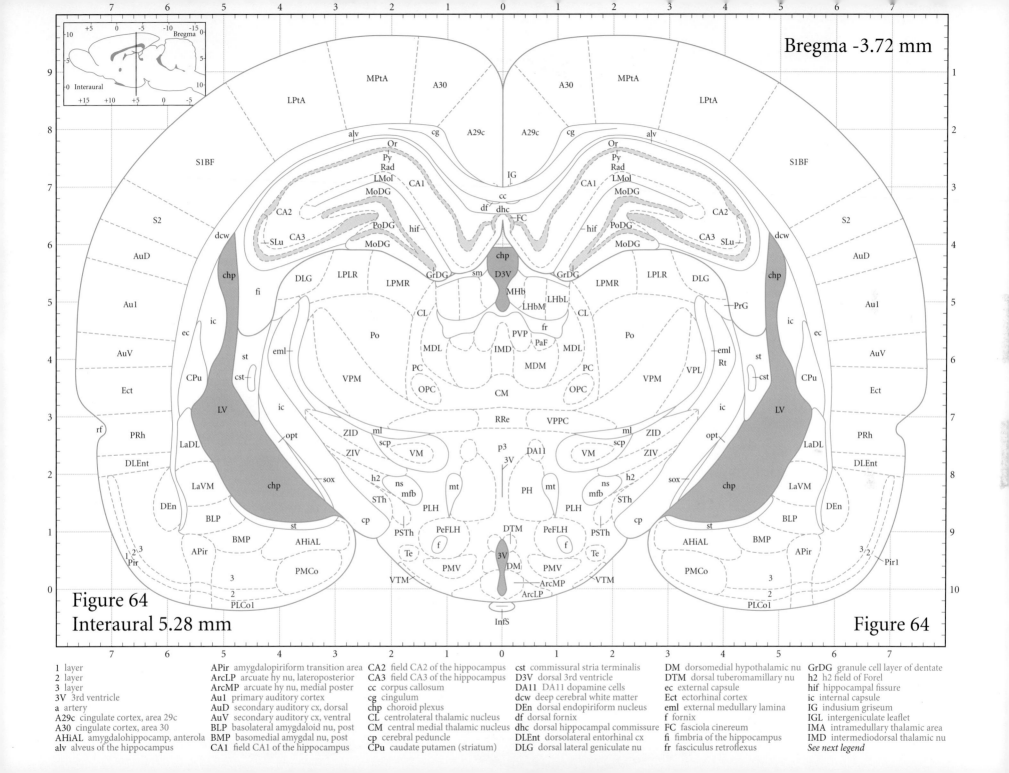

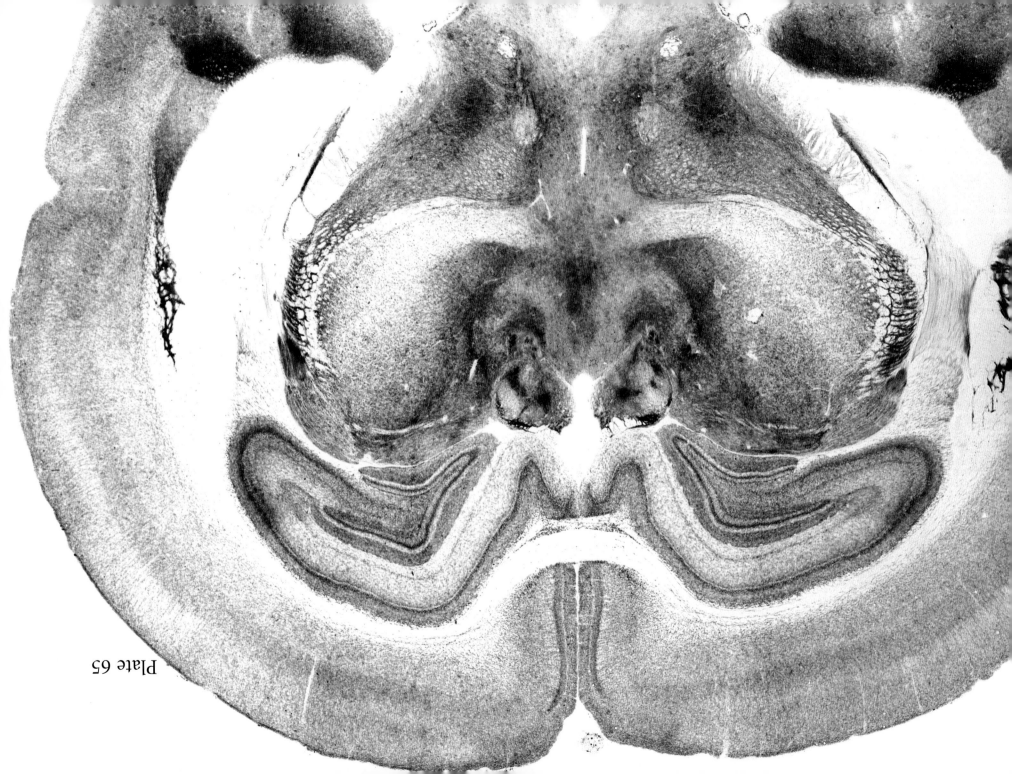

Plate 65

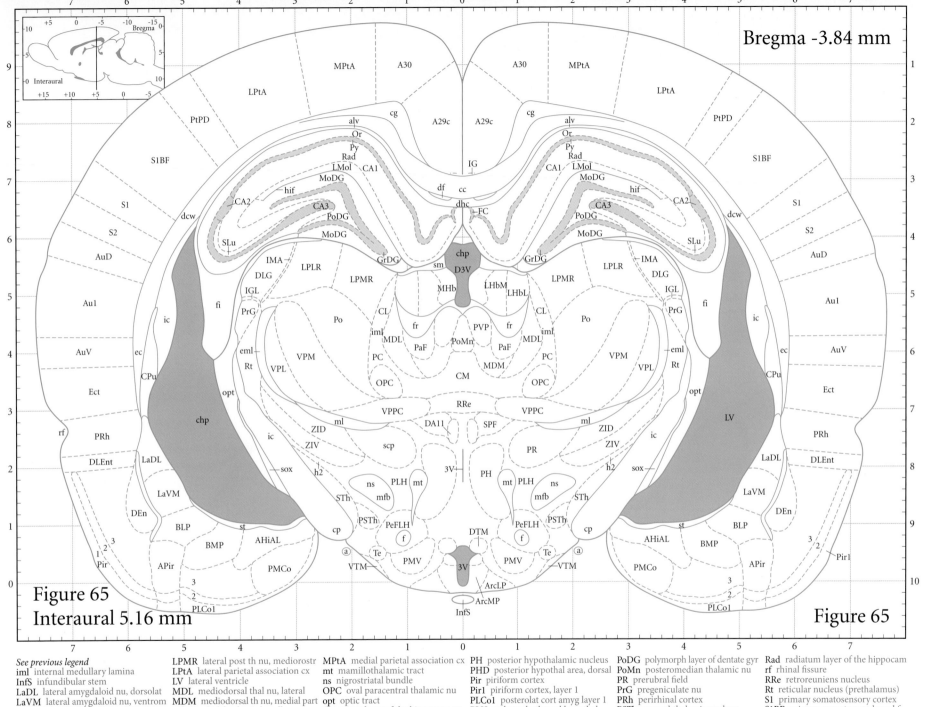

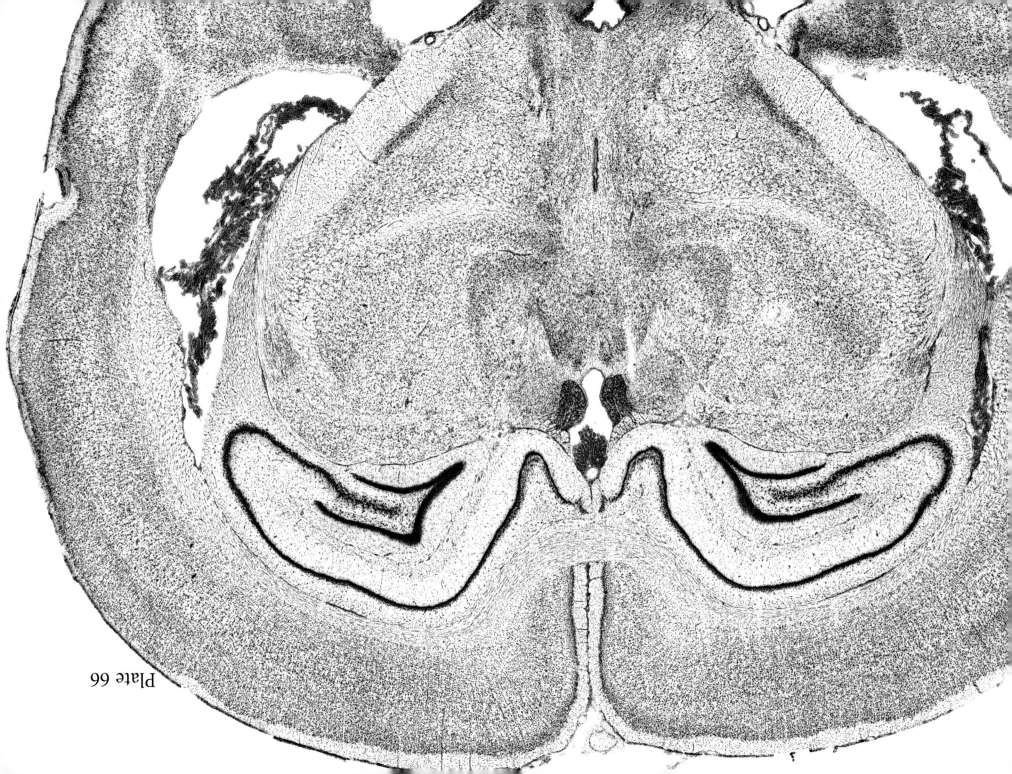

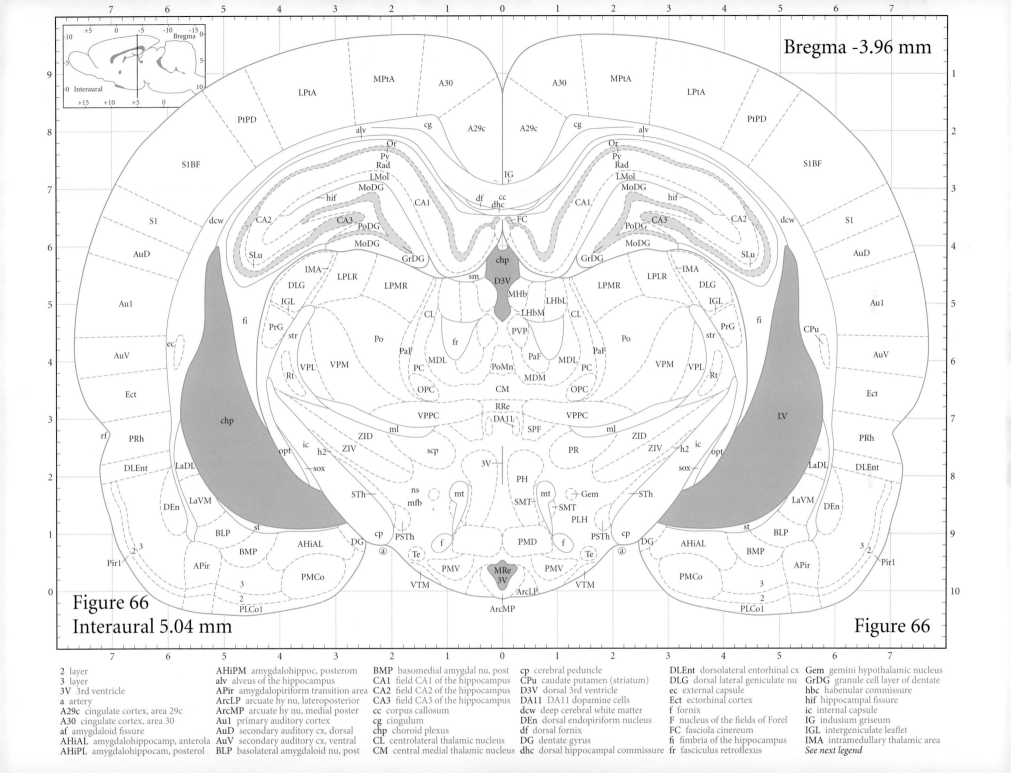

Figure 66 Interaural 5.04 mm Bregma -3.96 mm

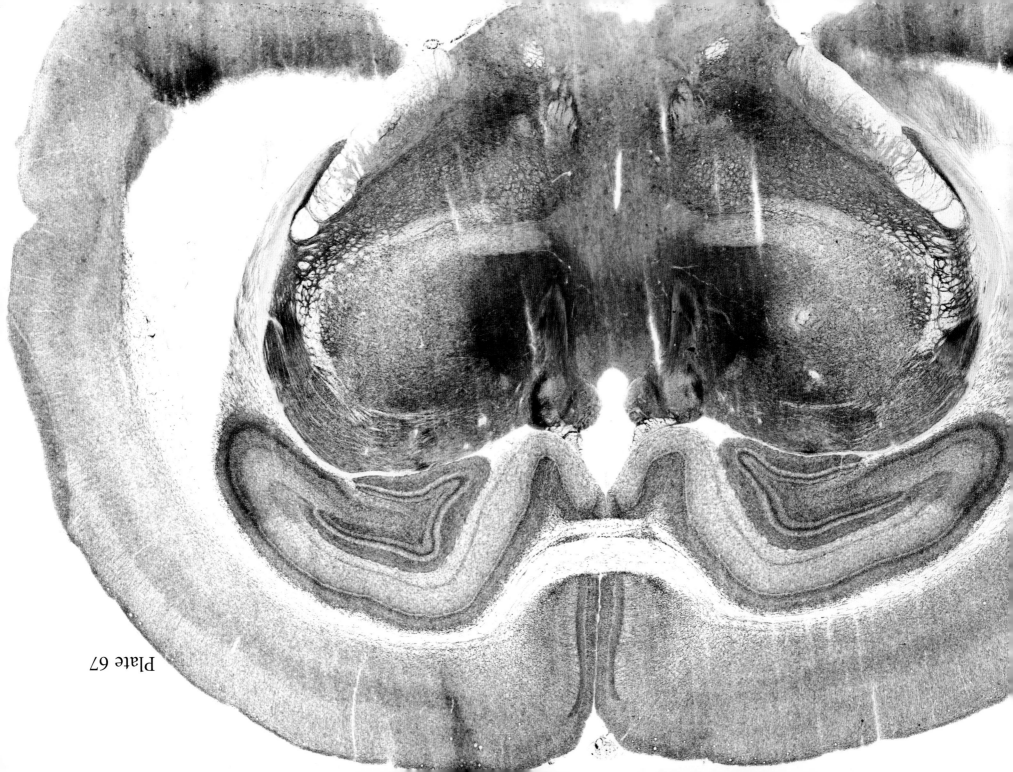

Plate 67

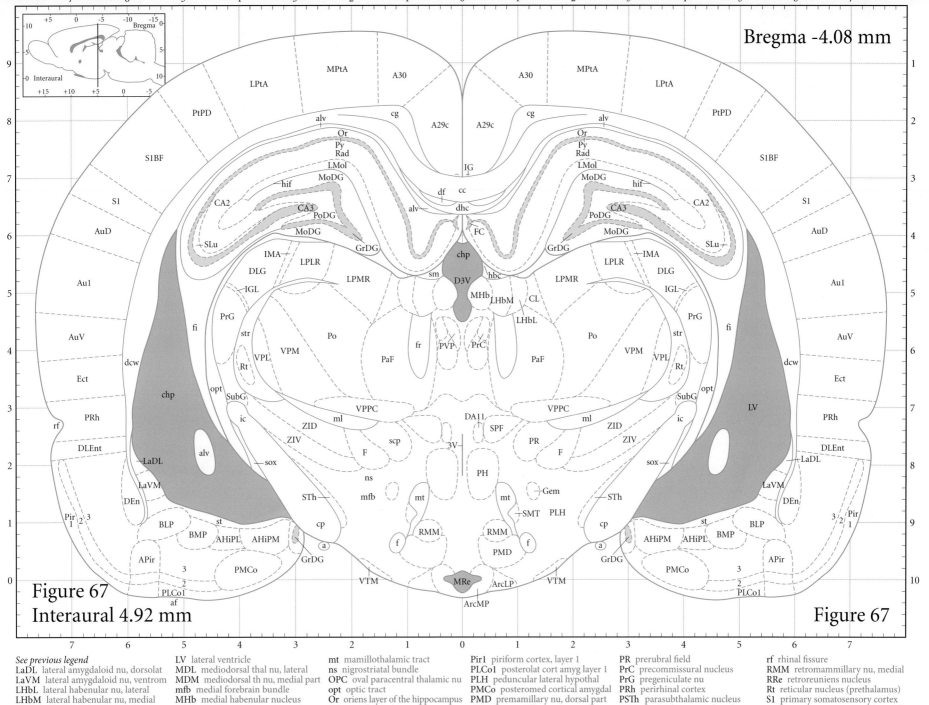

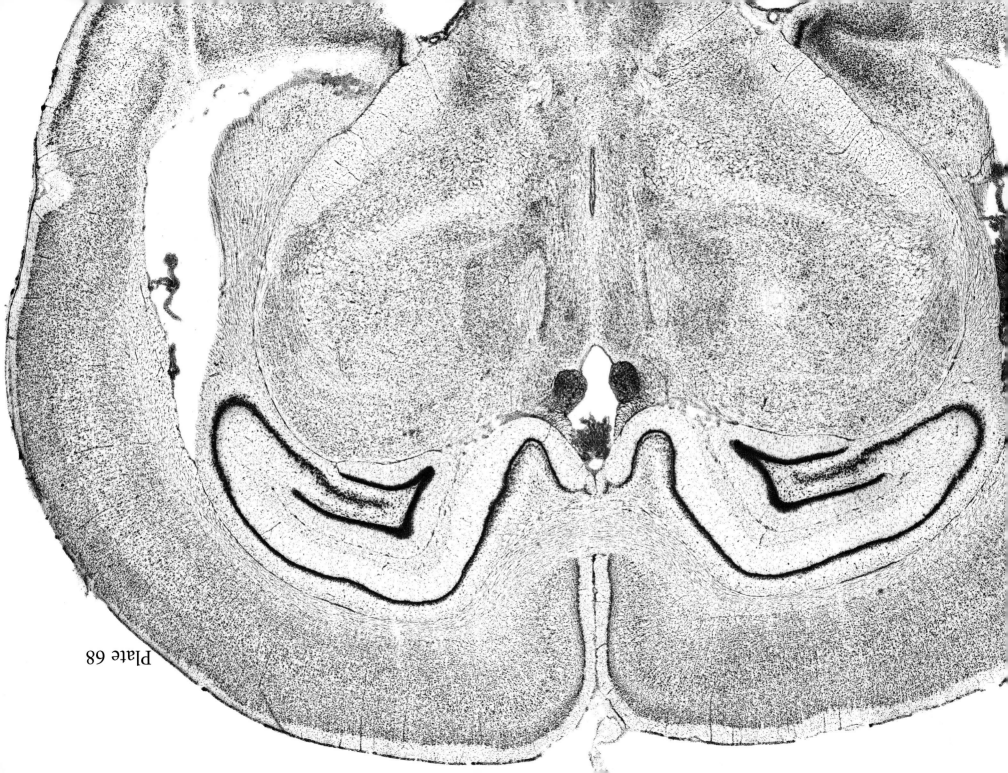

Plate 68

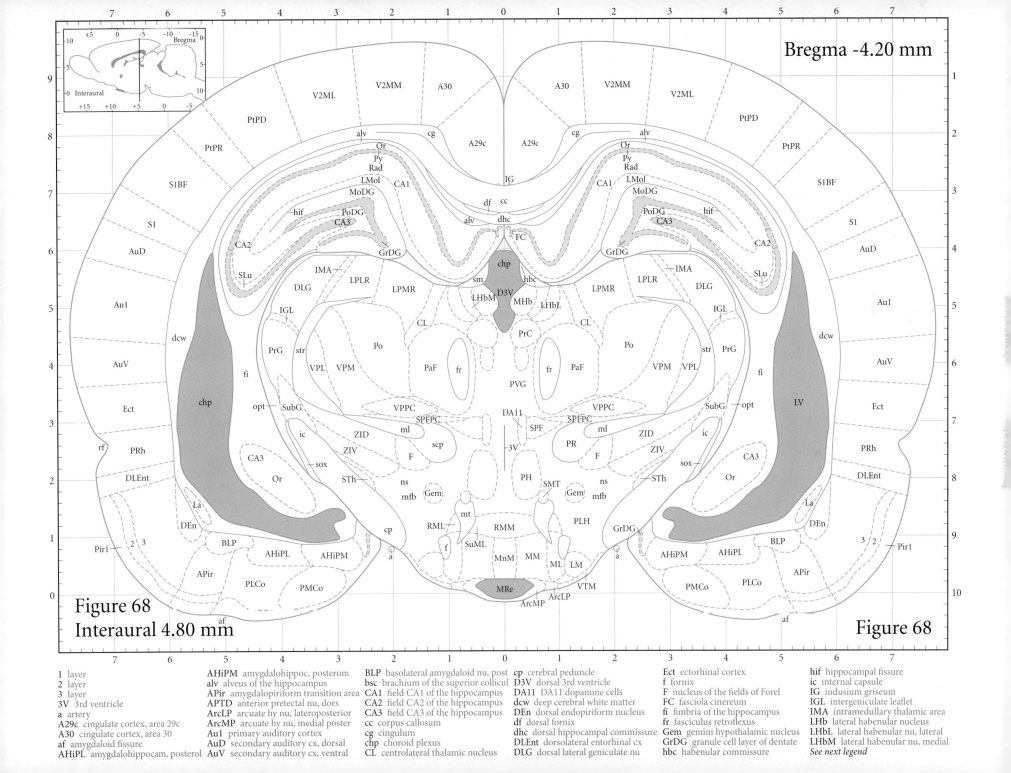

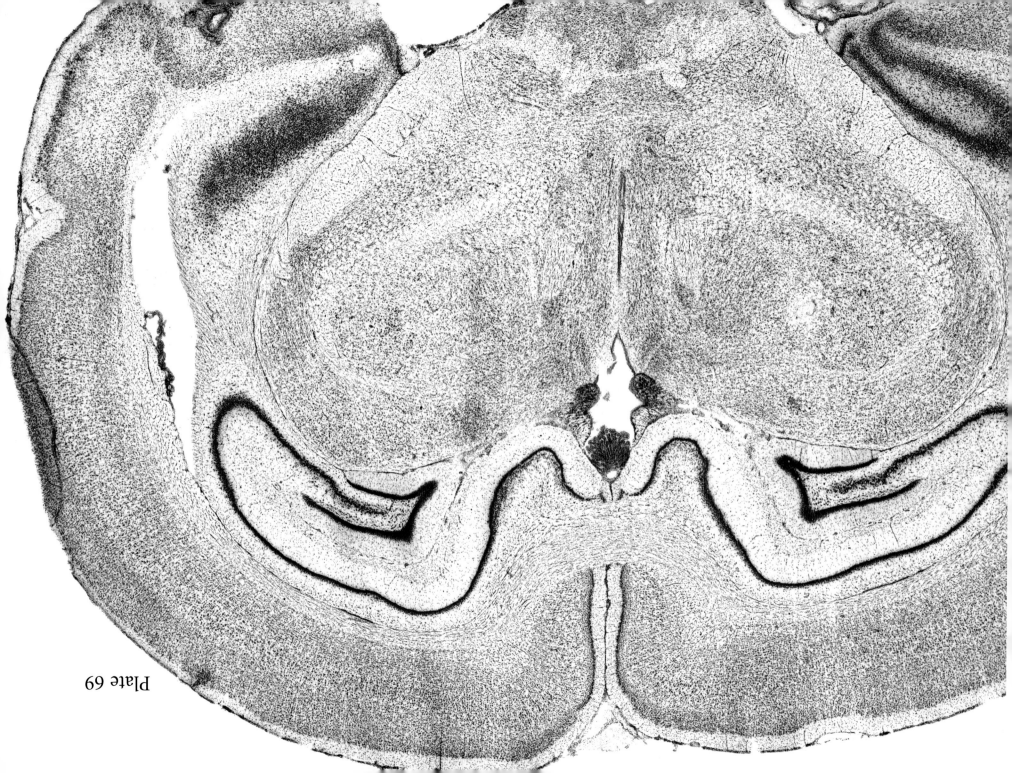

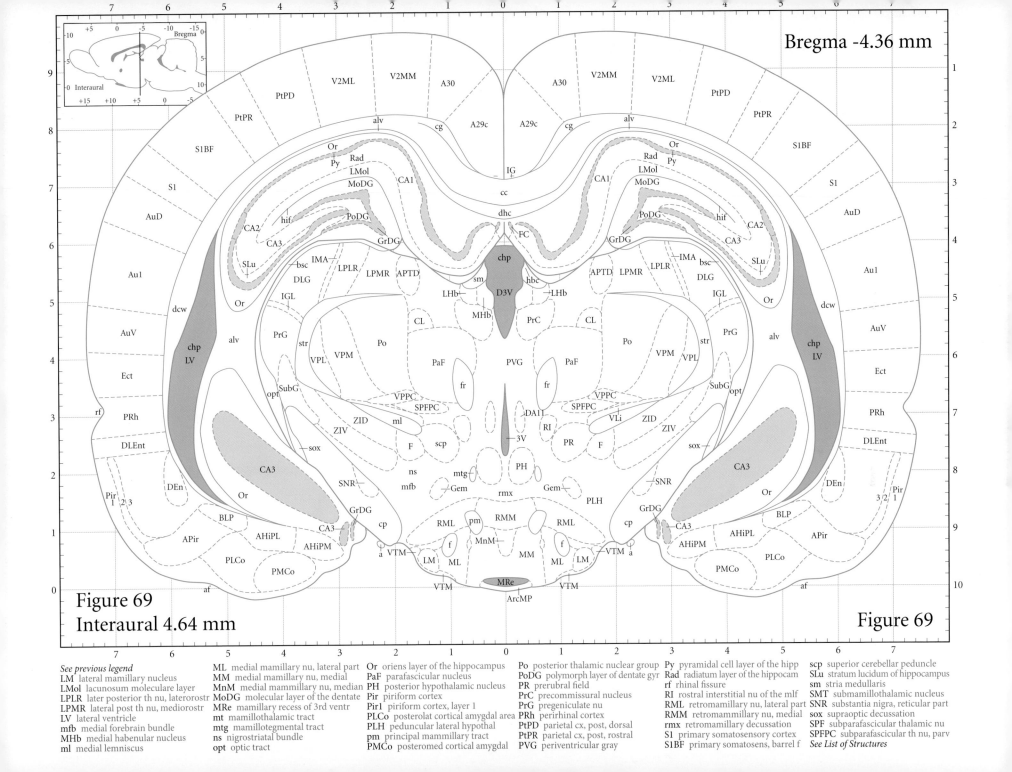

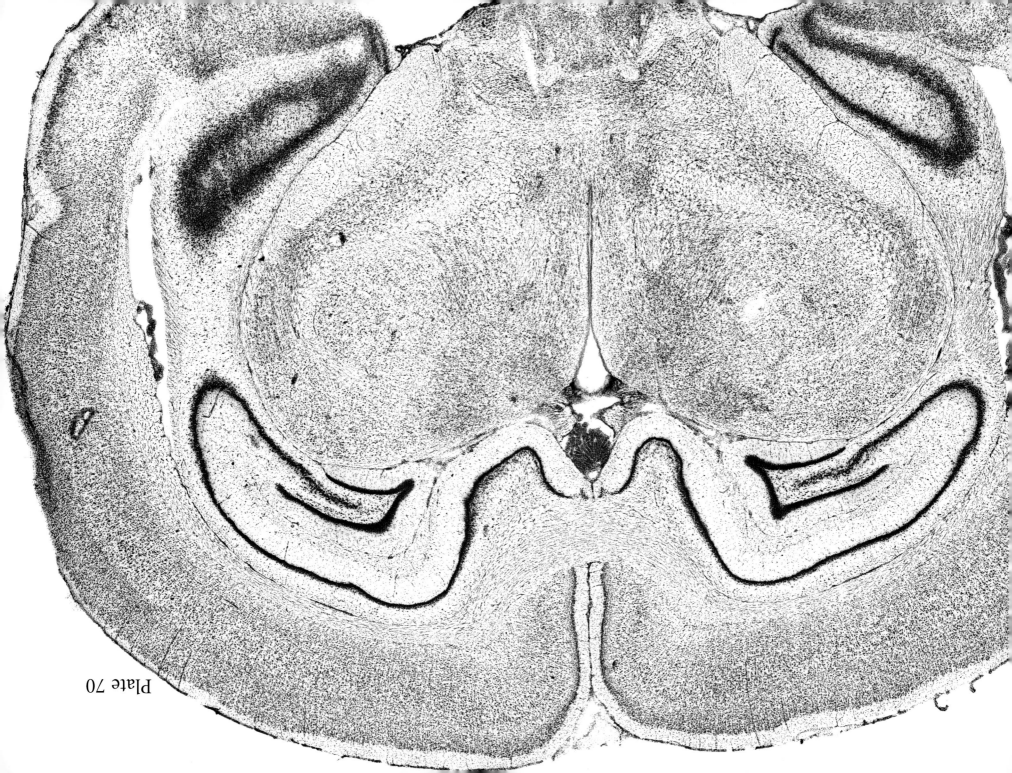

Plate 70

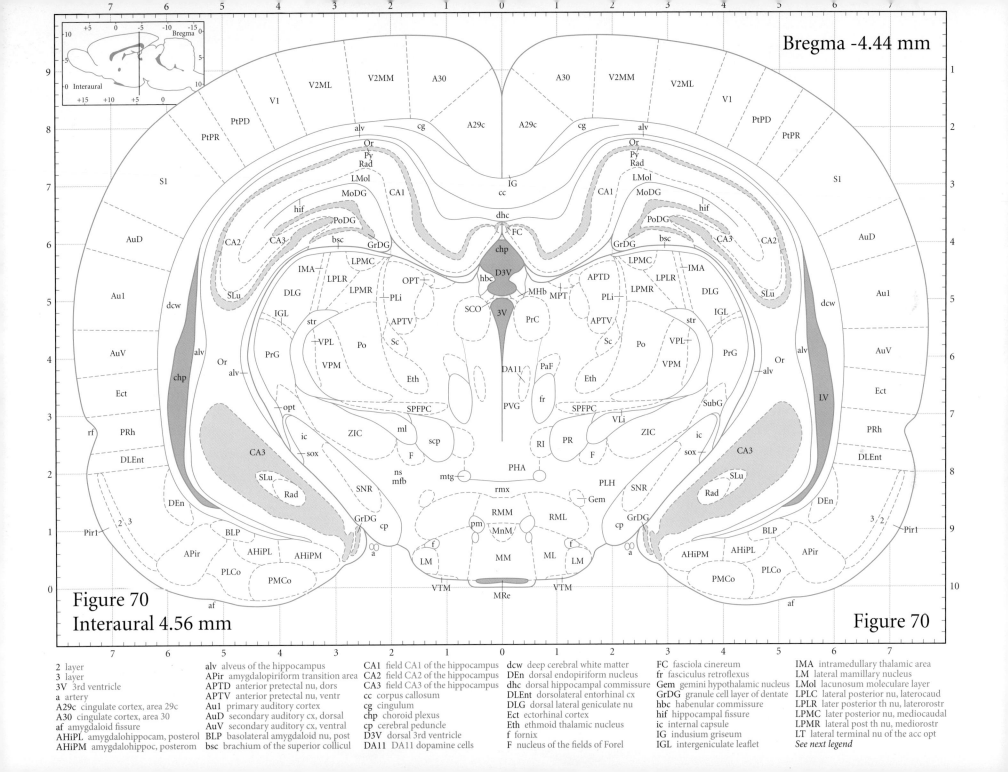

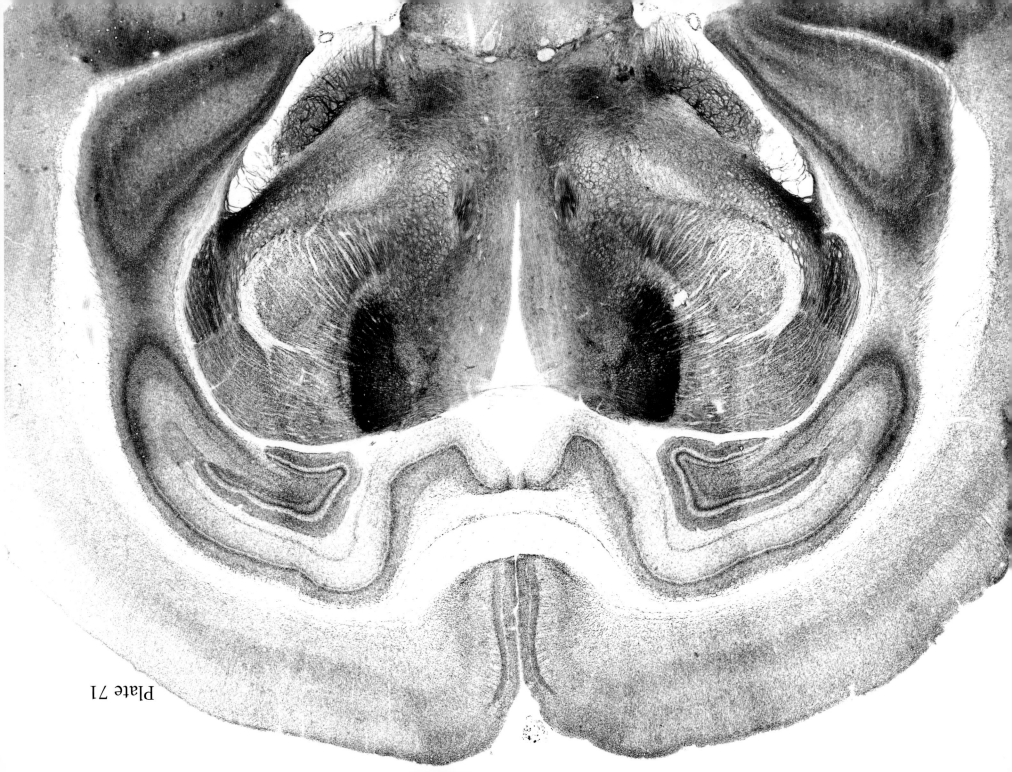

Plate 71

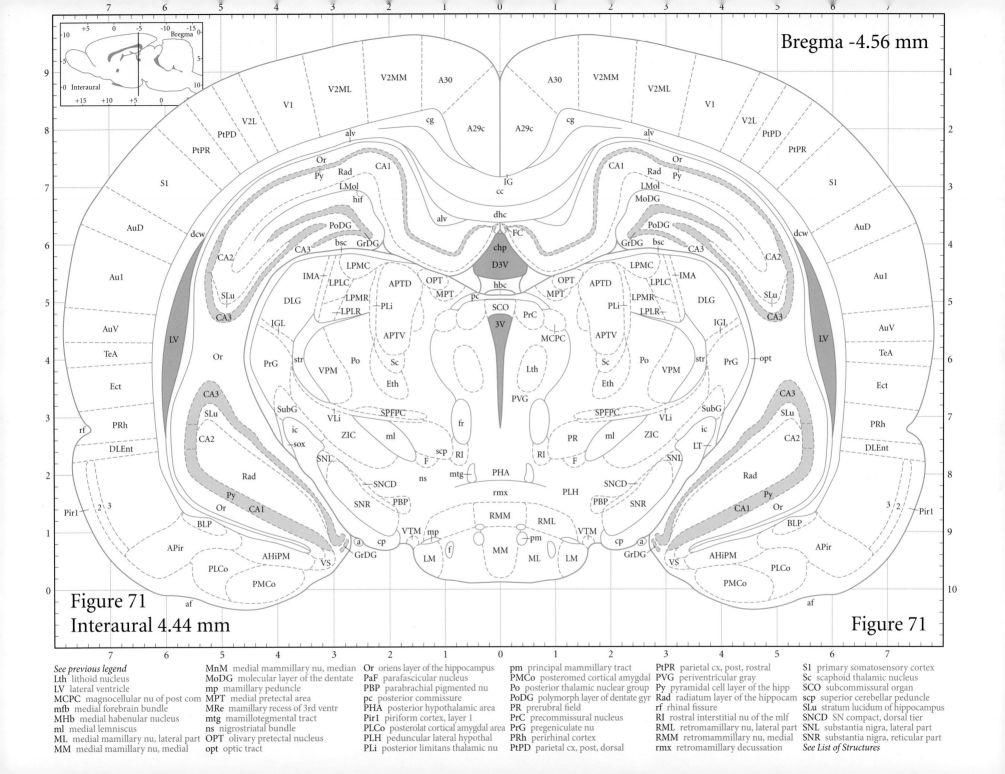

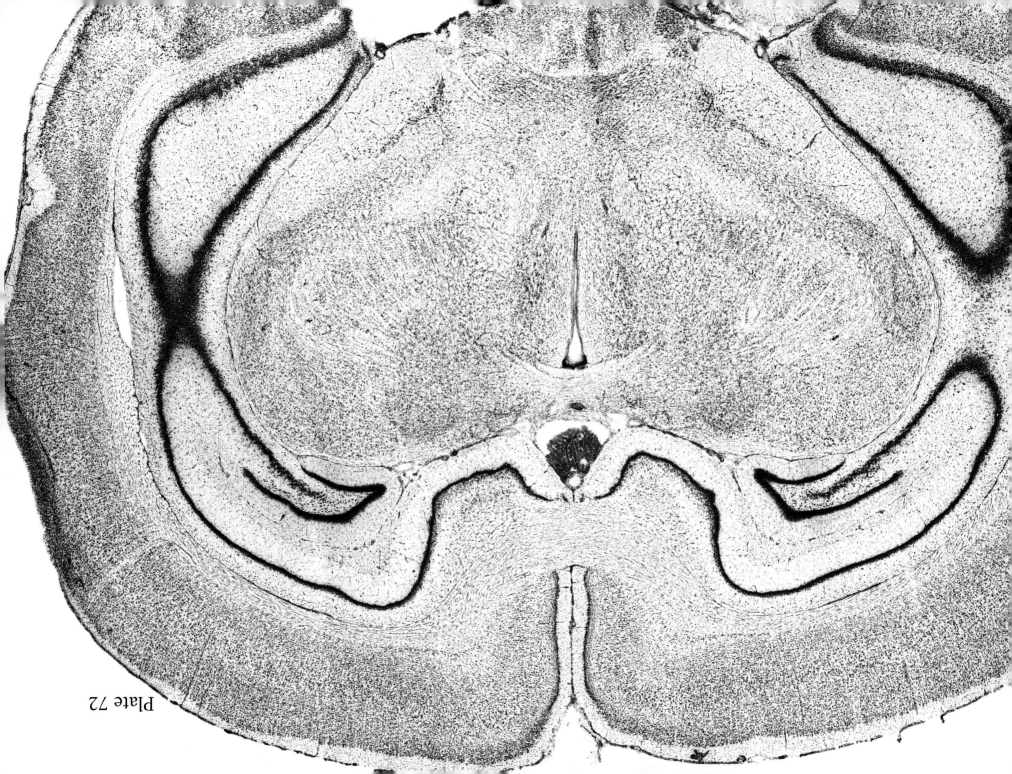

Plate 72

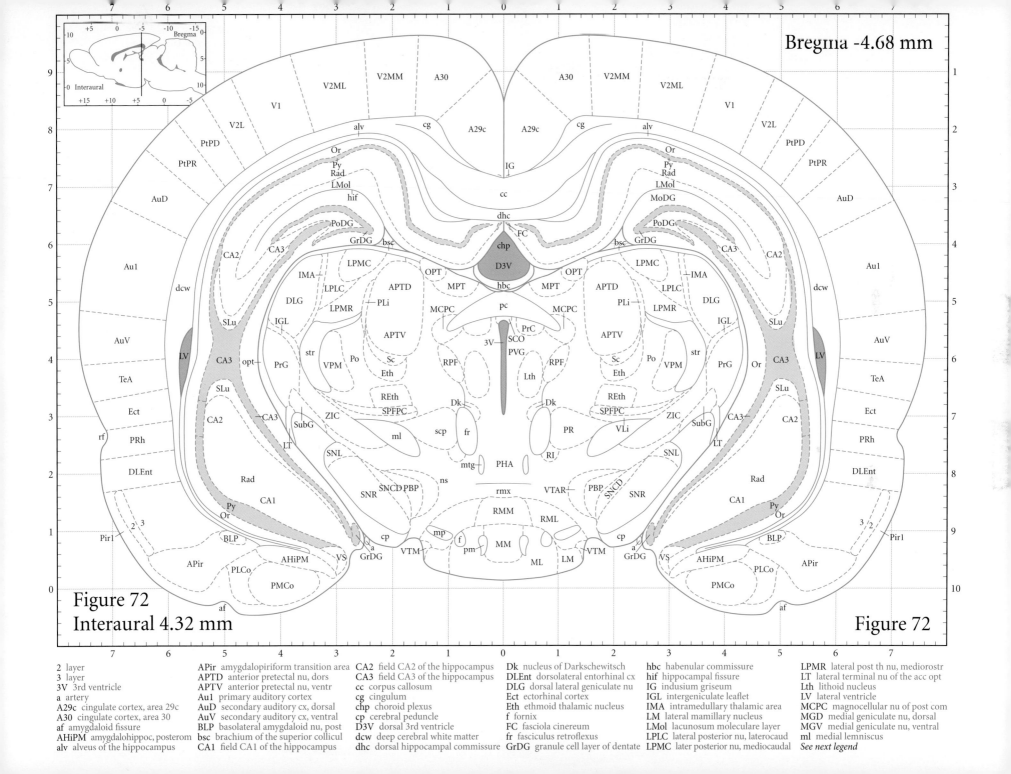

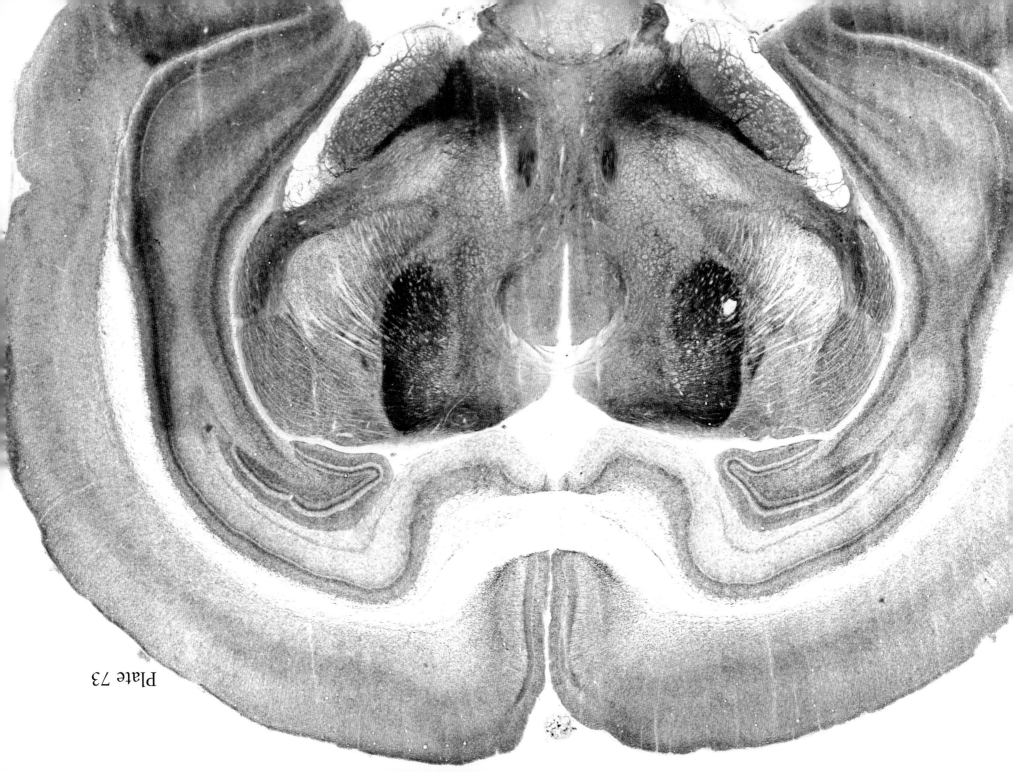

Plate 73

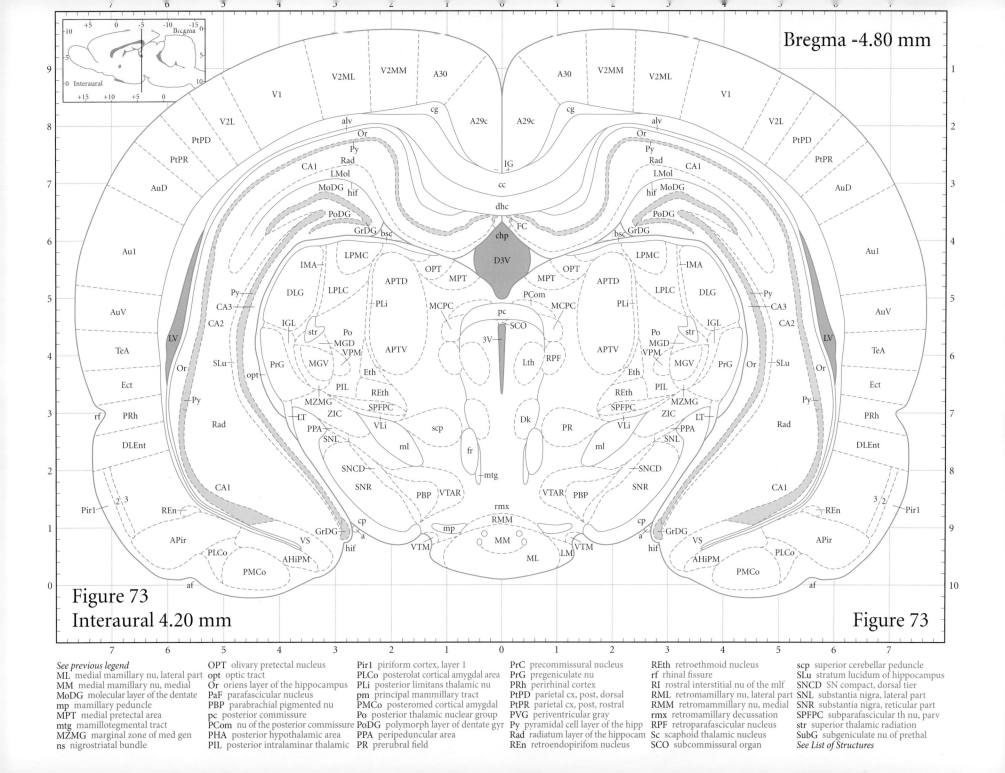

Figure 73
Interaural 4.20 mm
Bregma -4.80 mm

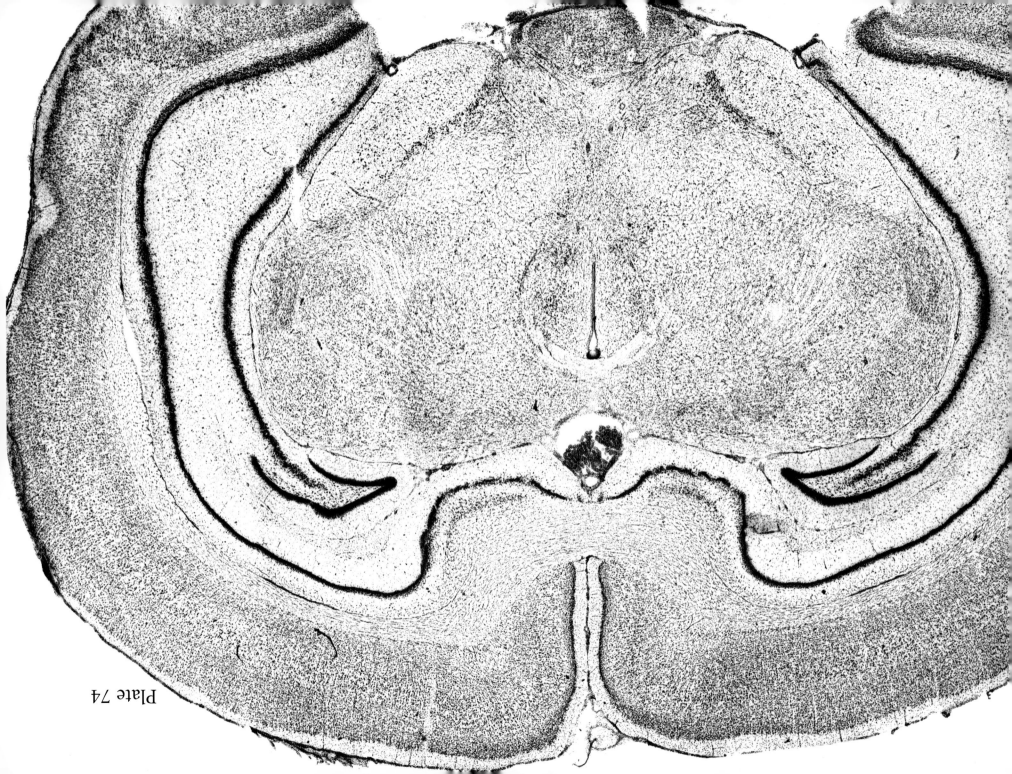

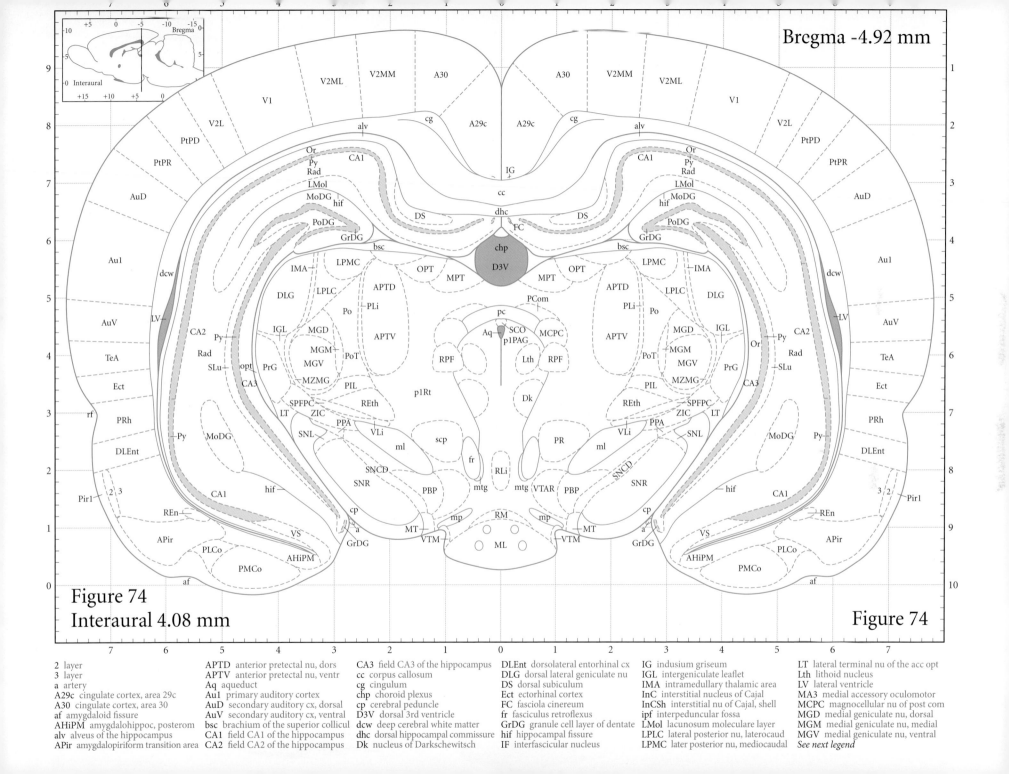

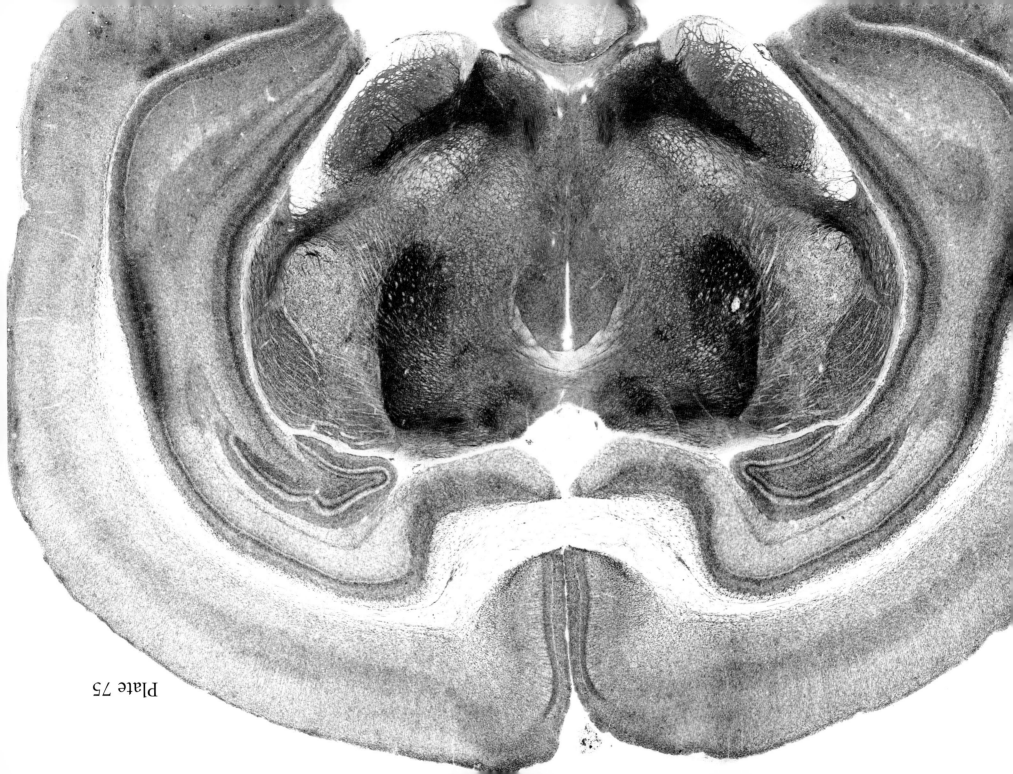

Plate 75

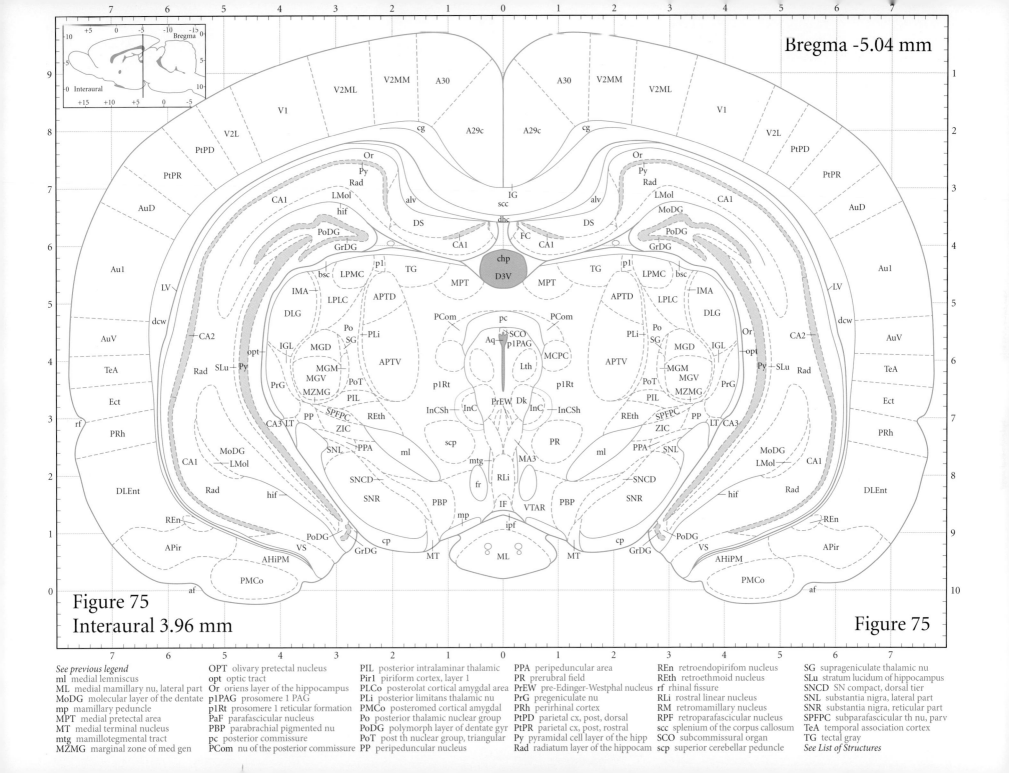

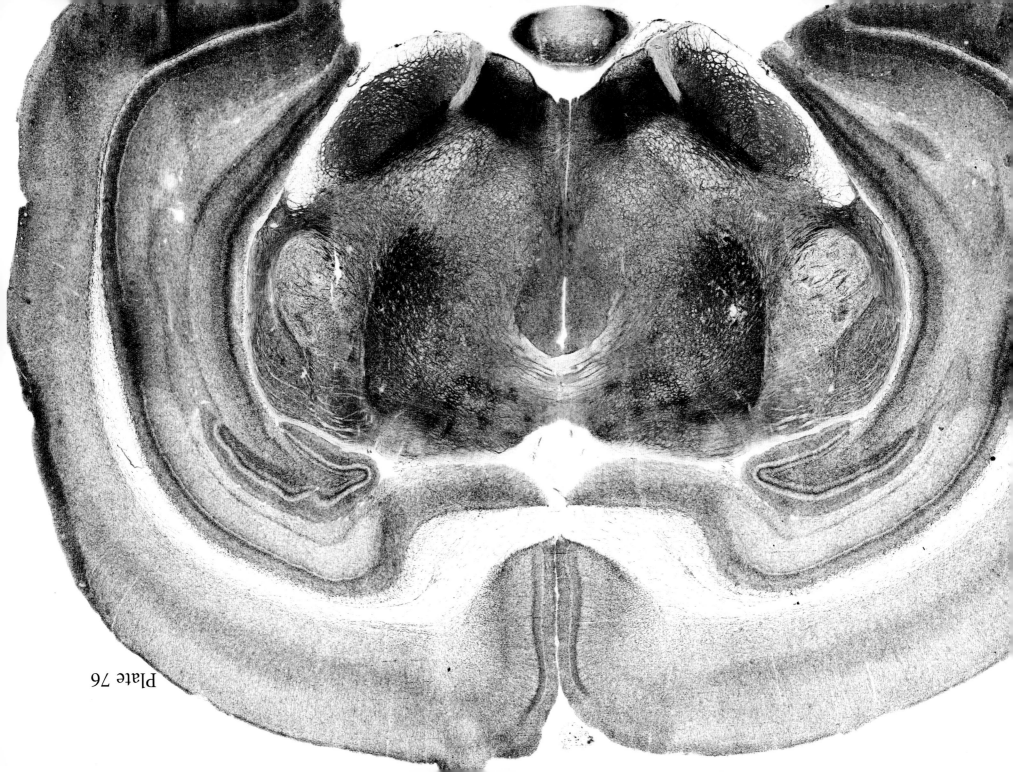

Plate 76

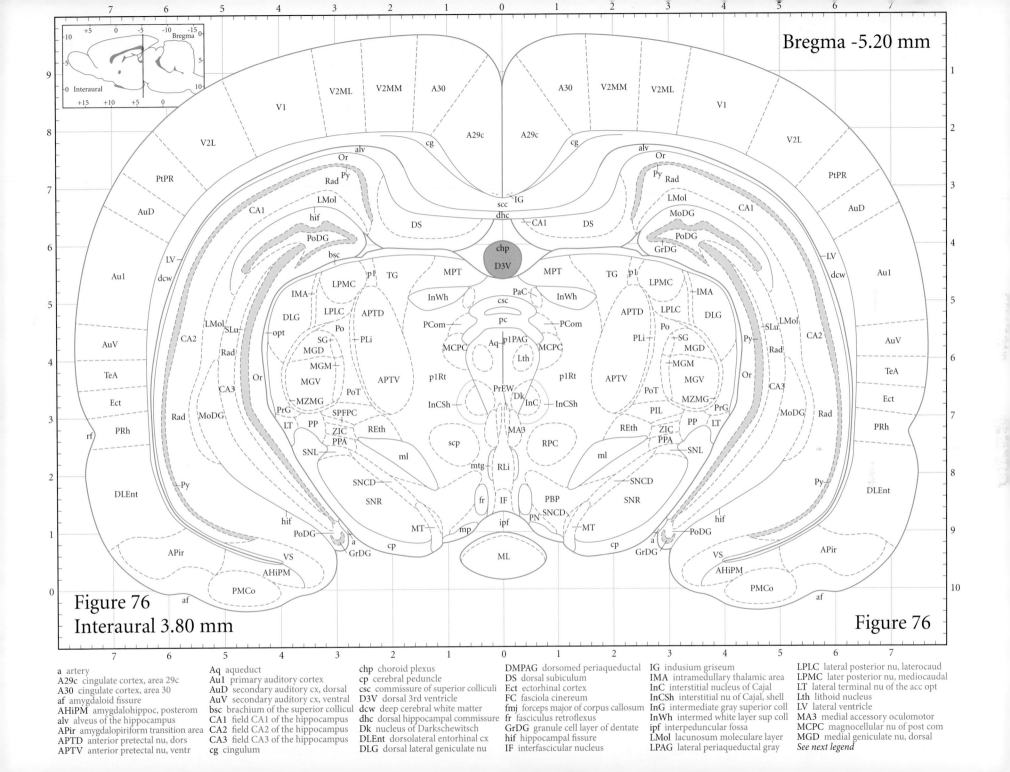

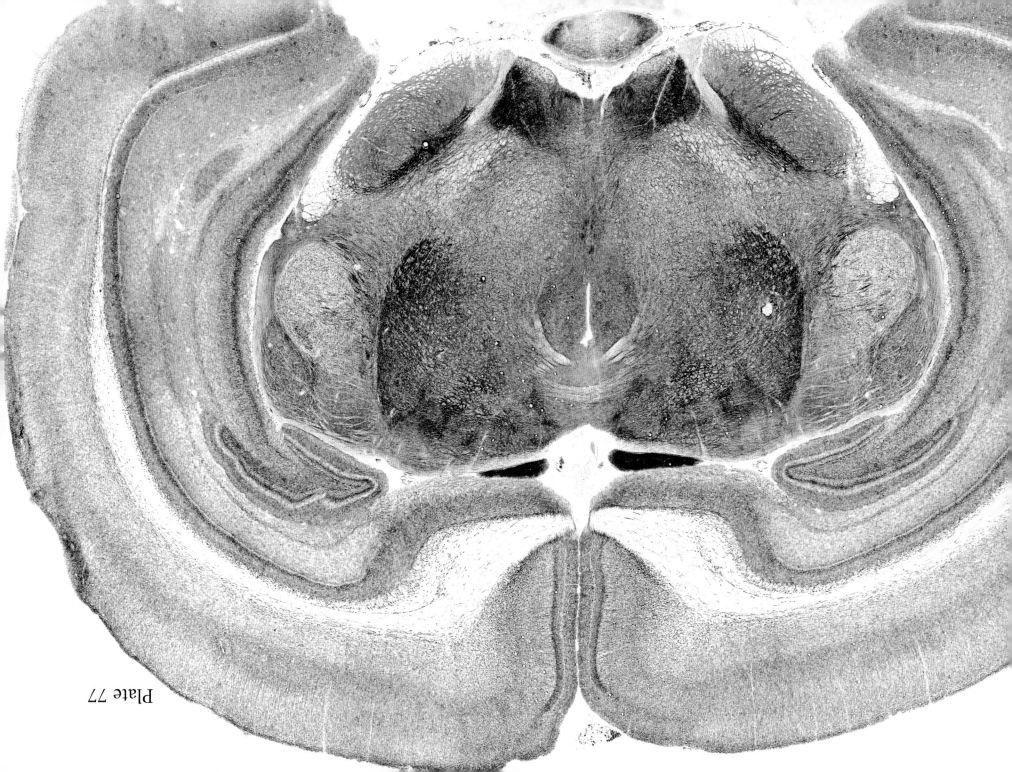

Plate 77

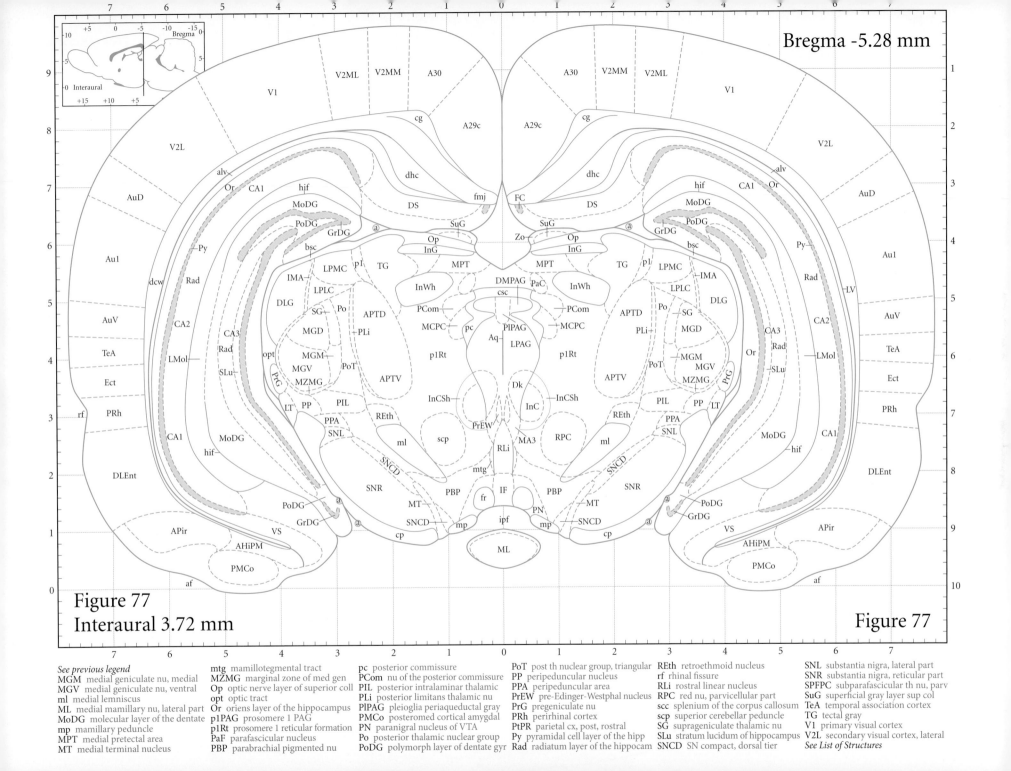

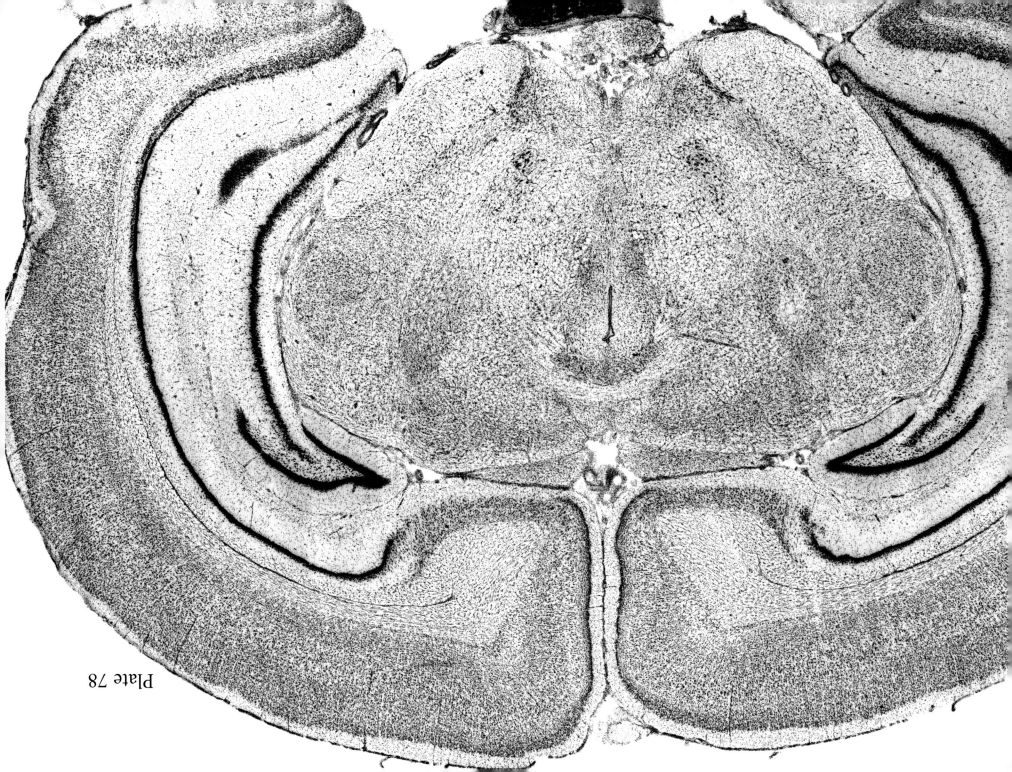

Plate 78

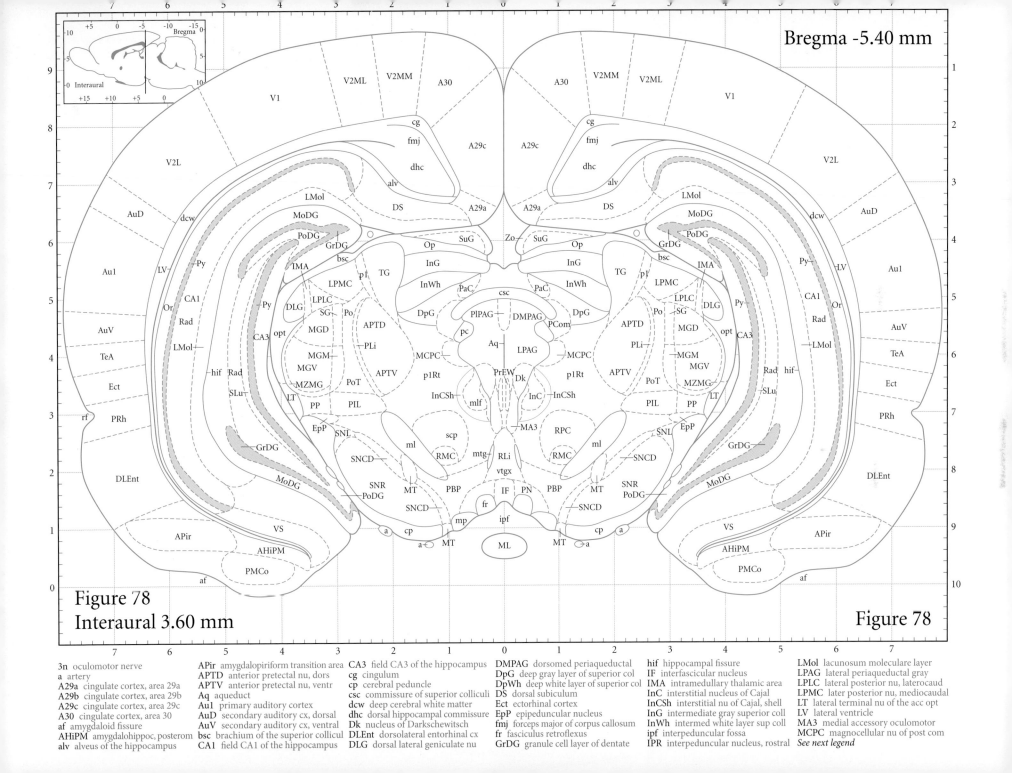

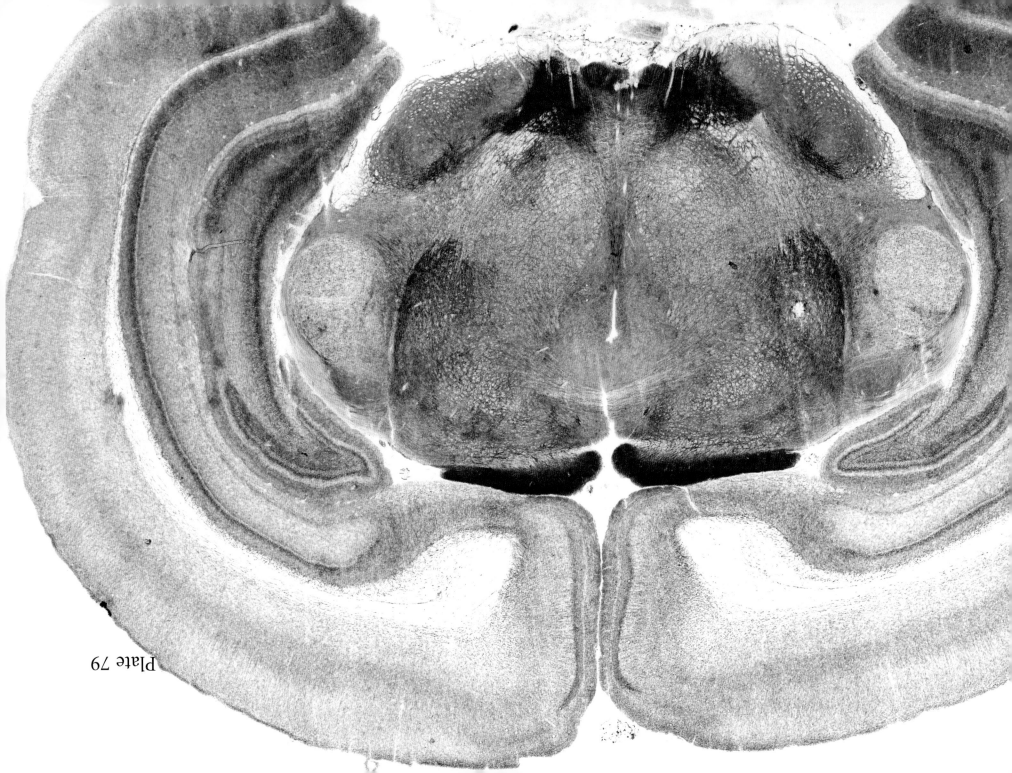

Plate 79

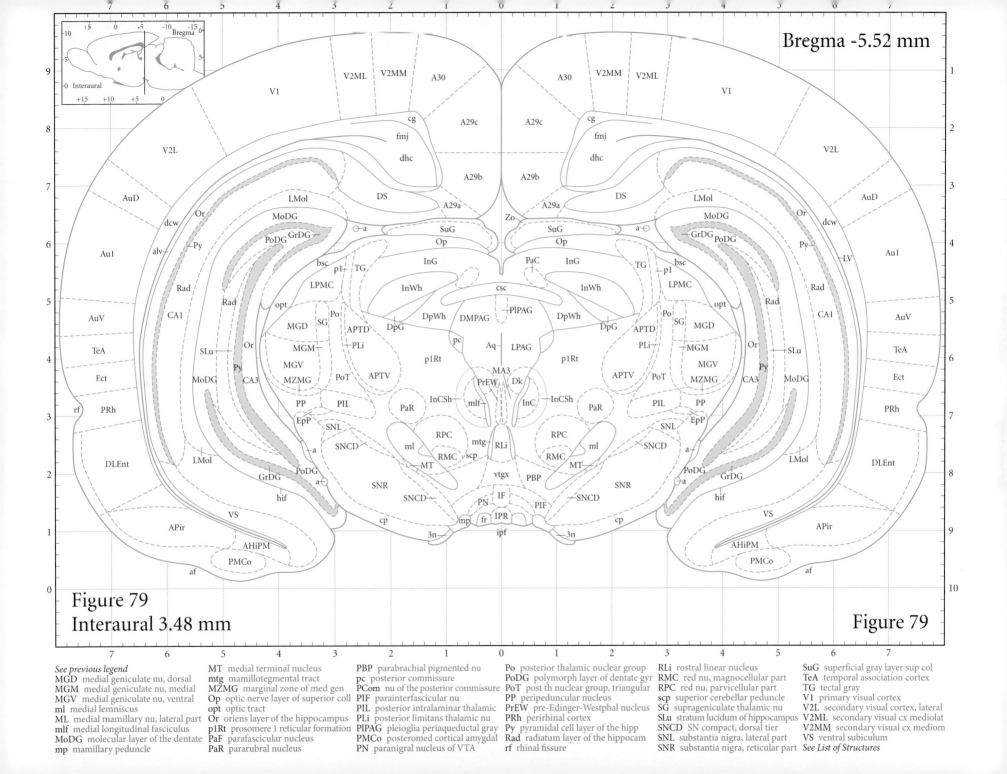

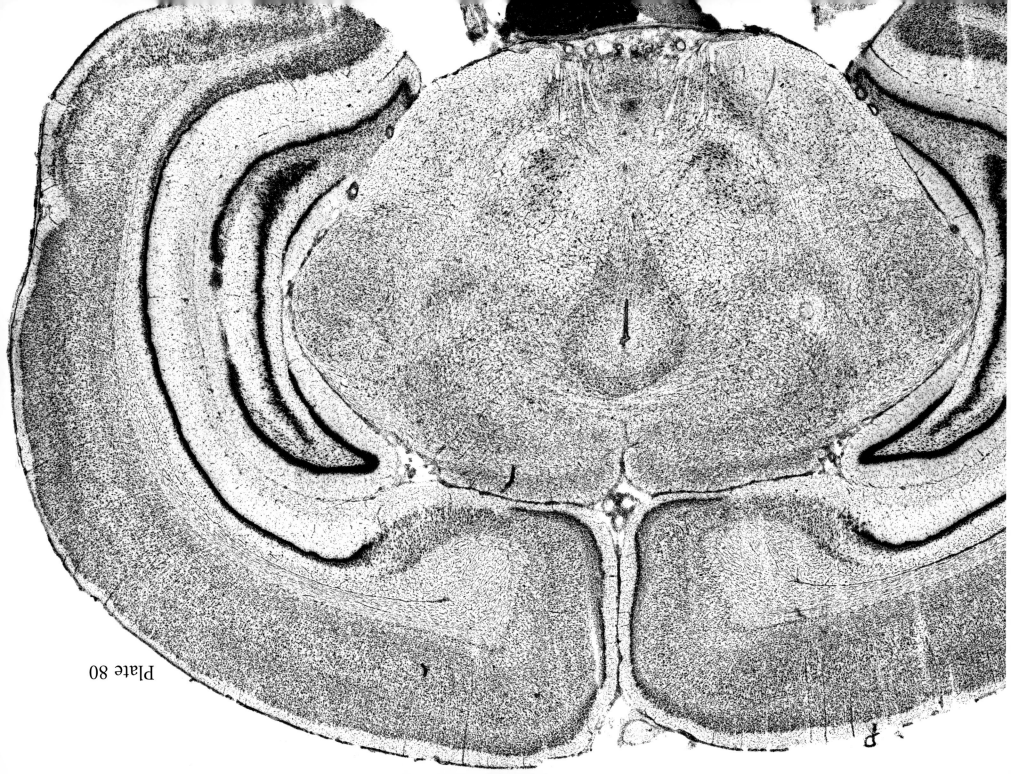

Plate 80

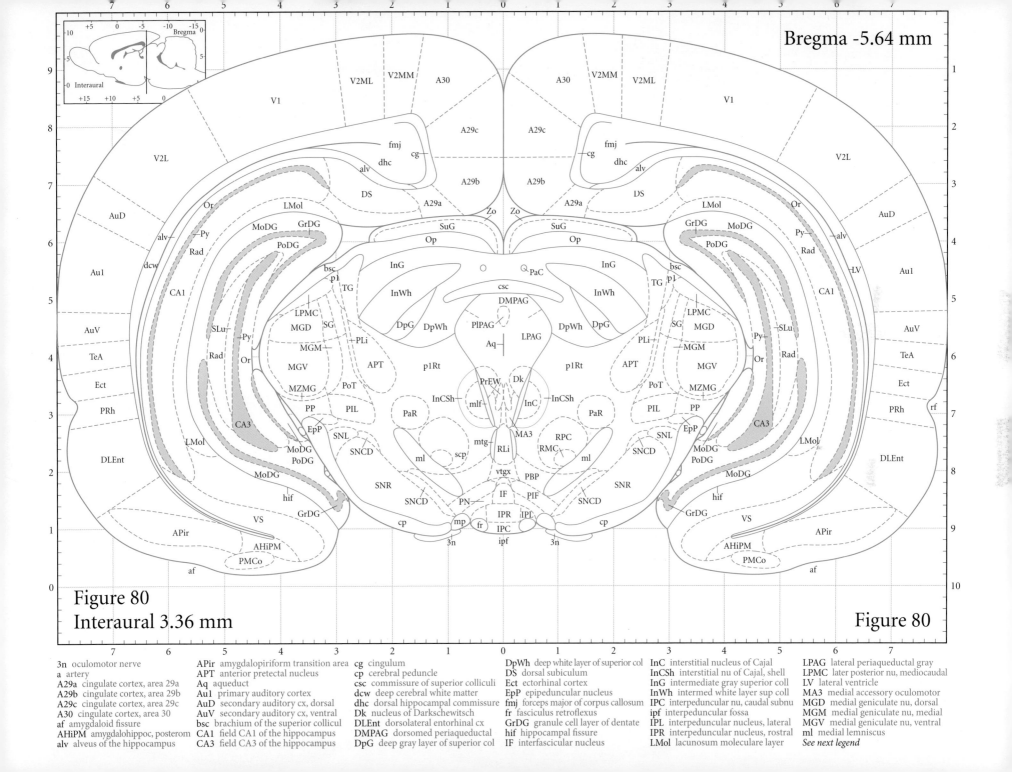

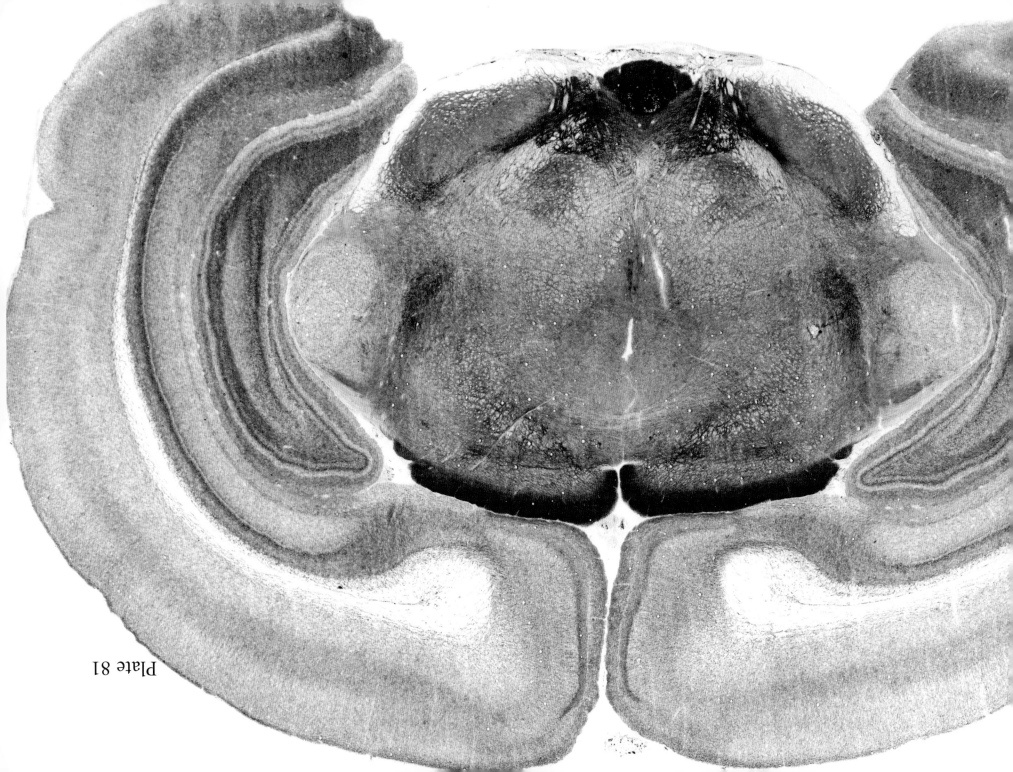

Plate 81

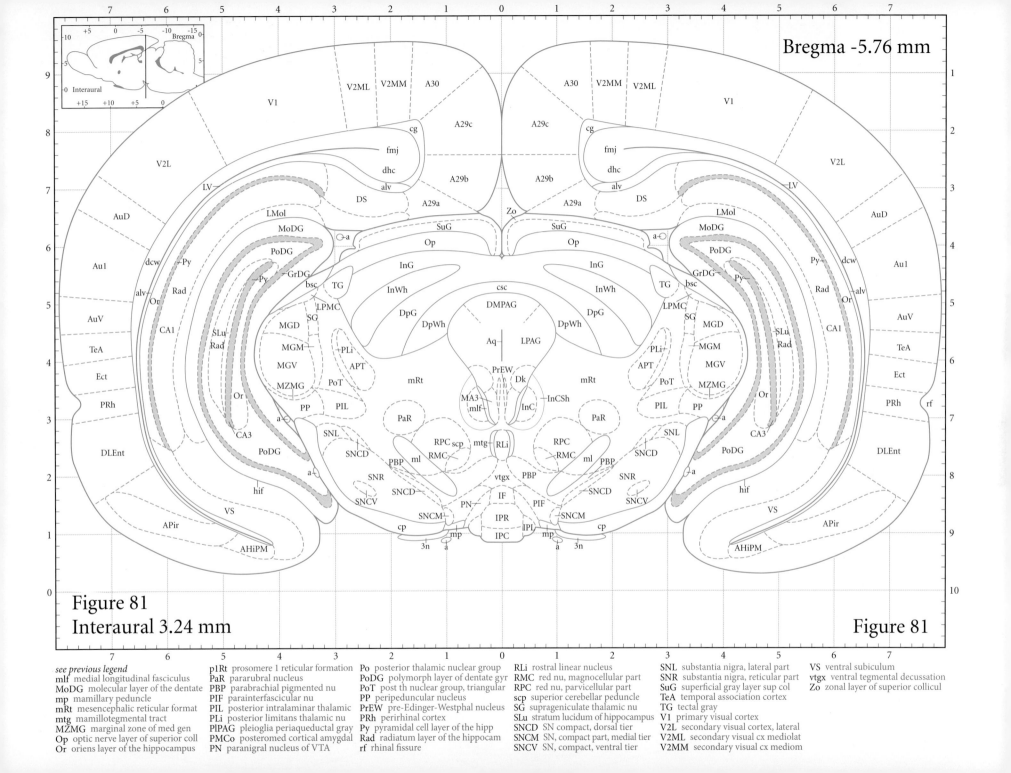

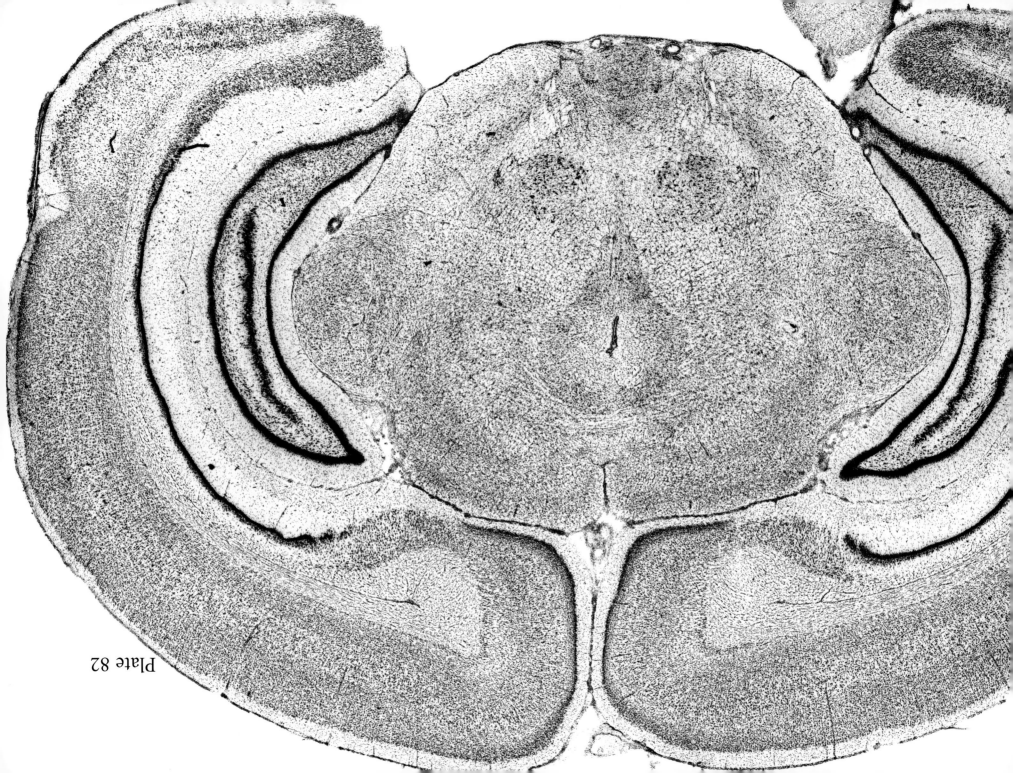

Plate 82

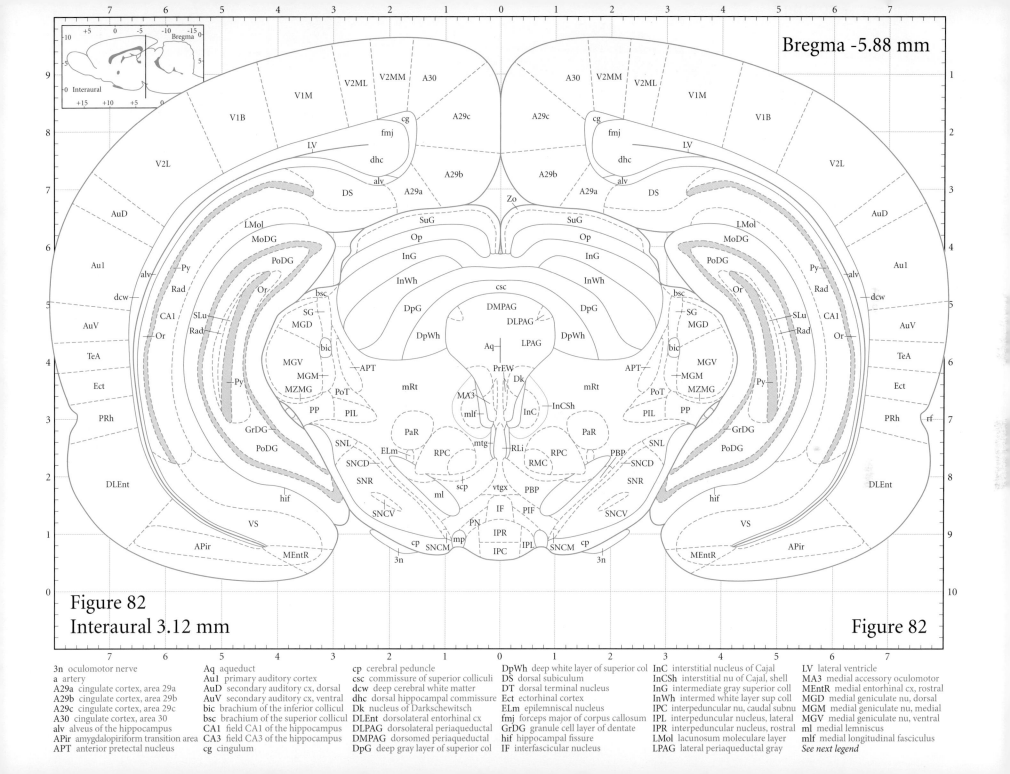

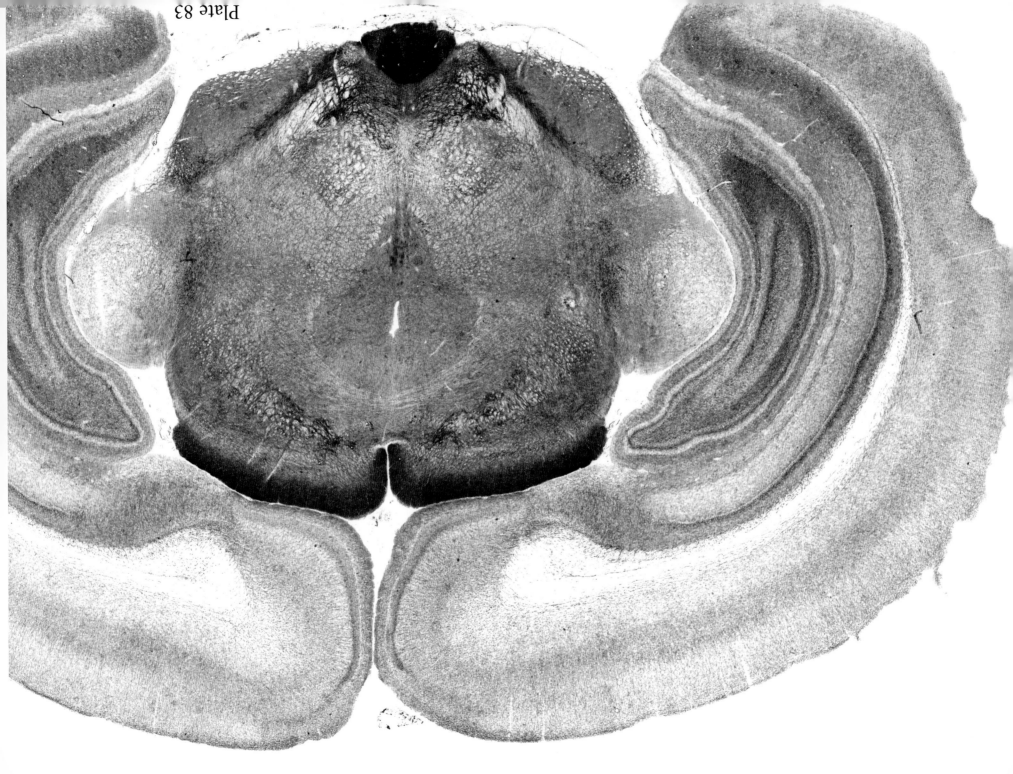
Plate 83

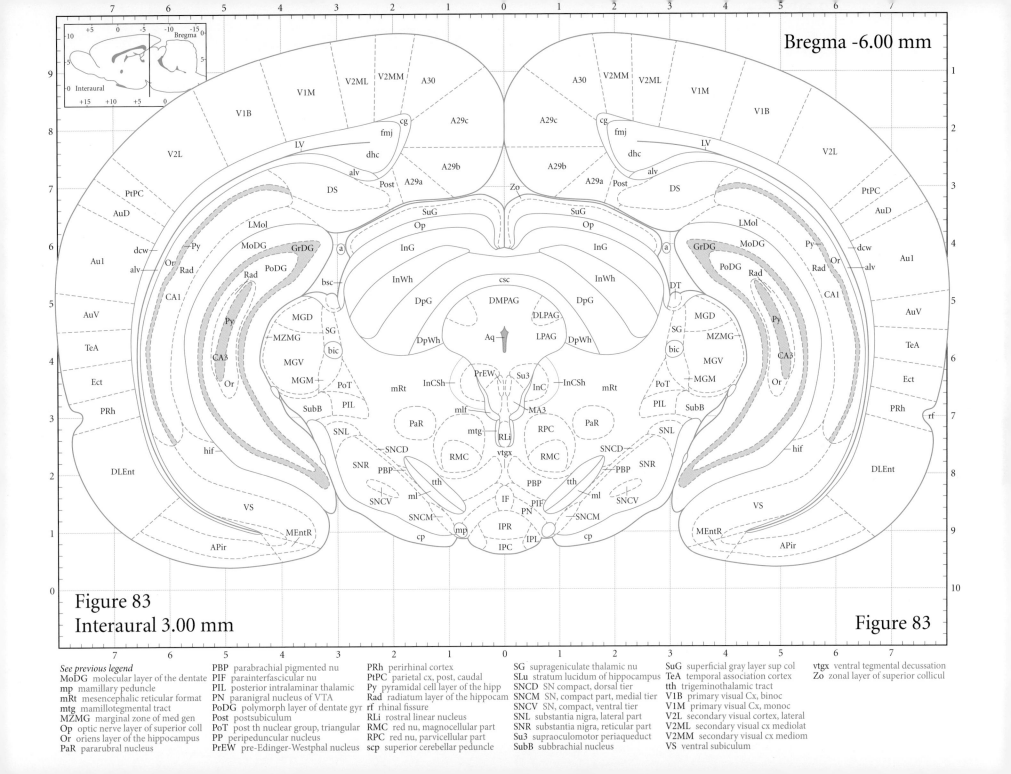

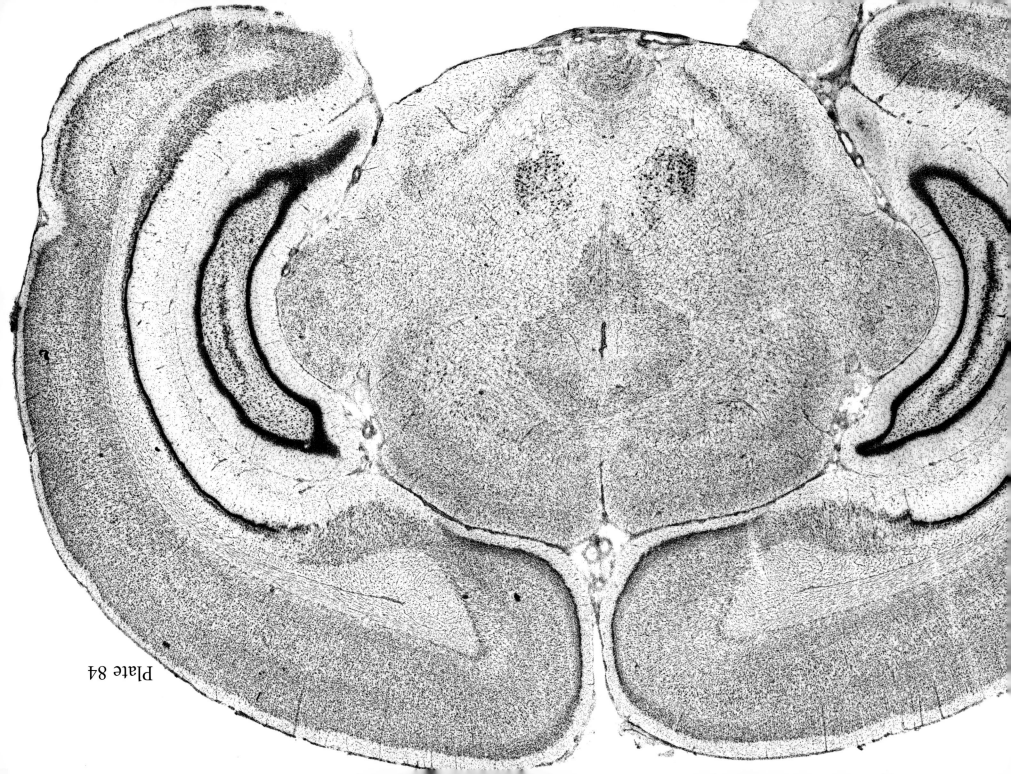

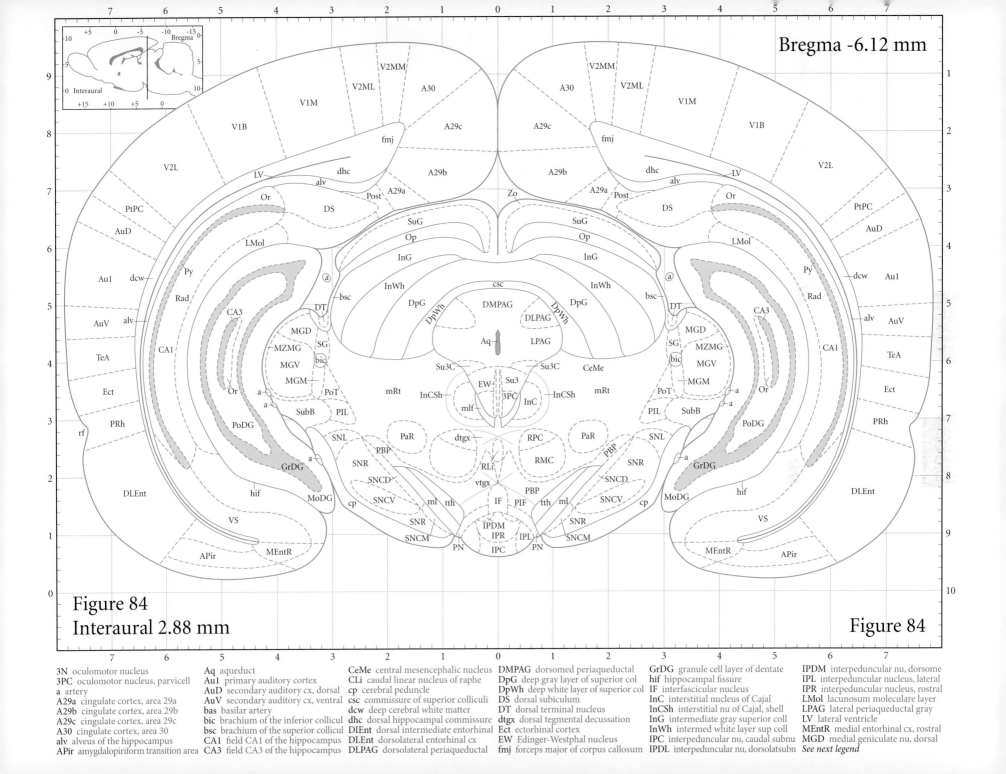

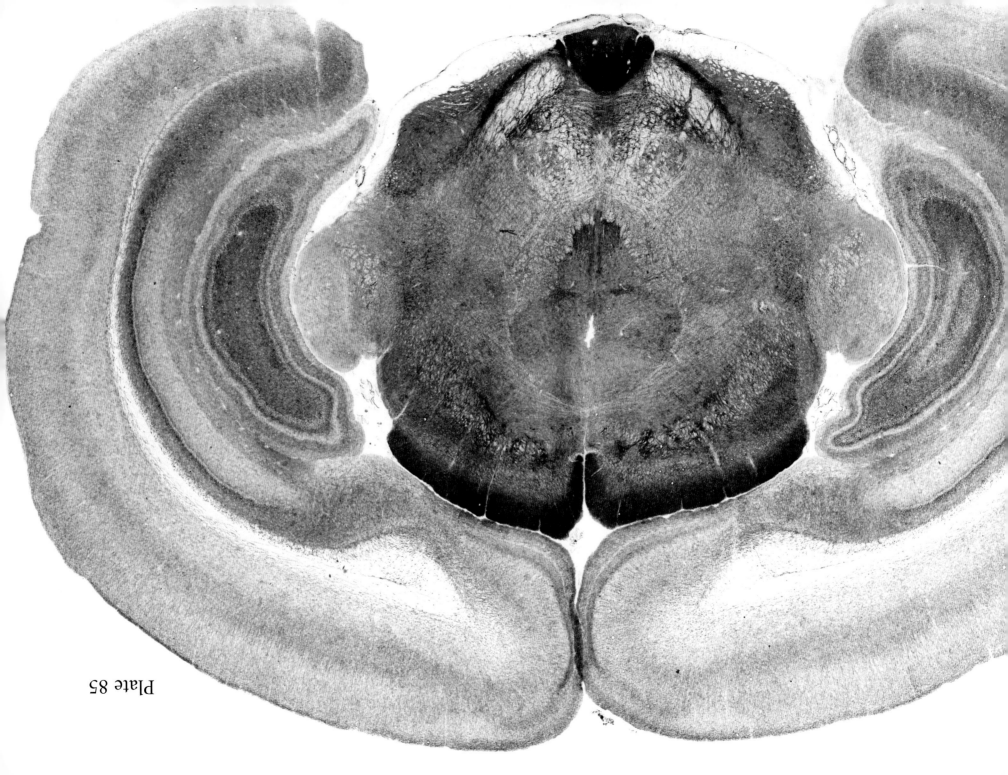

Plate 85

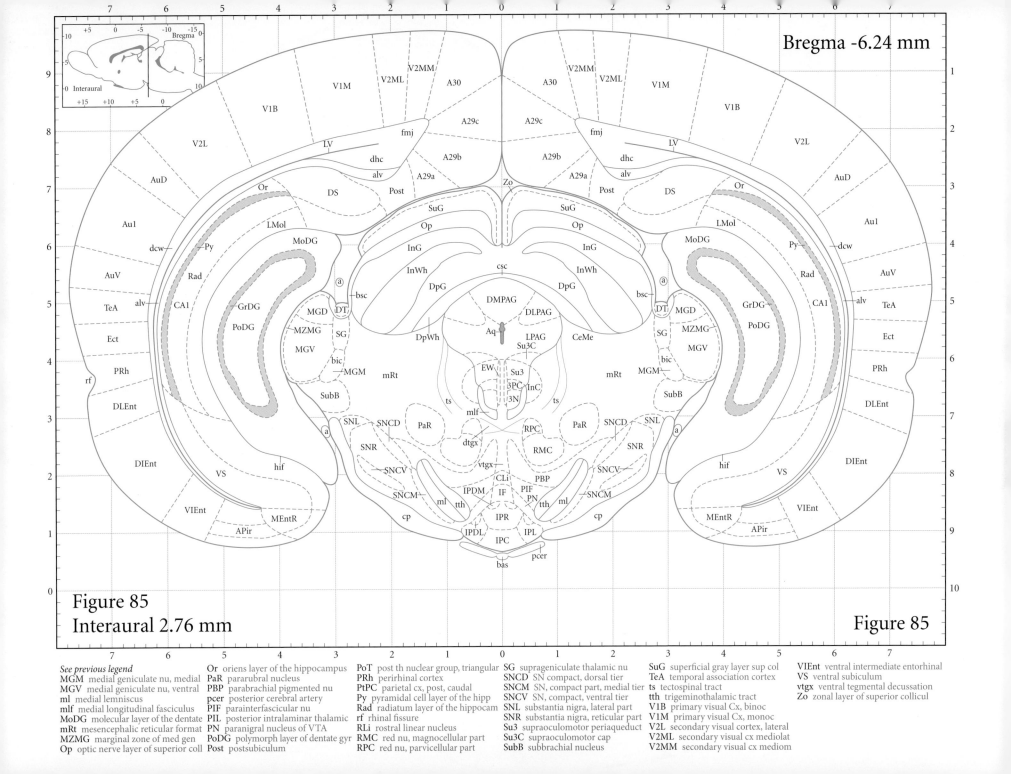

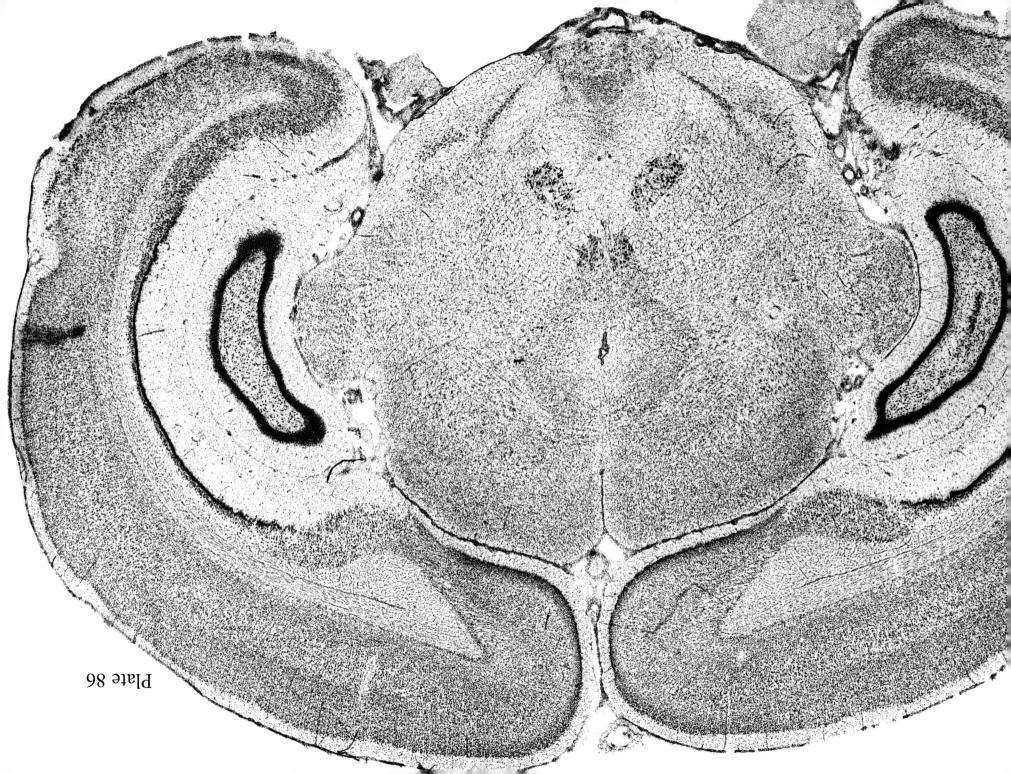

Plate 86

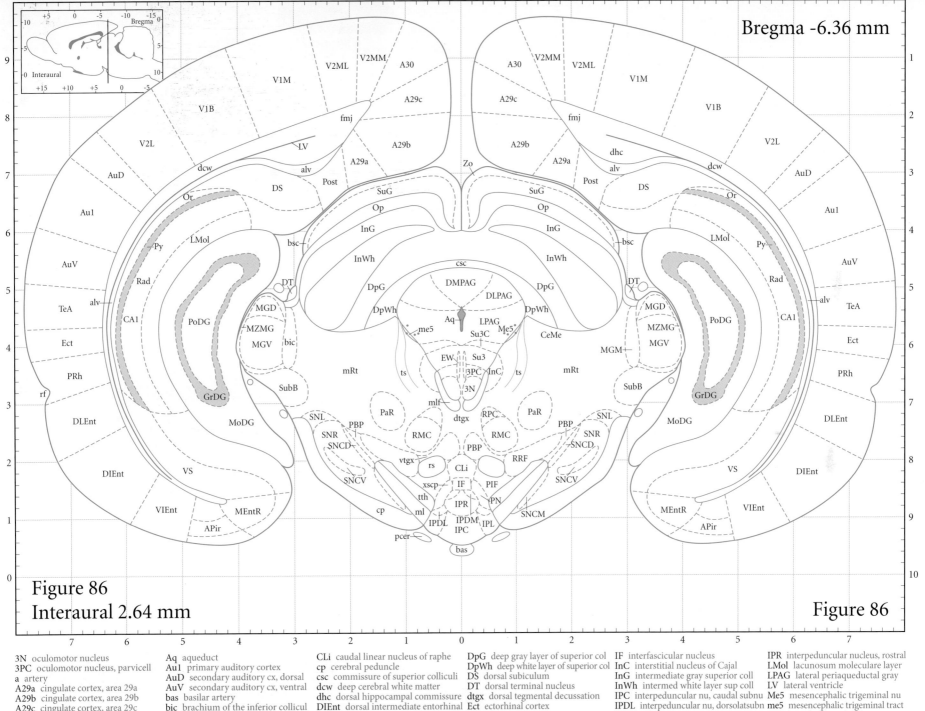

Figure 86
Interaural 2.64 mm
Bregma -6.36 mm

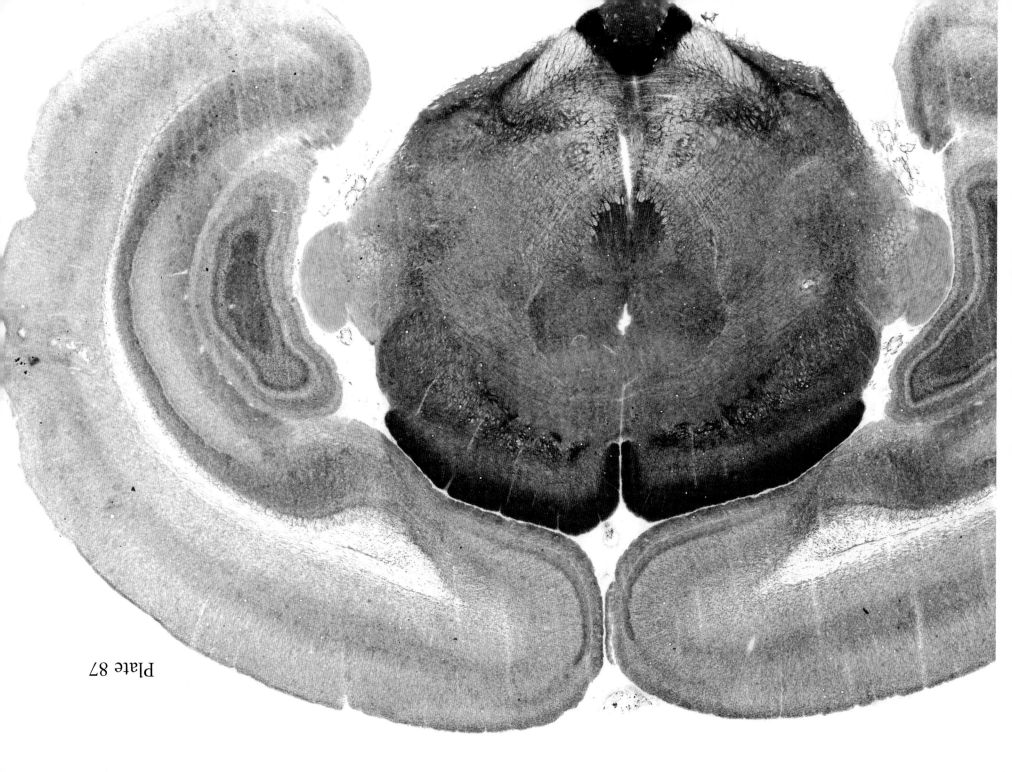

Plate 87

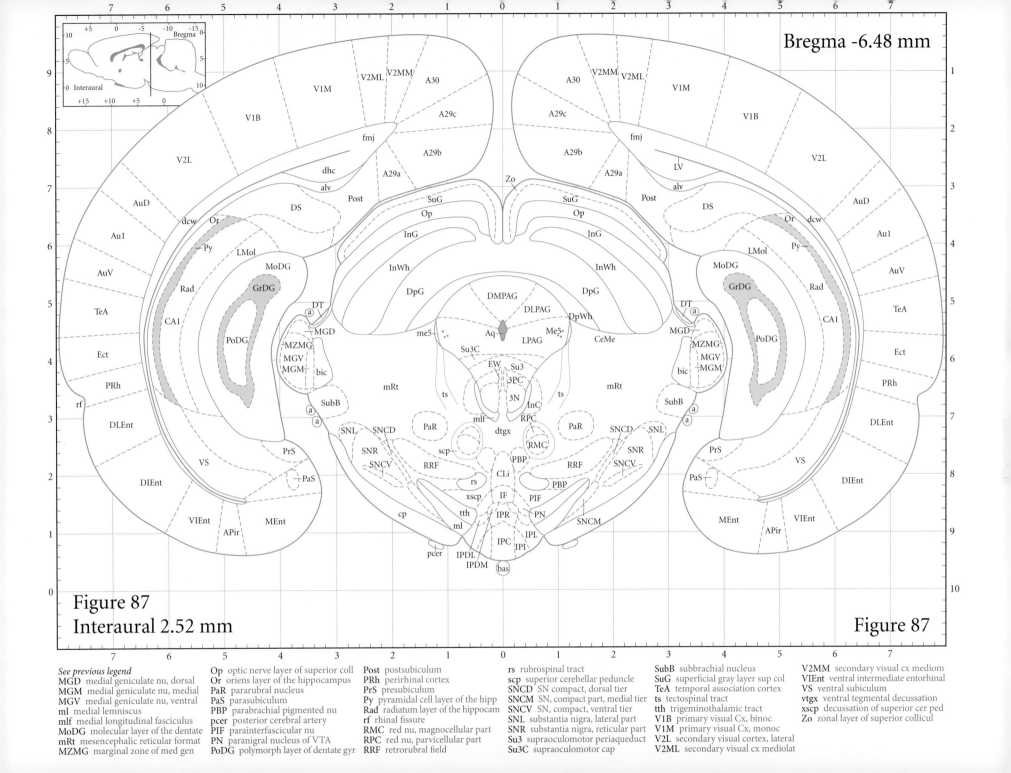

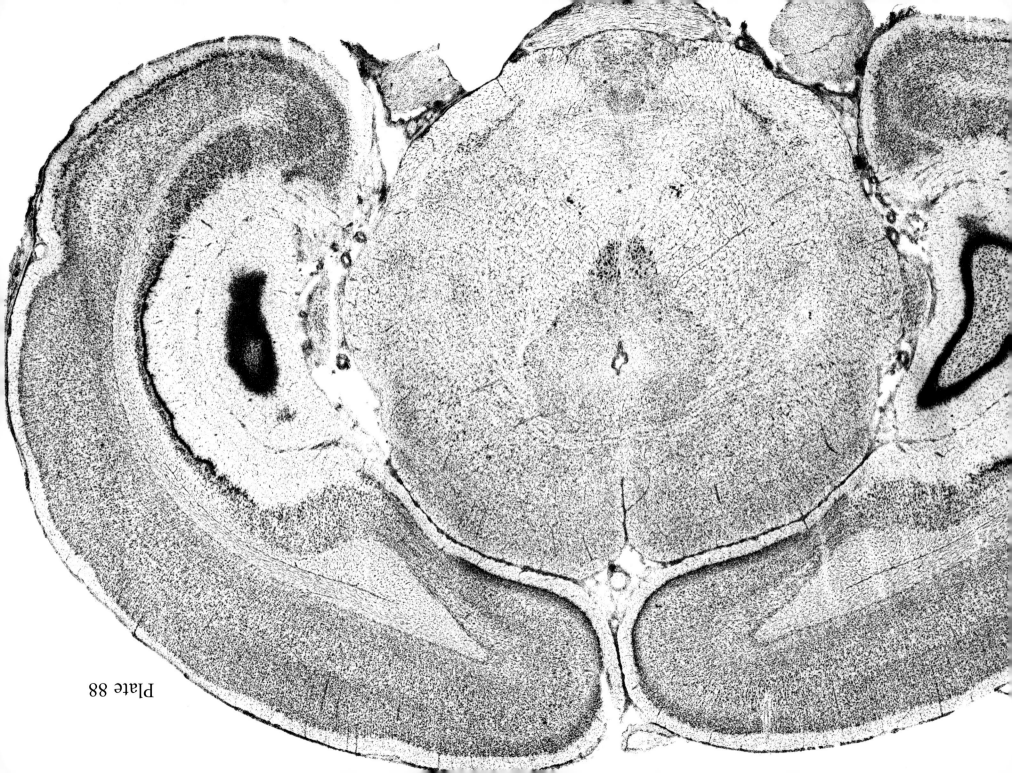

Plate 88

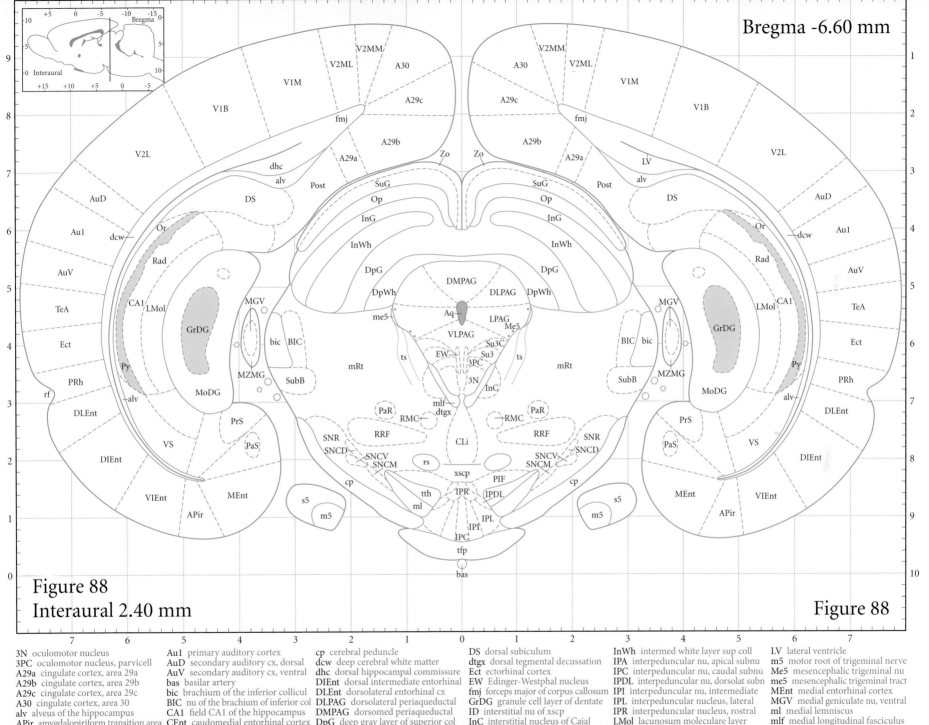

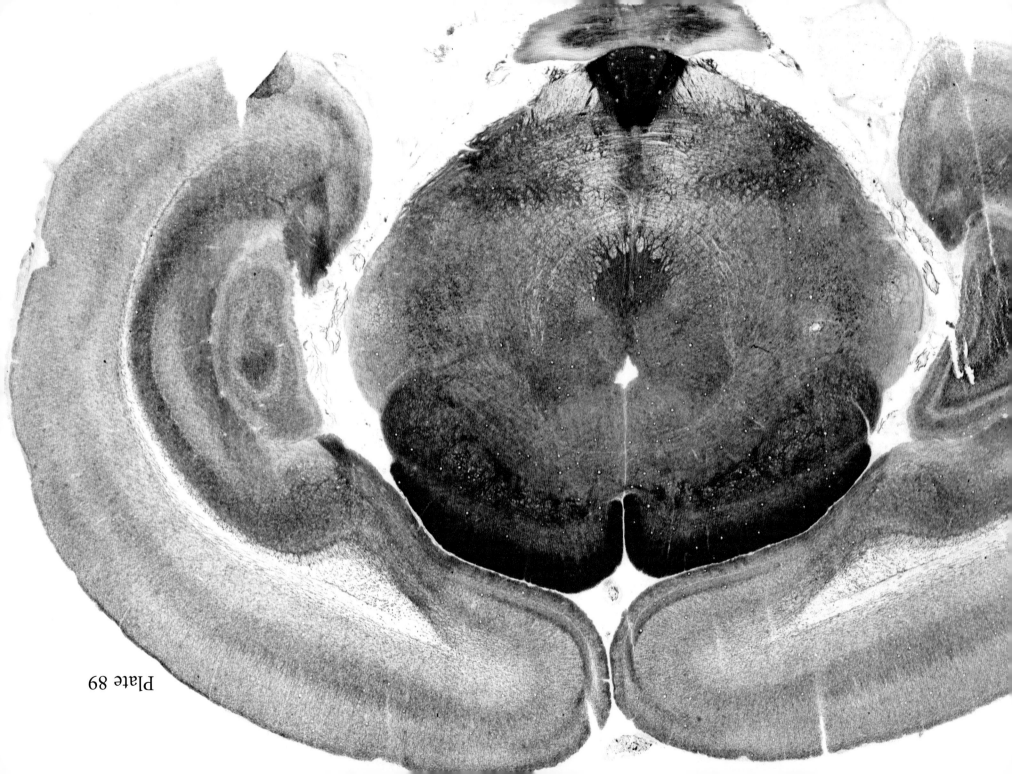

Plate 89

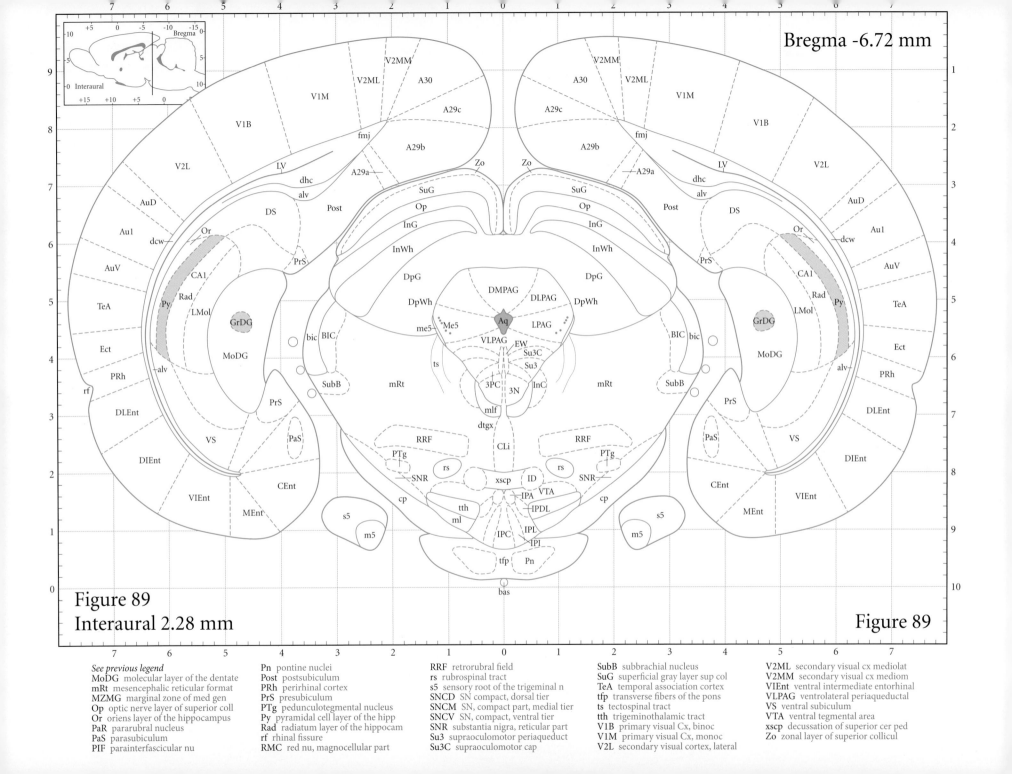

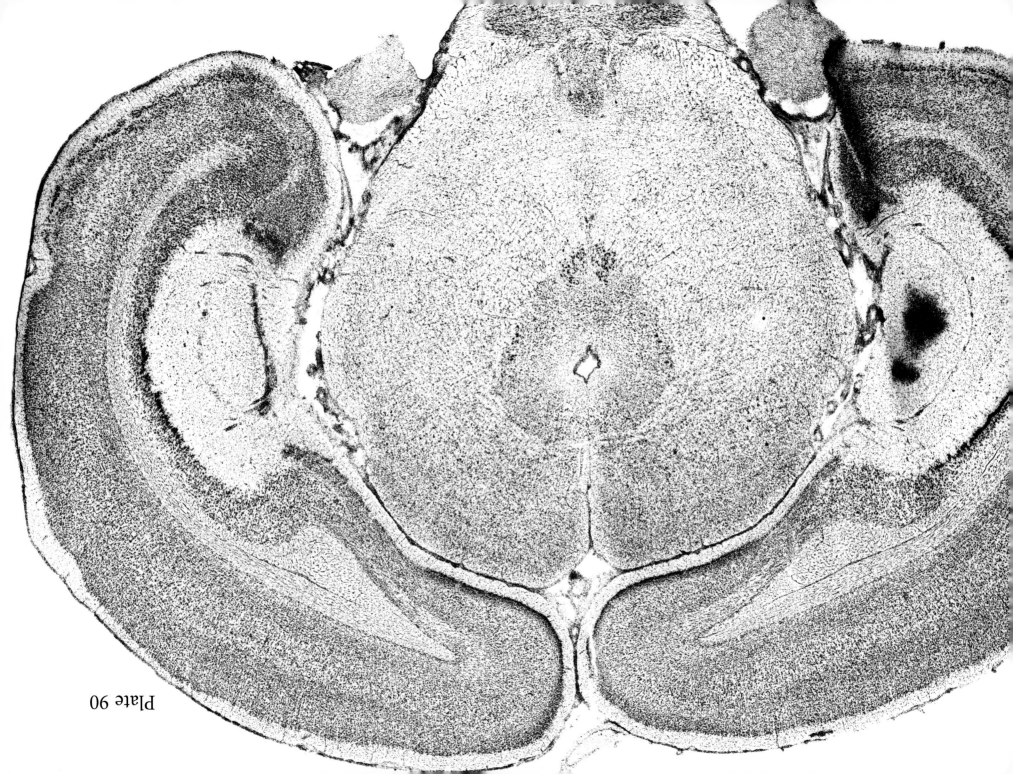
Plate 90

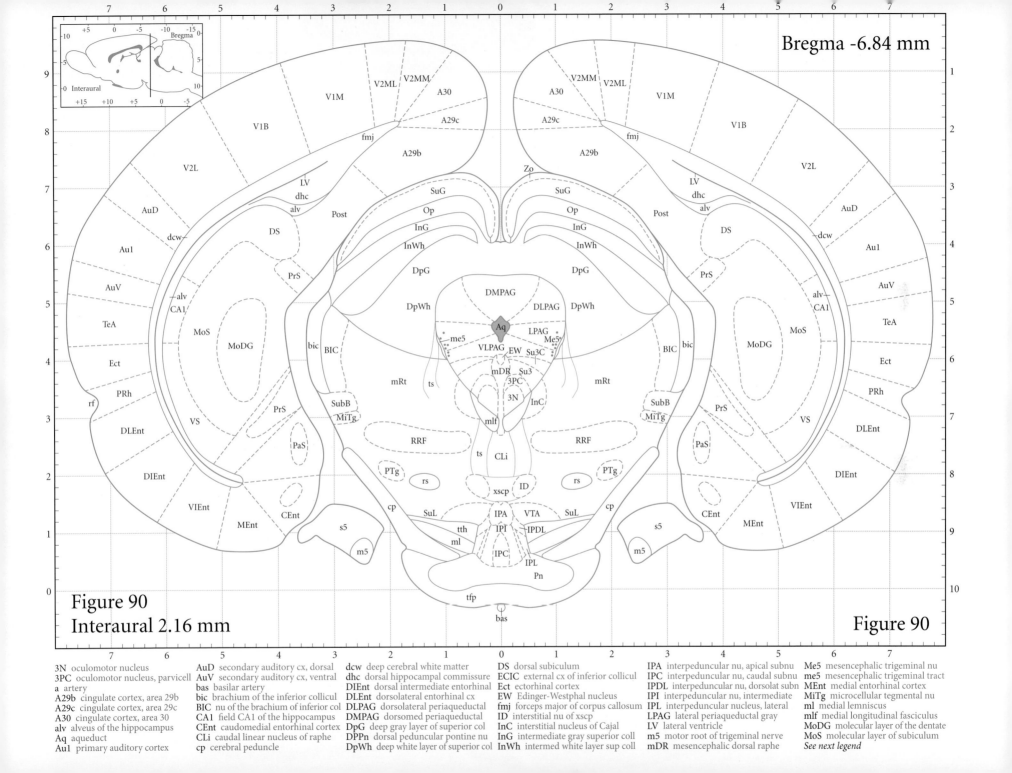

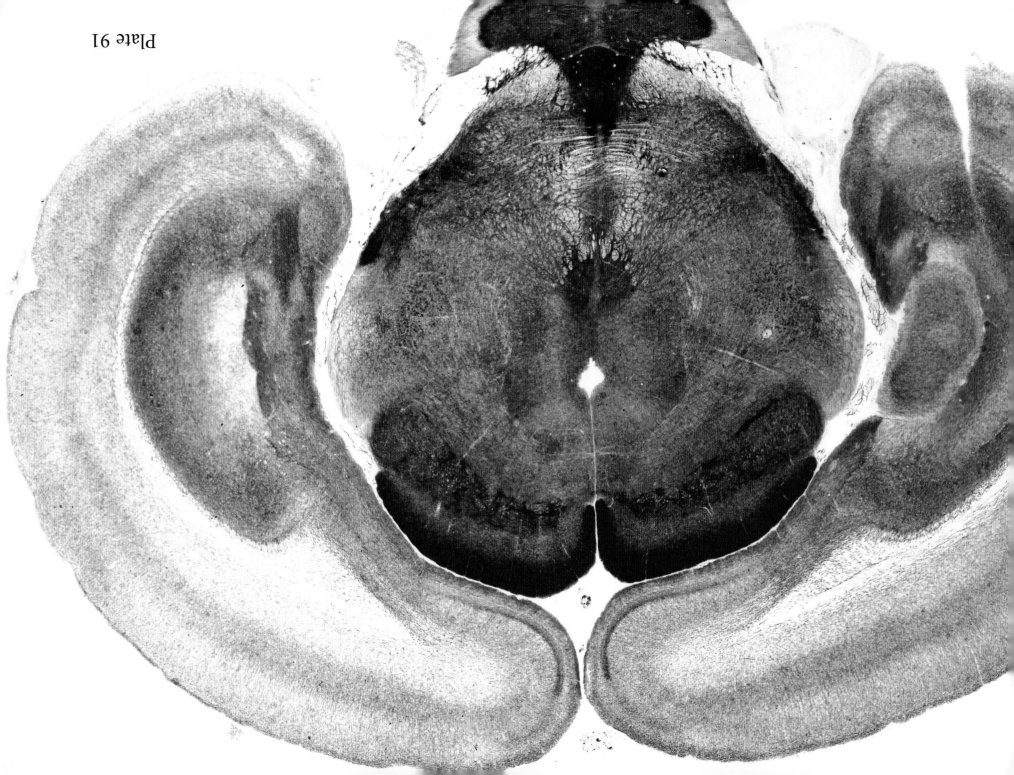

Plate 91

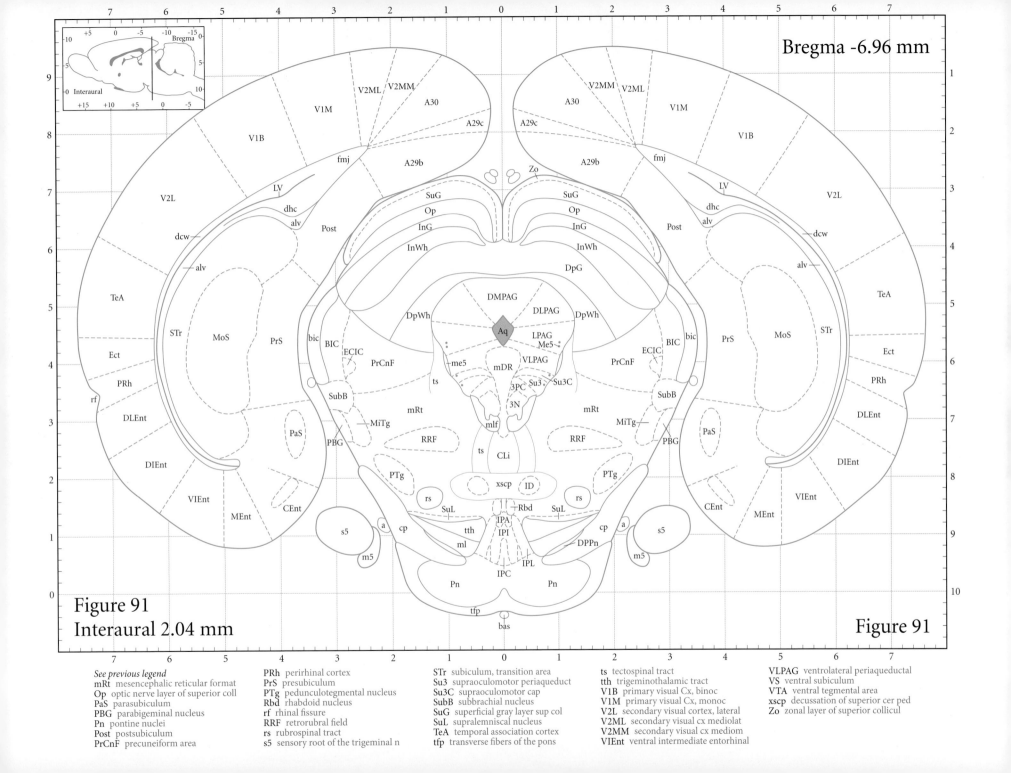

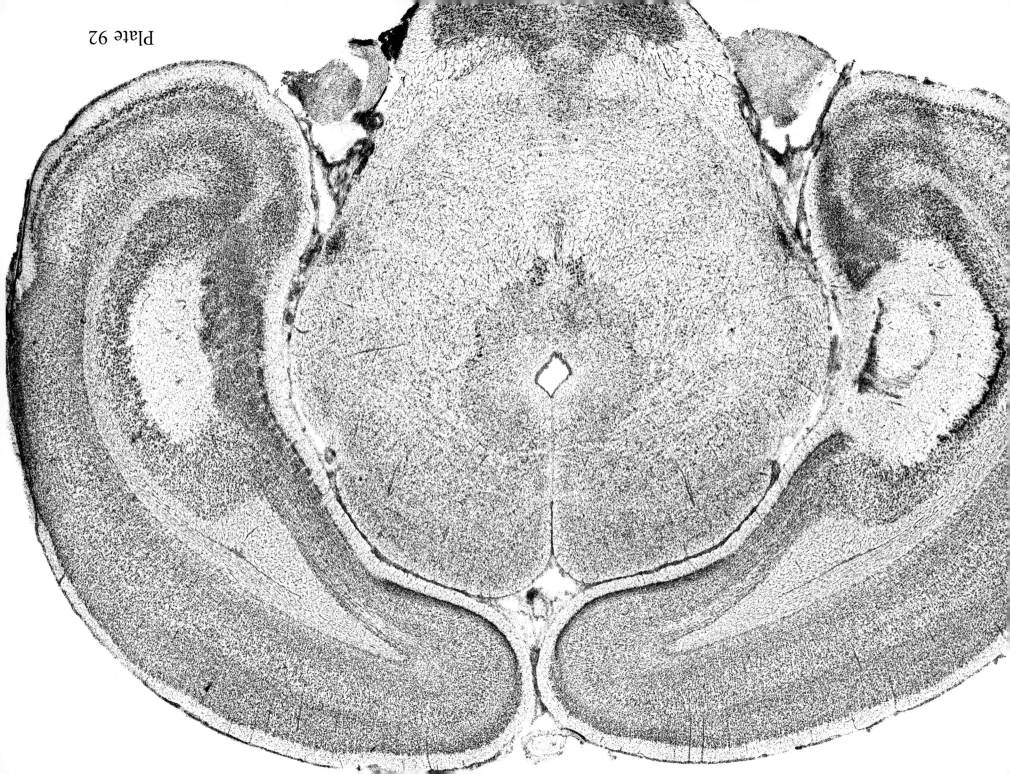

Plate 92

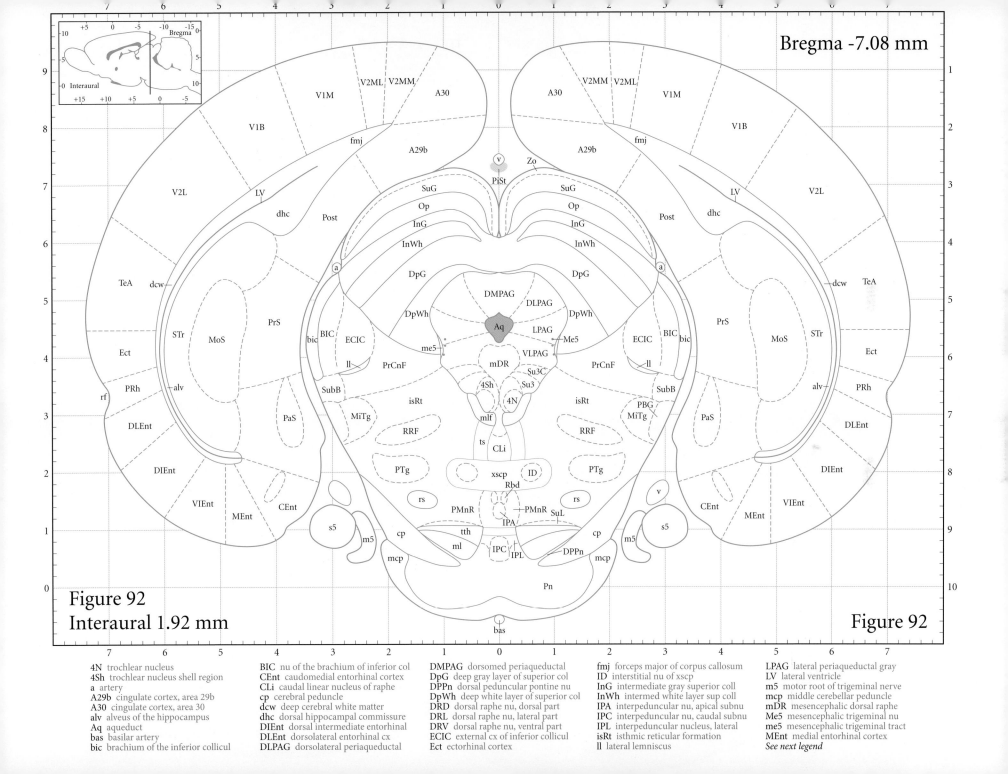

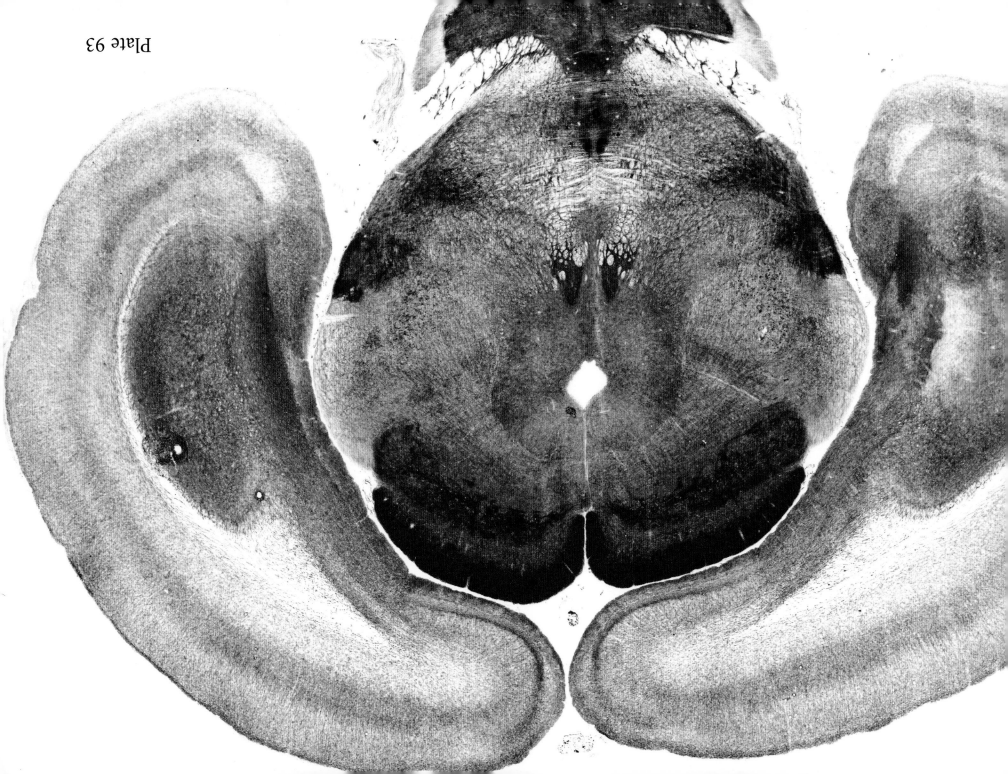

Plate 93

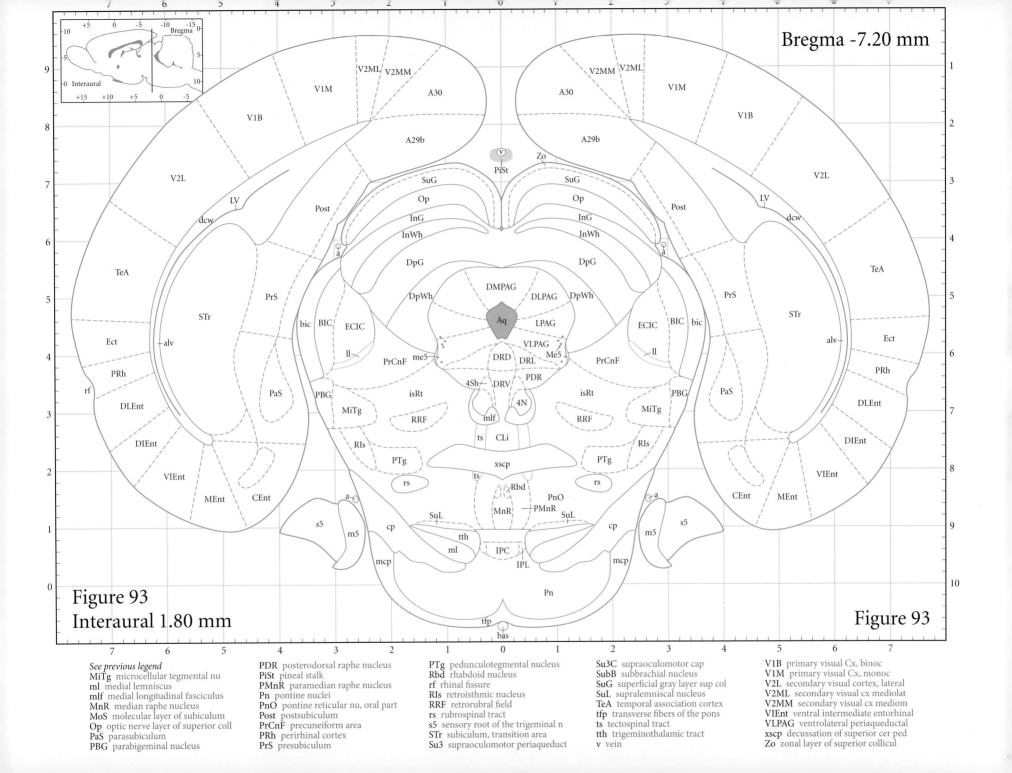

Figure 93
Interaural 1.80 mm
Bregma -7.20 mm

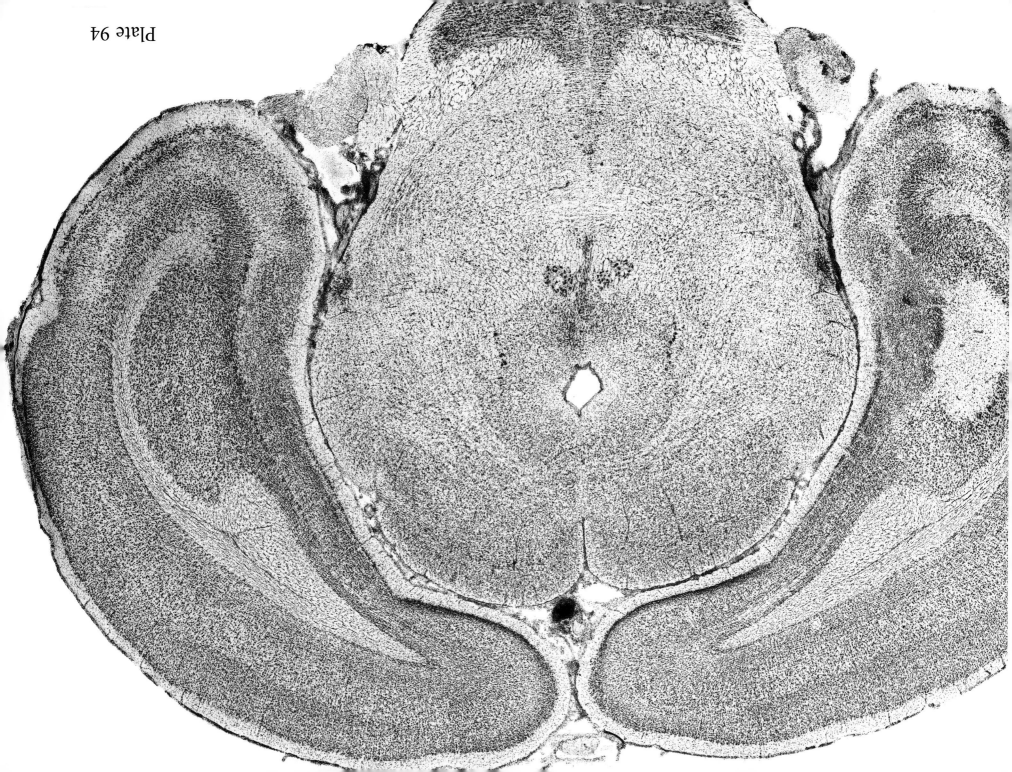
Plate 94

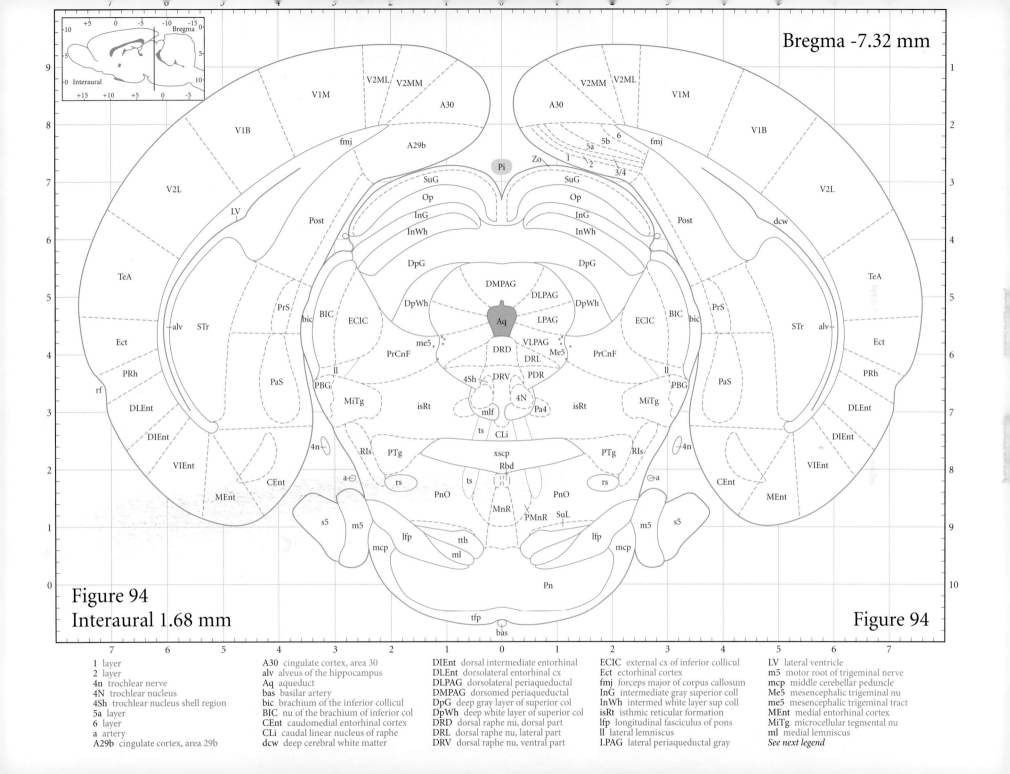

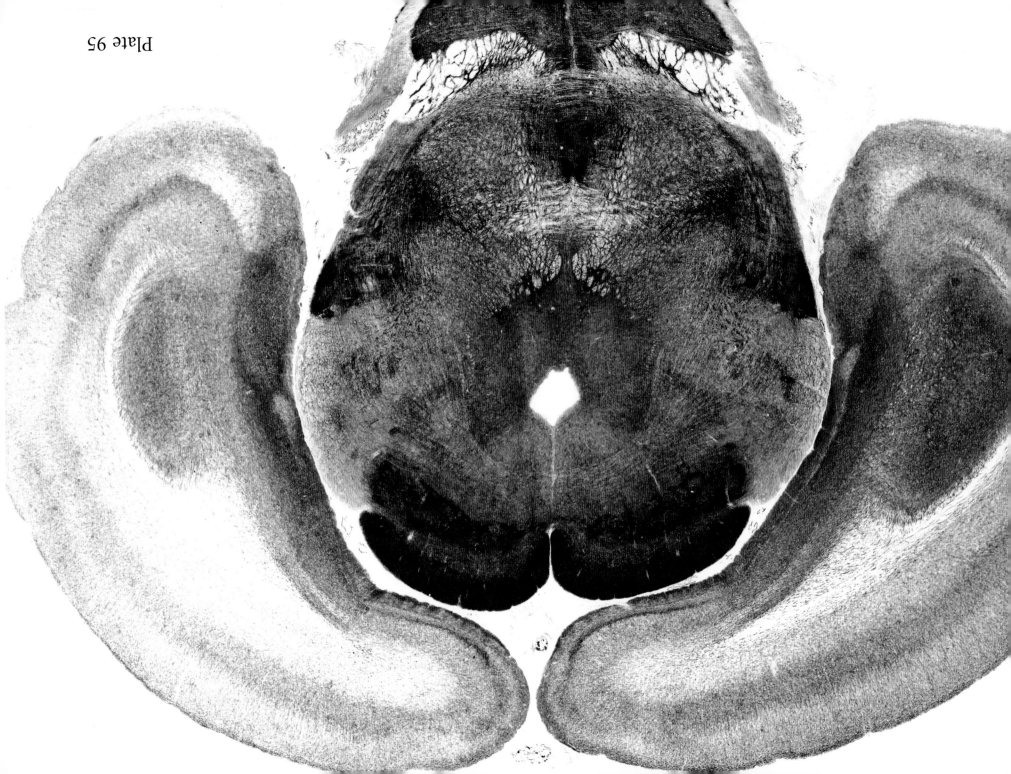

Plate 95

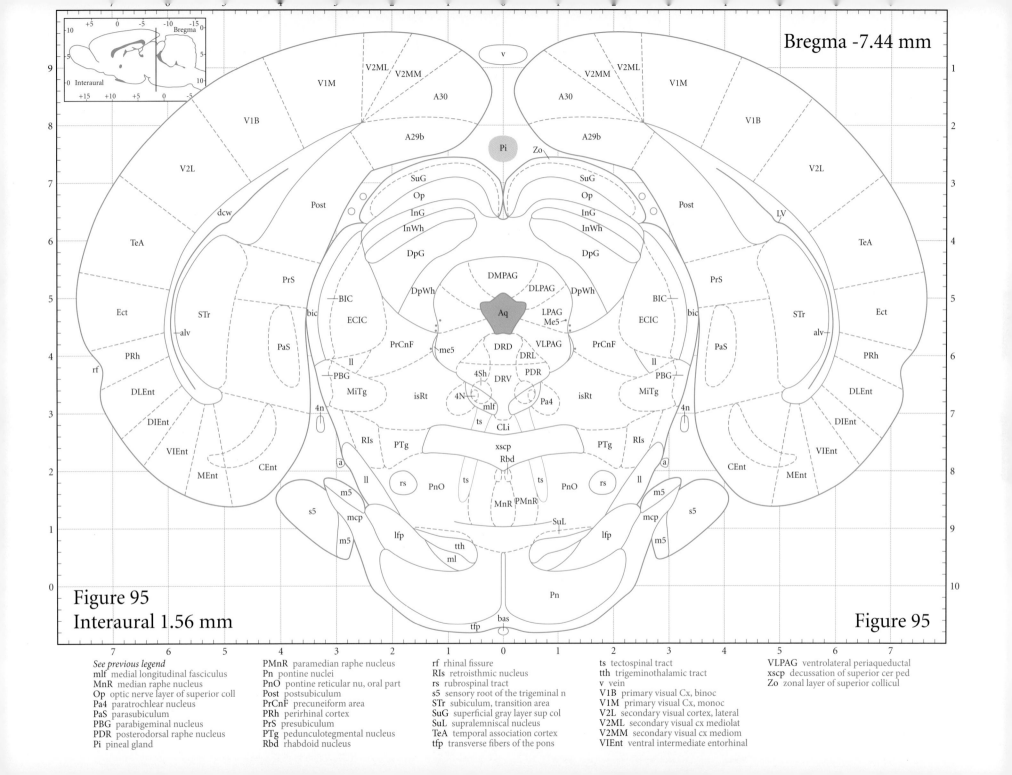

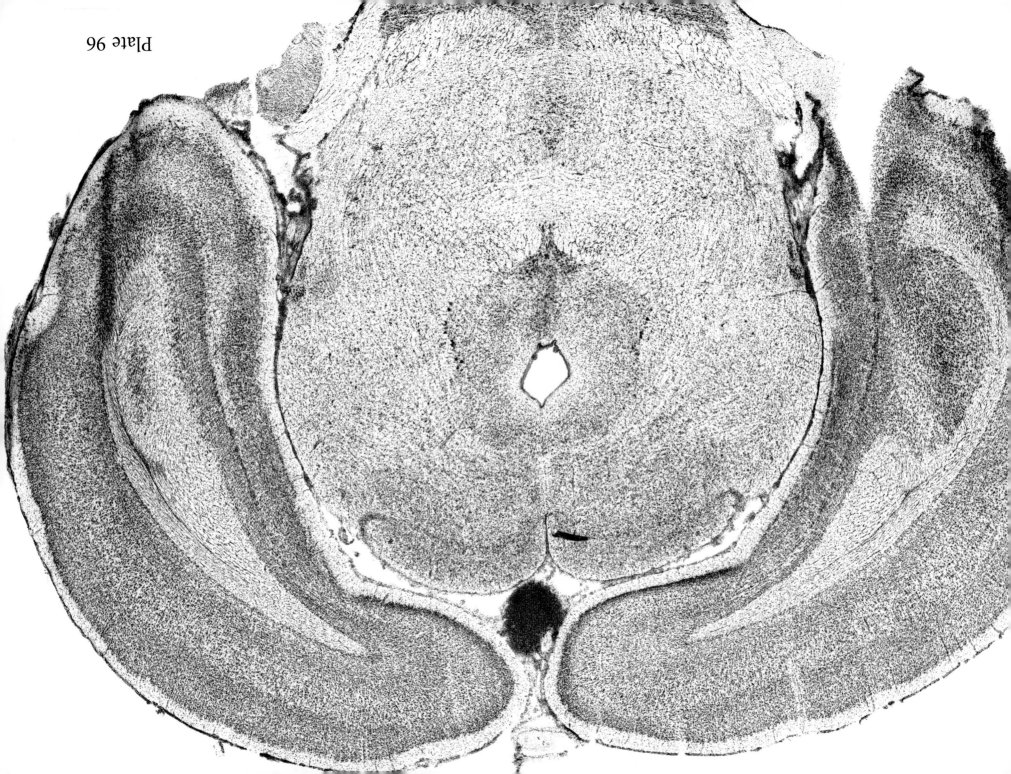

Plate 96

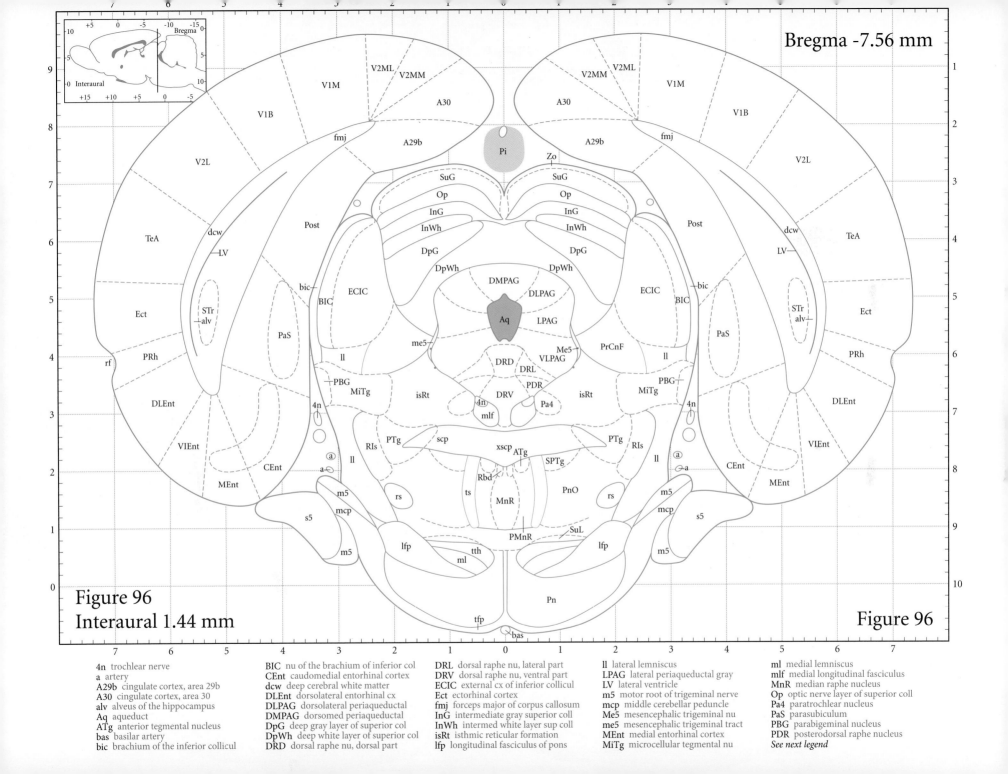

Plate 97

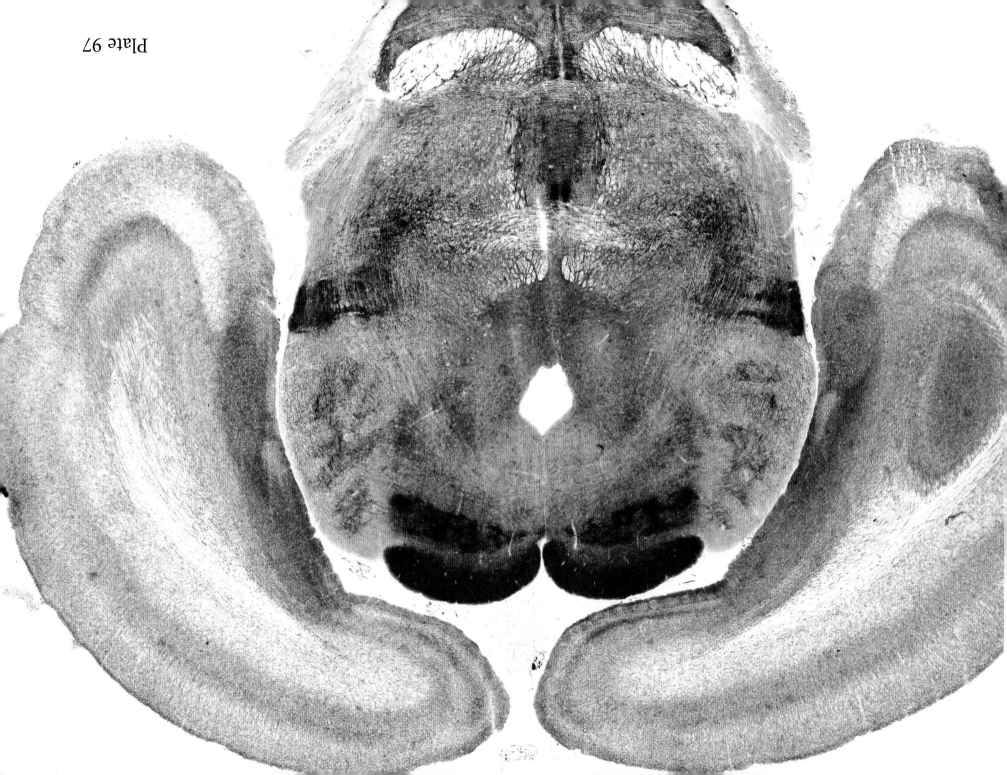

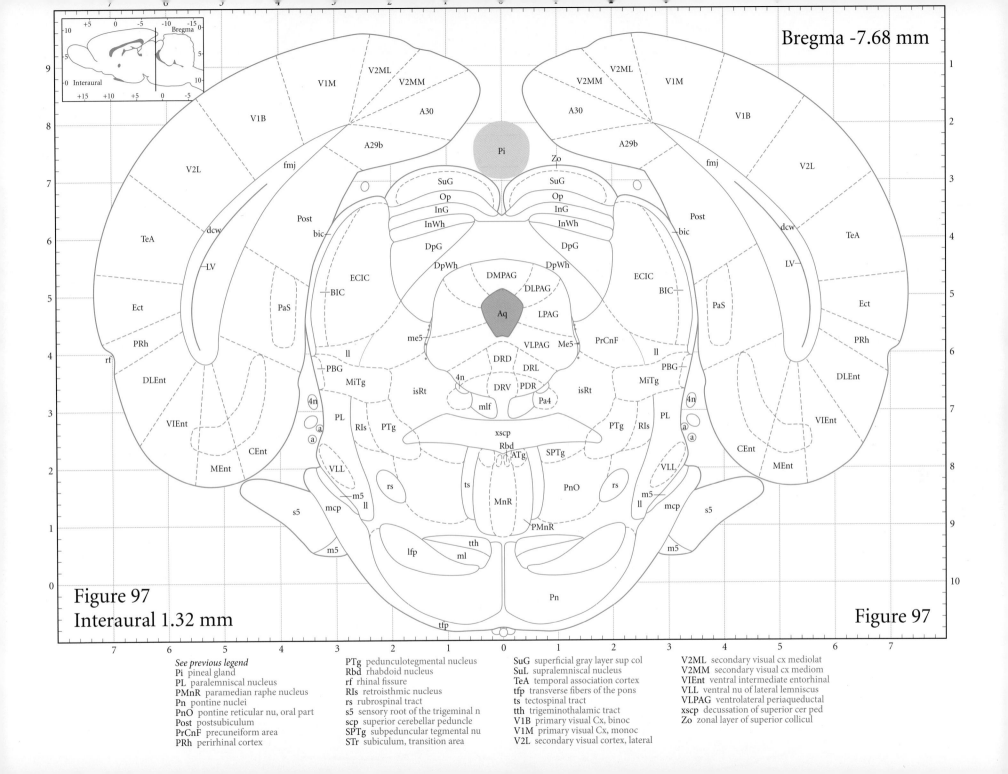

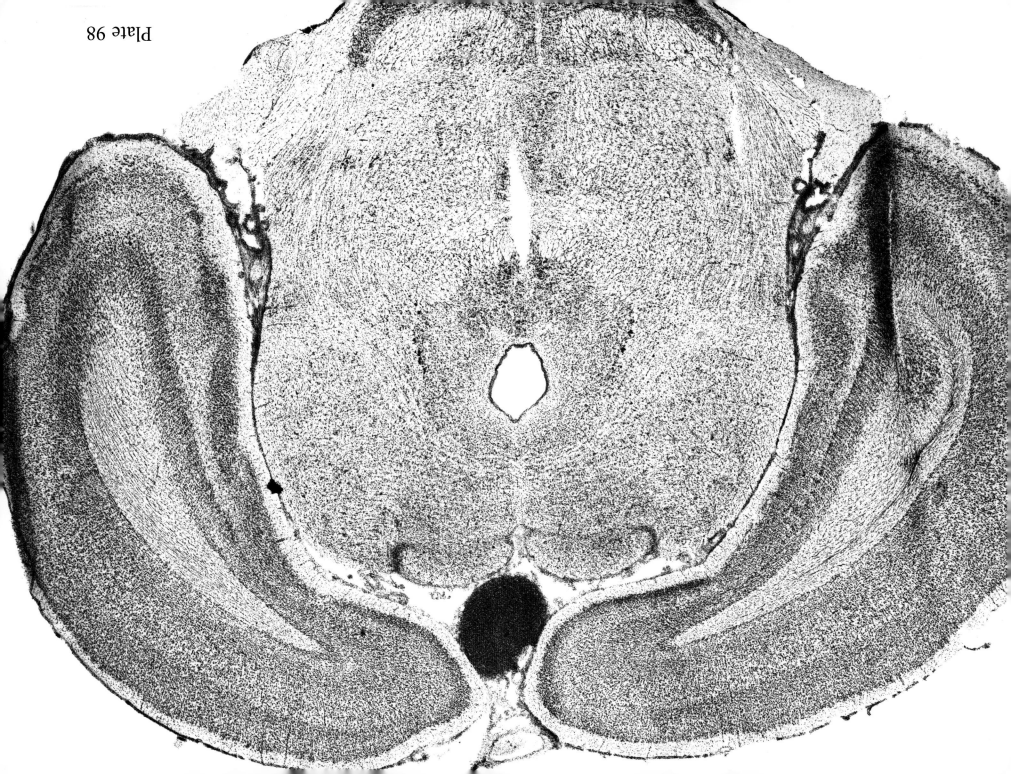

Plate 98

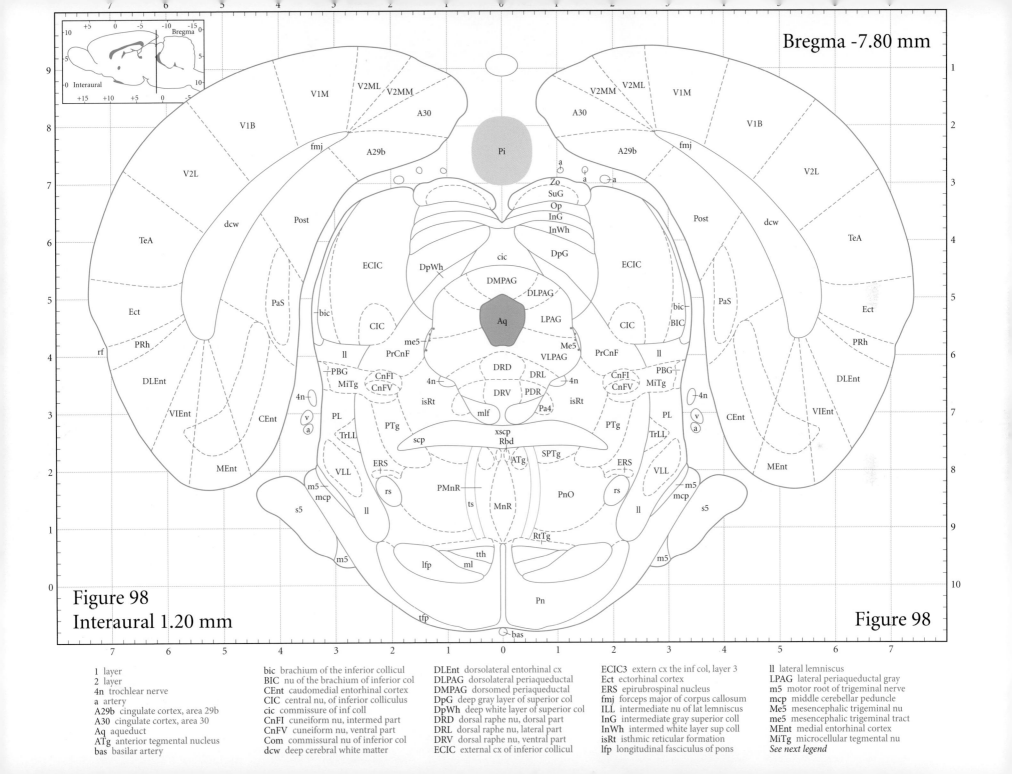

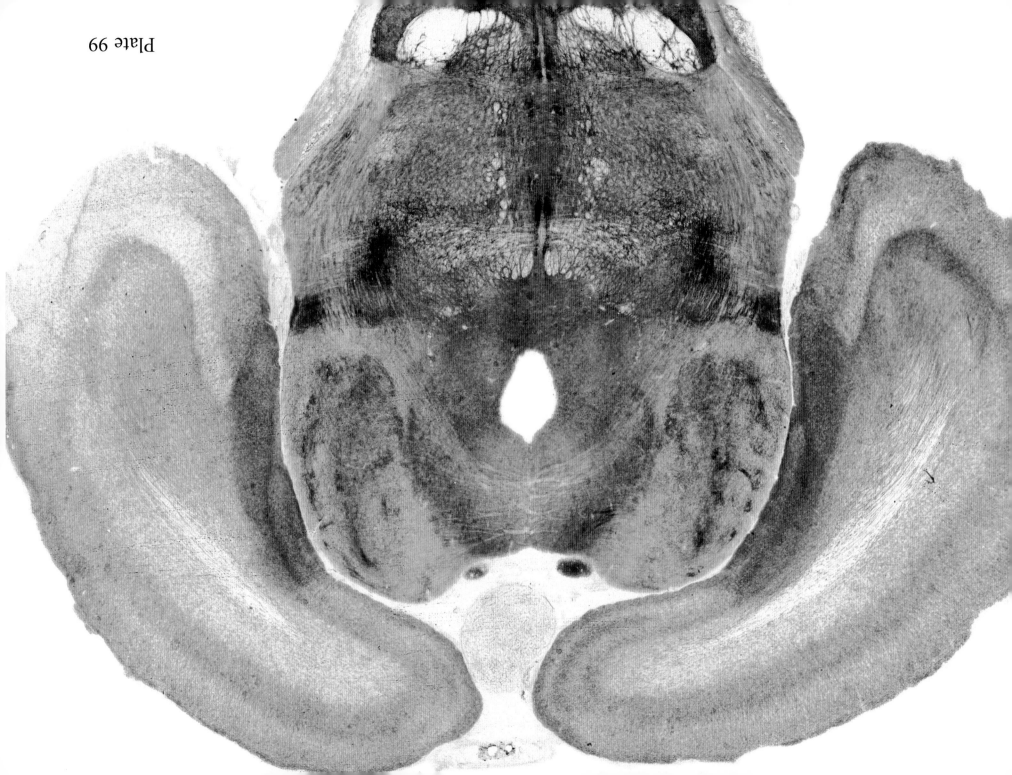

Plate 99

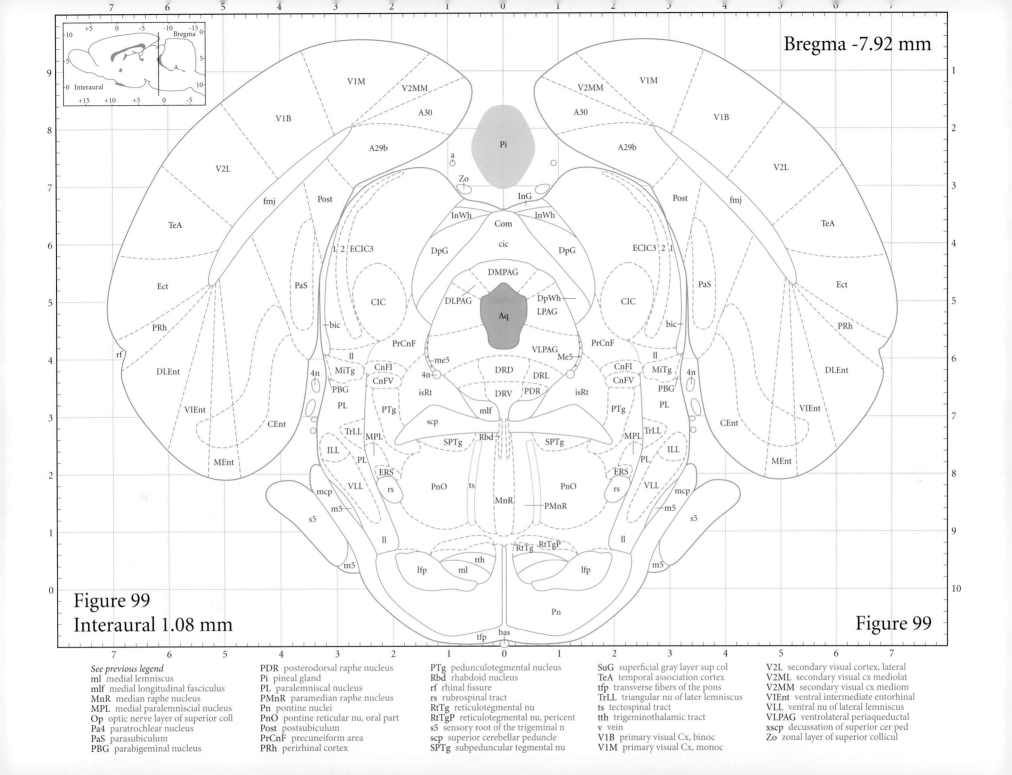

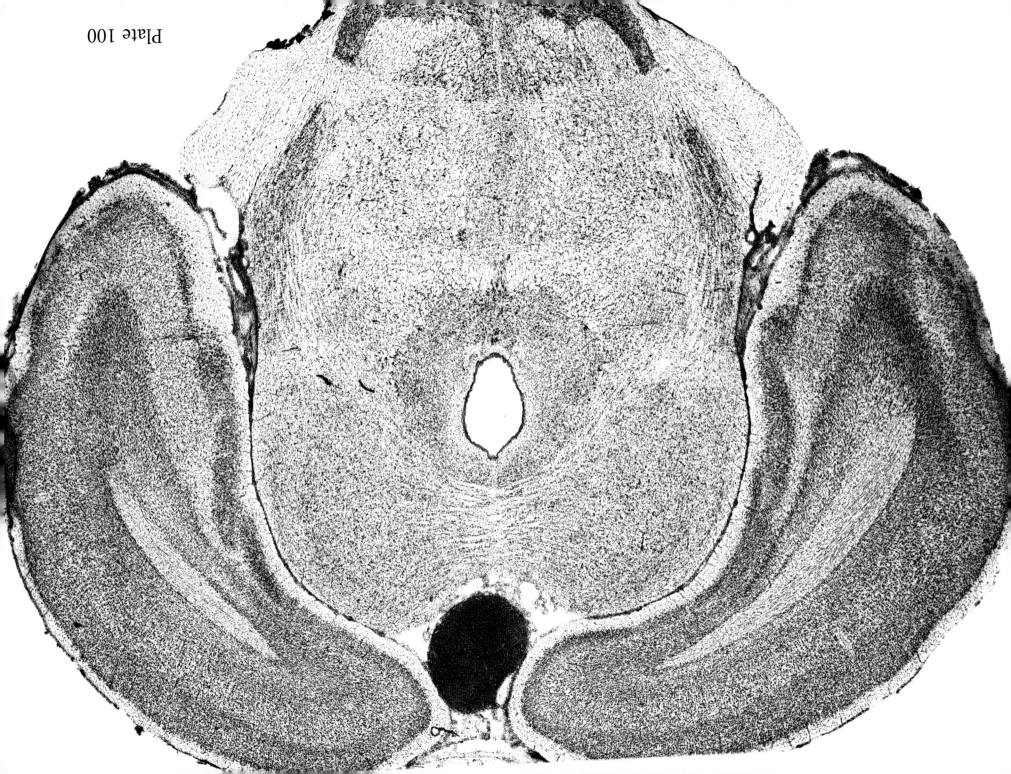

Plate 100

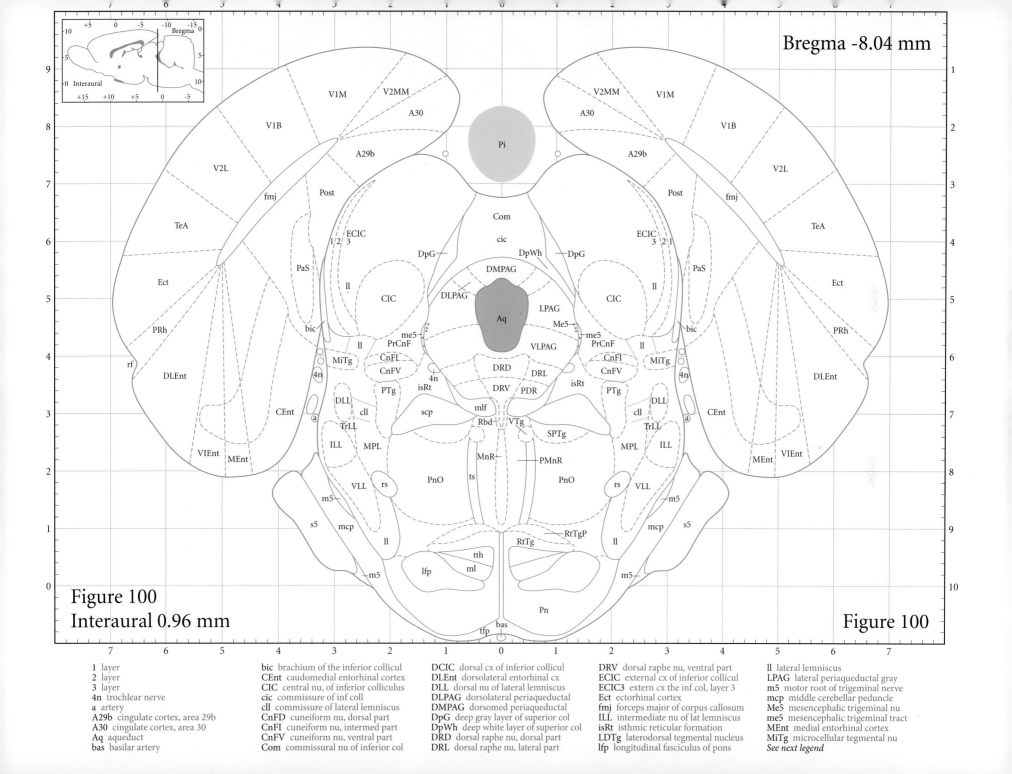

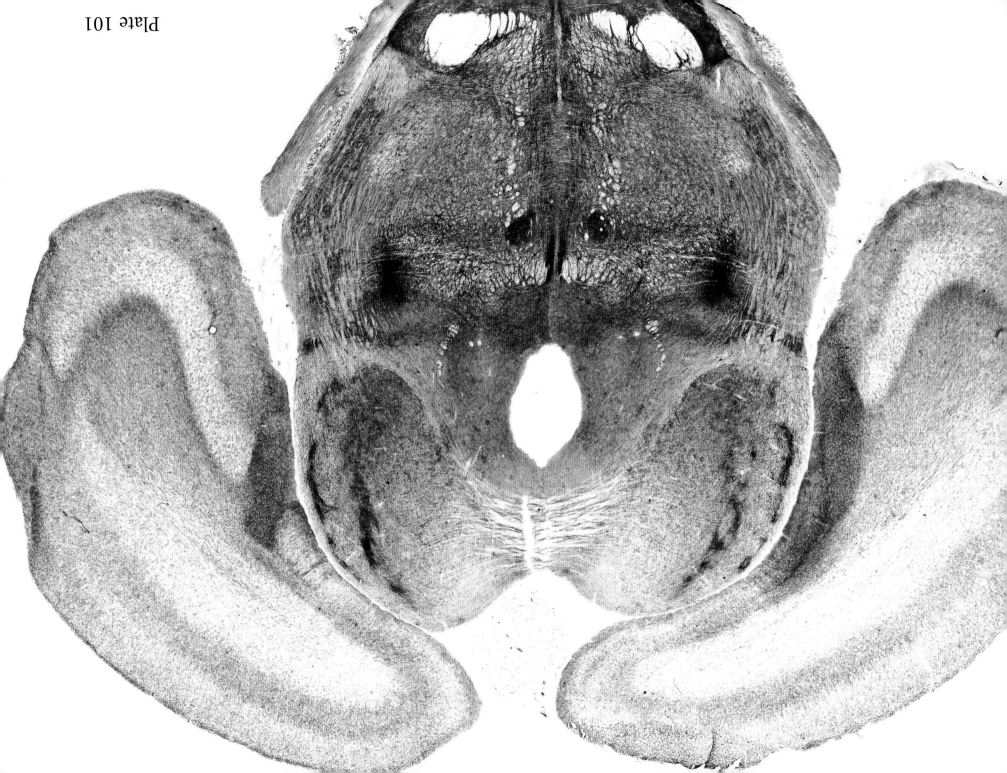
Plate 101

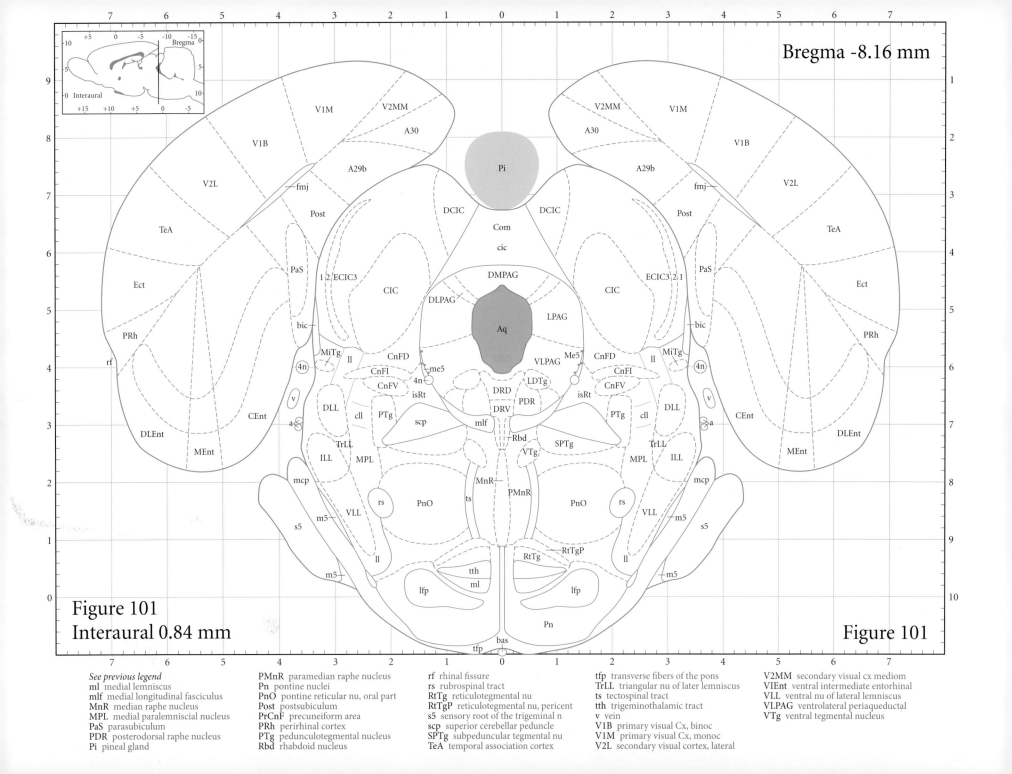

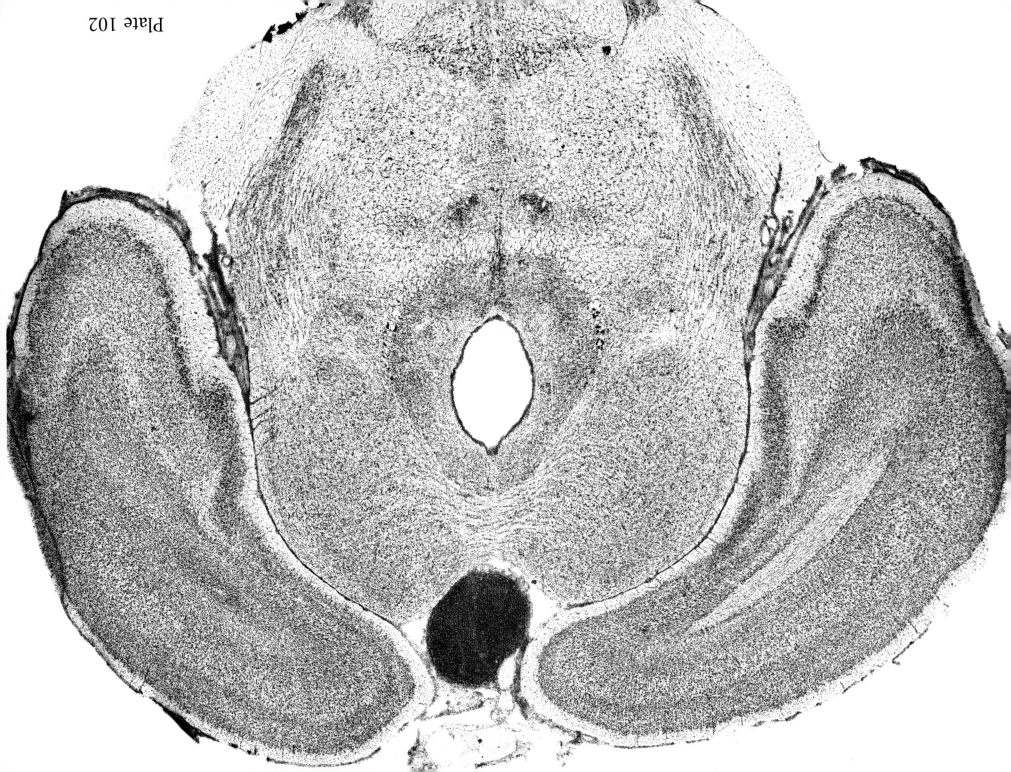

Plate 102

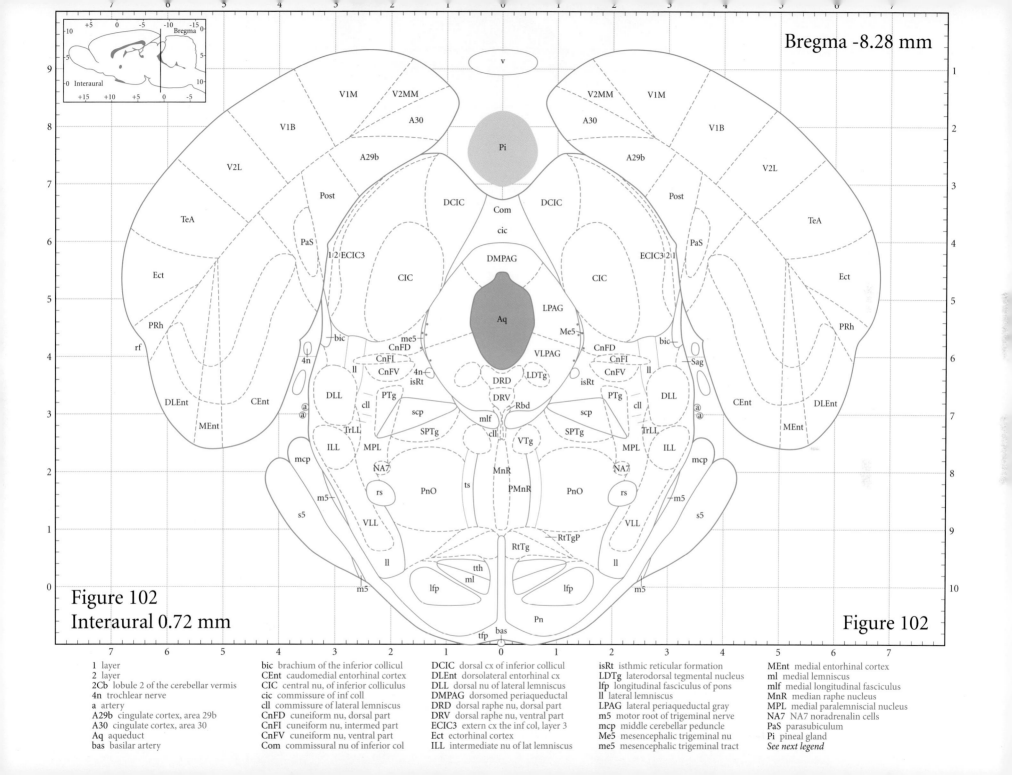

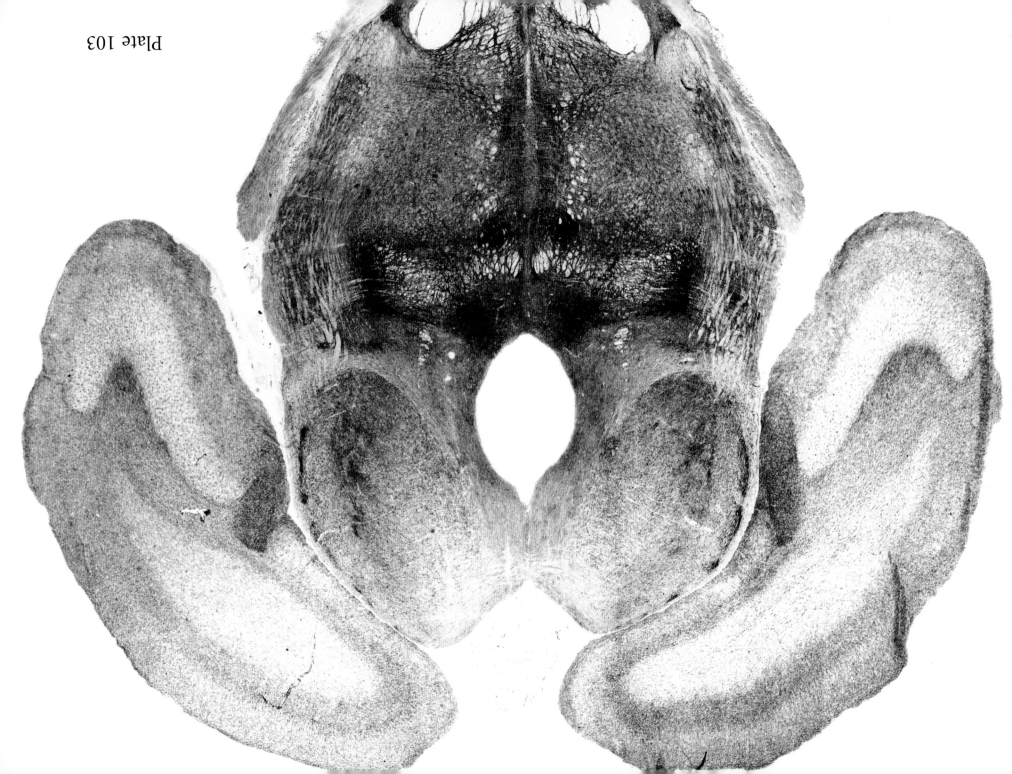

Plate 103

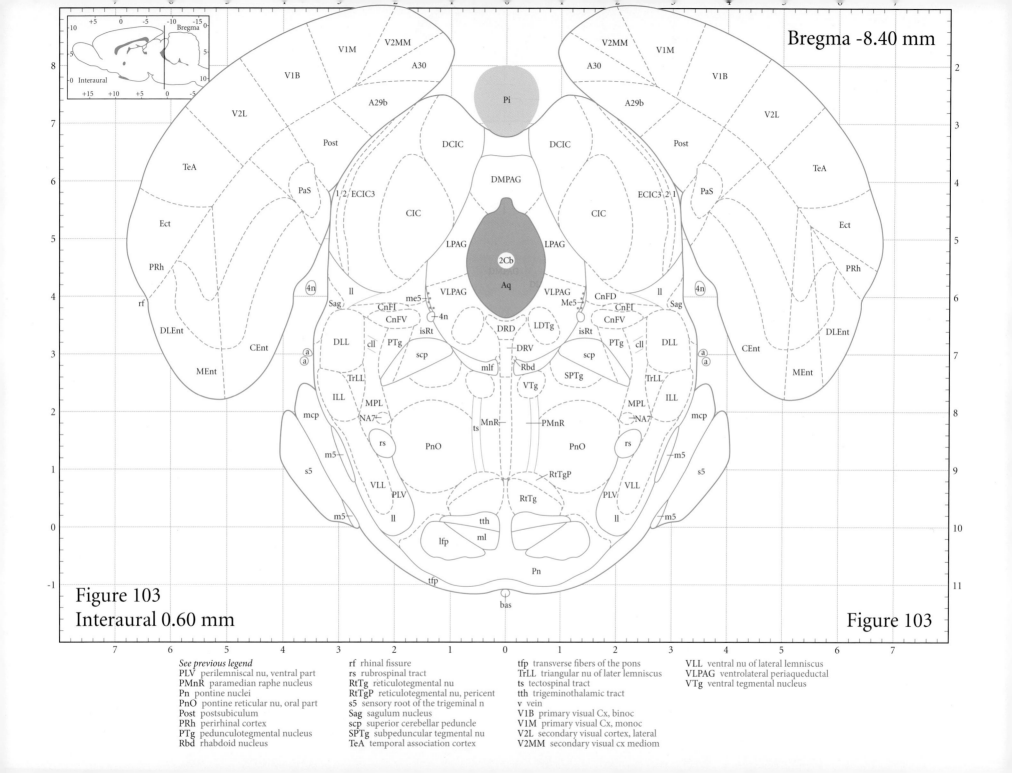

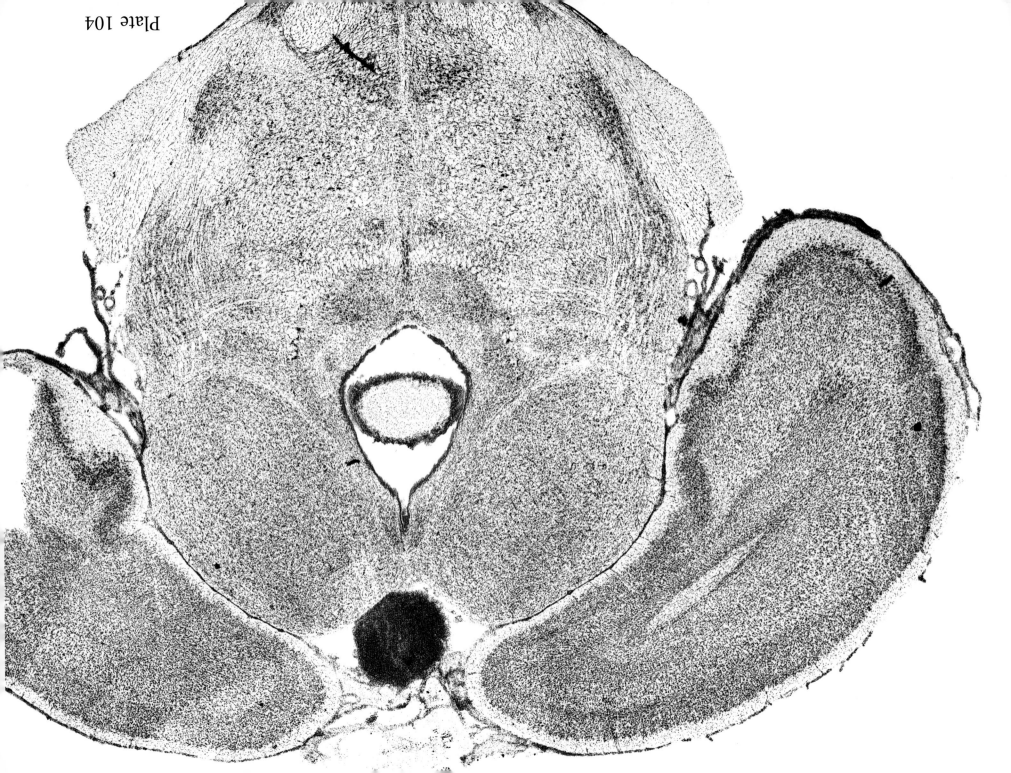

Plate 104

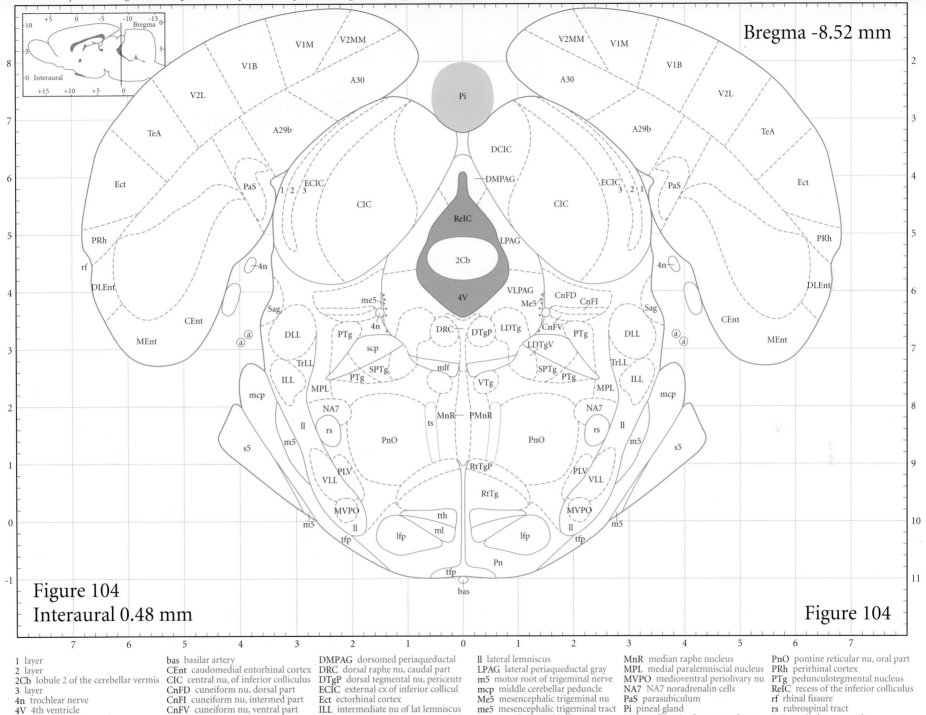

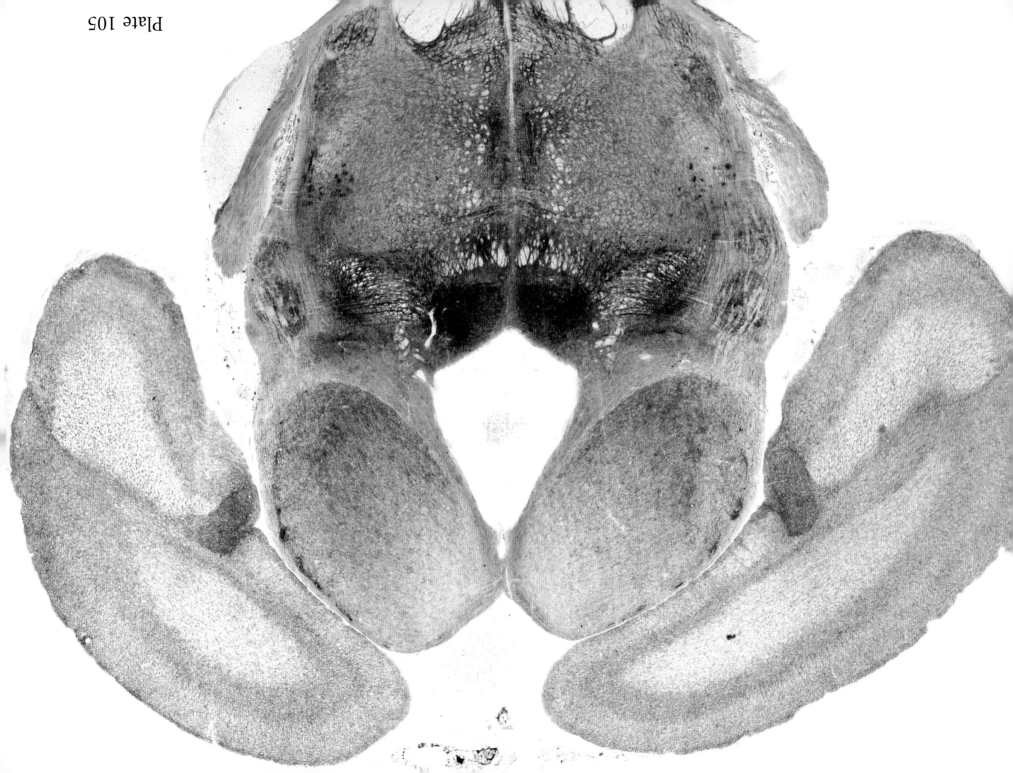

Plate 105

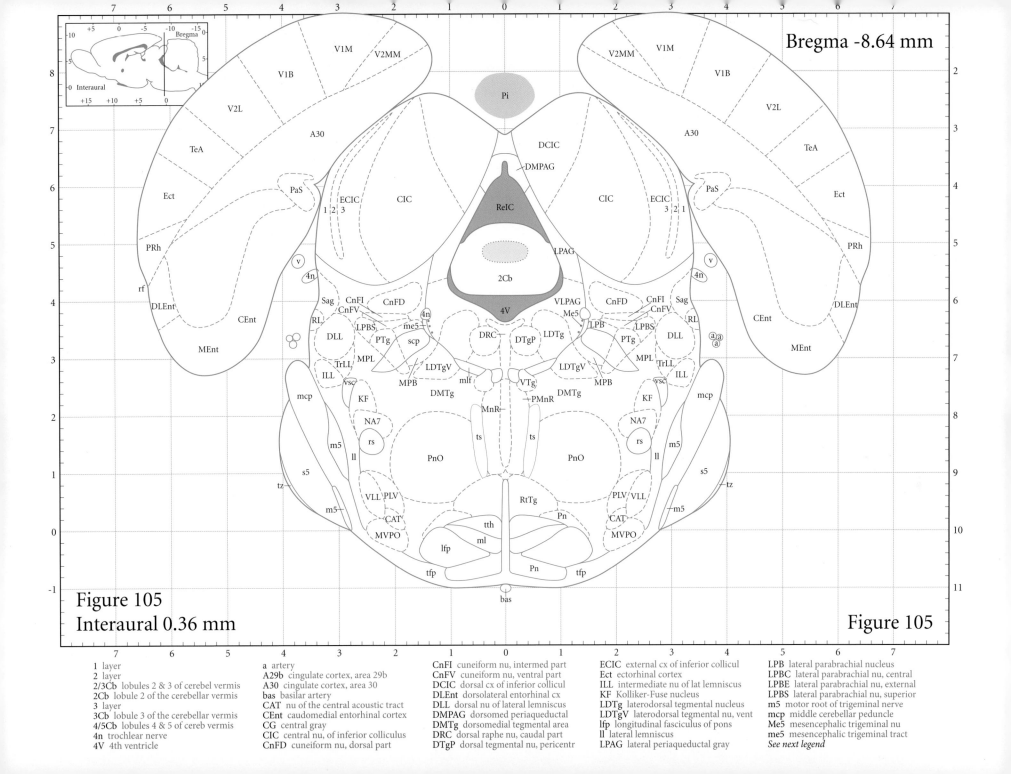

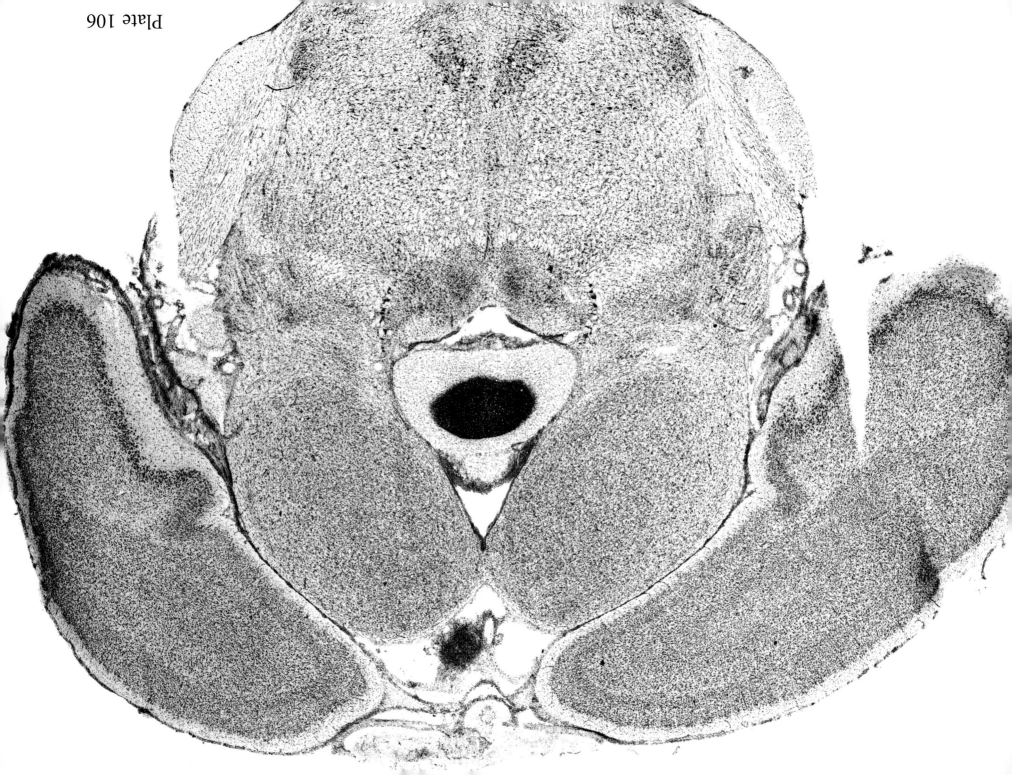

Plate 106

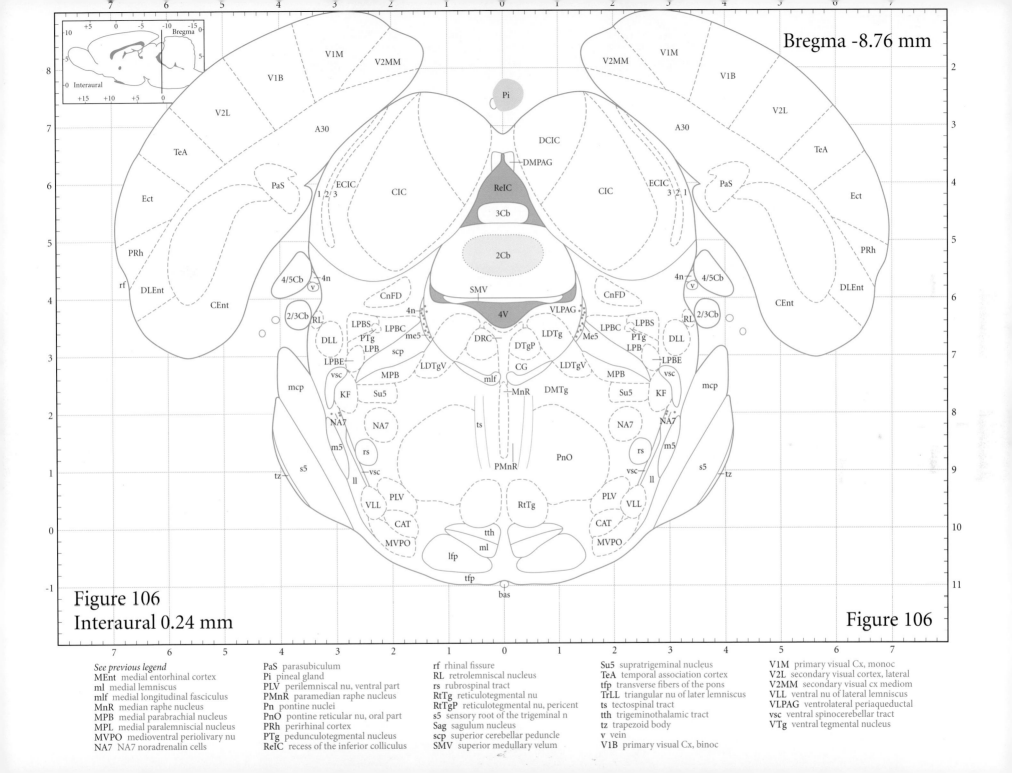

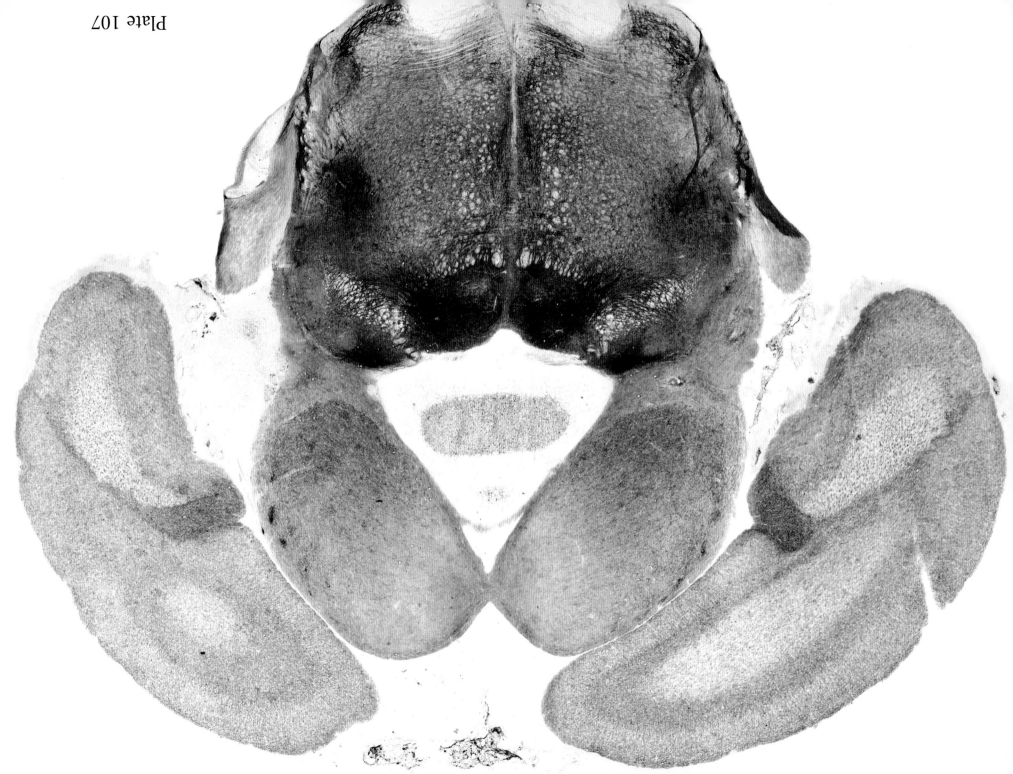

Plate 107

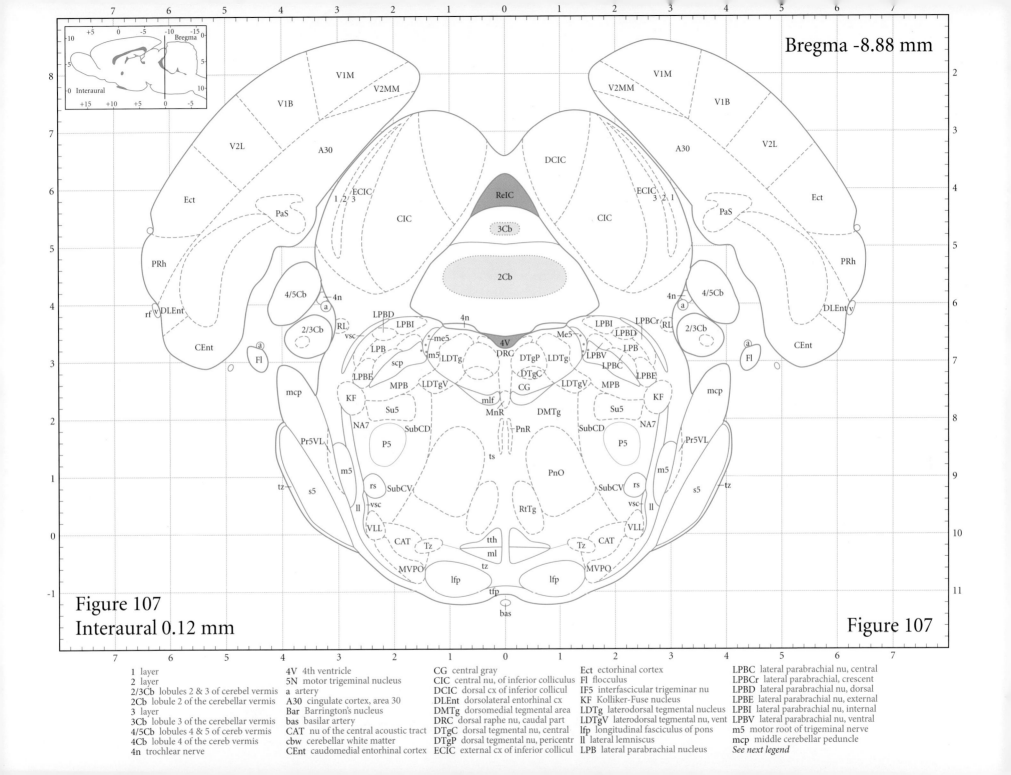

Figure 107
Interaural 0.12 mm
Bregma -8.88 mm

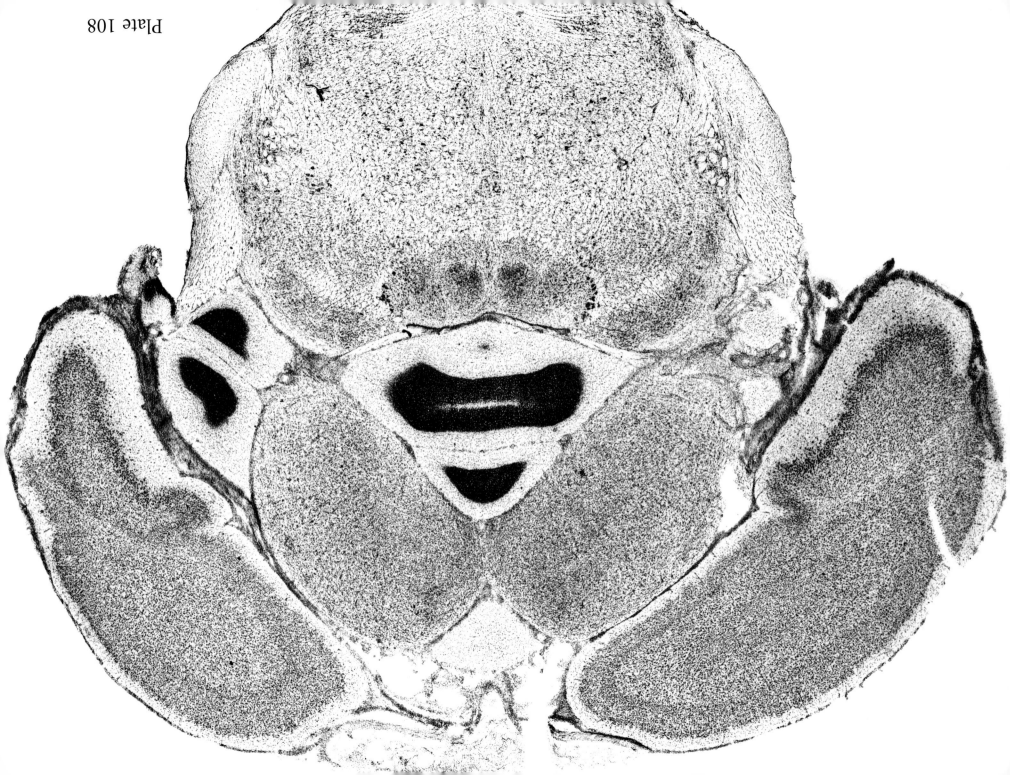

Plate 108

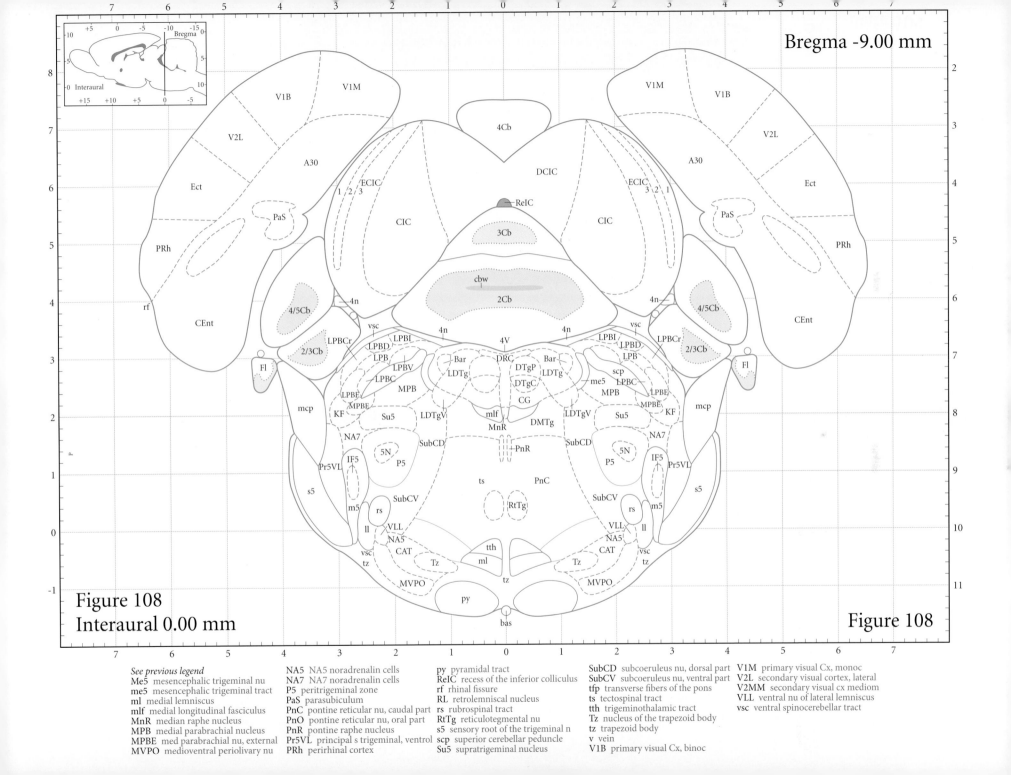

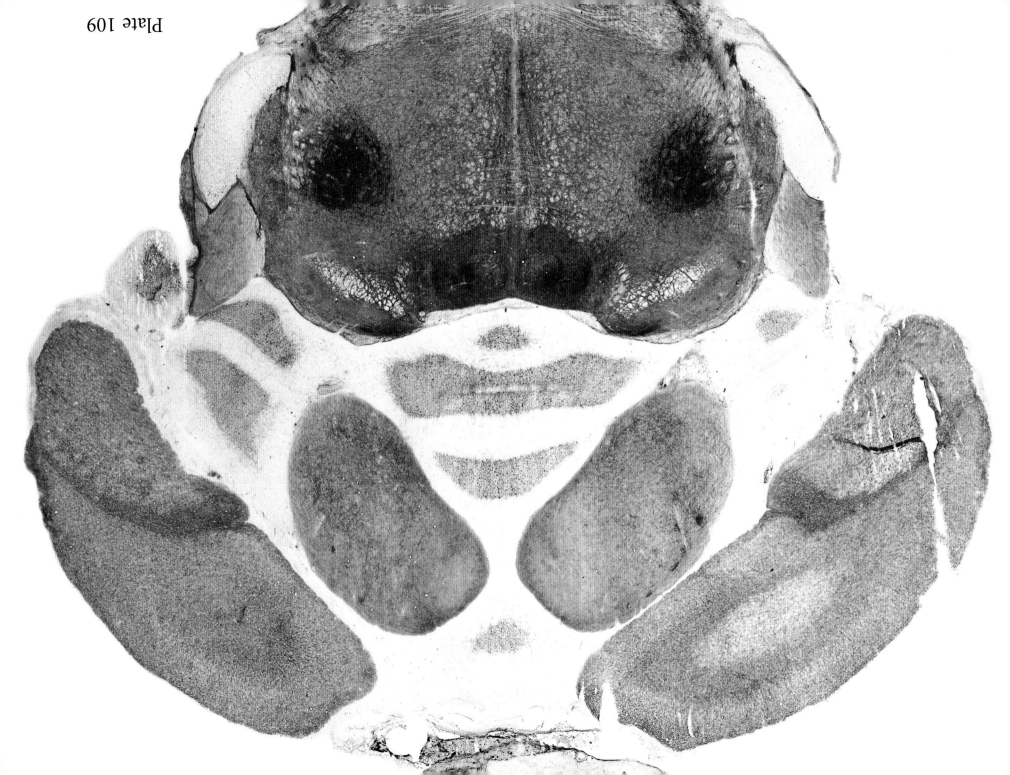

Plate 109

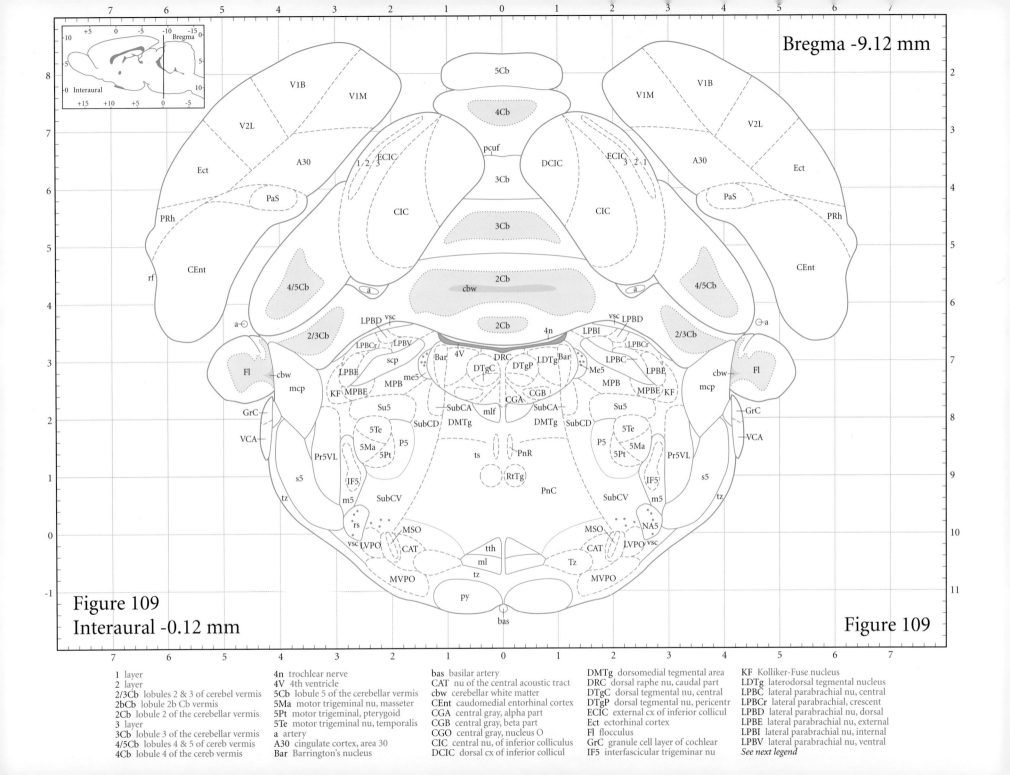

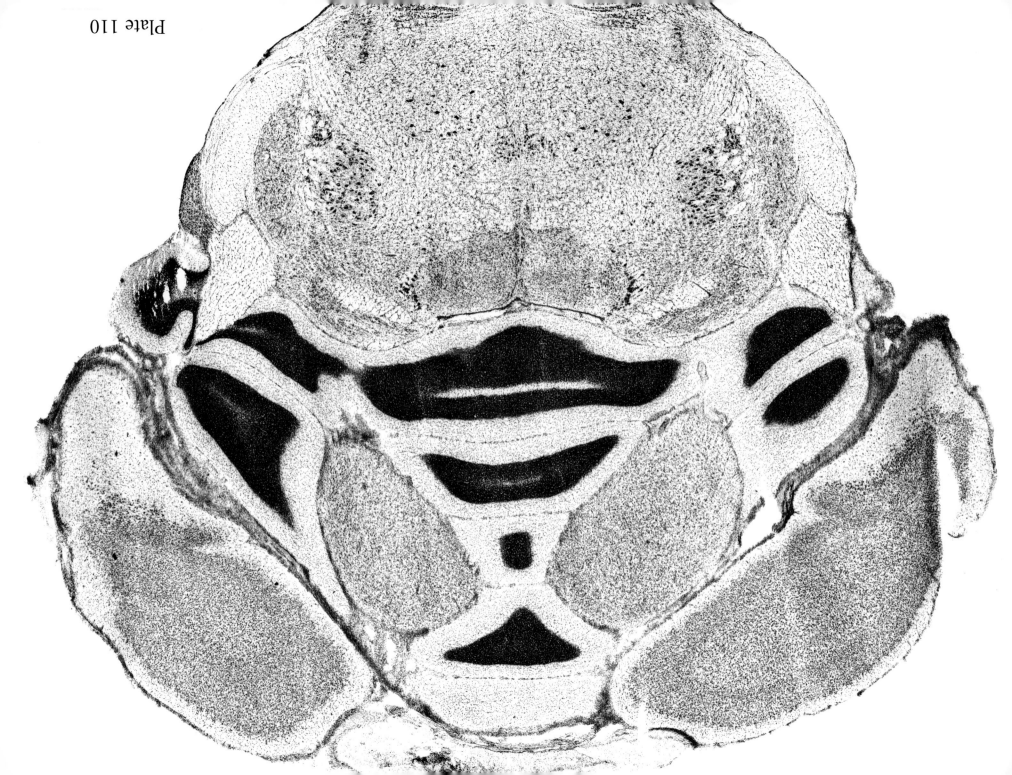

Plate 110

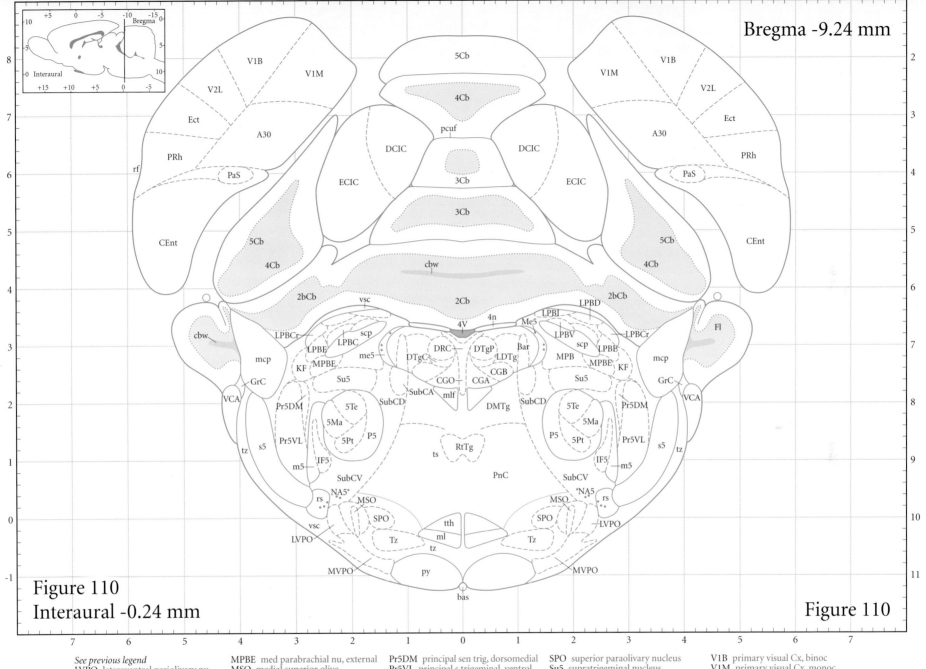

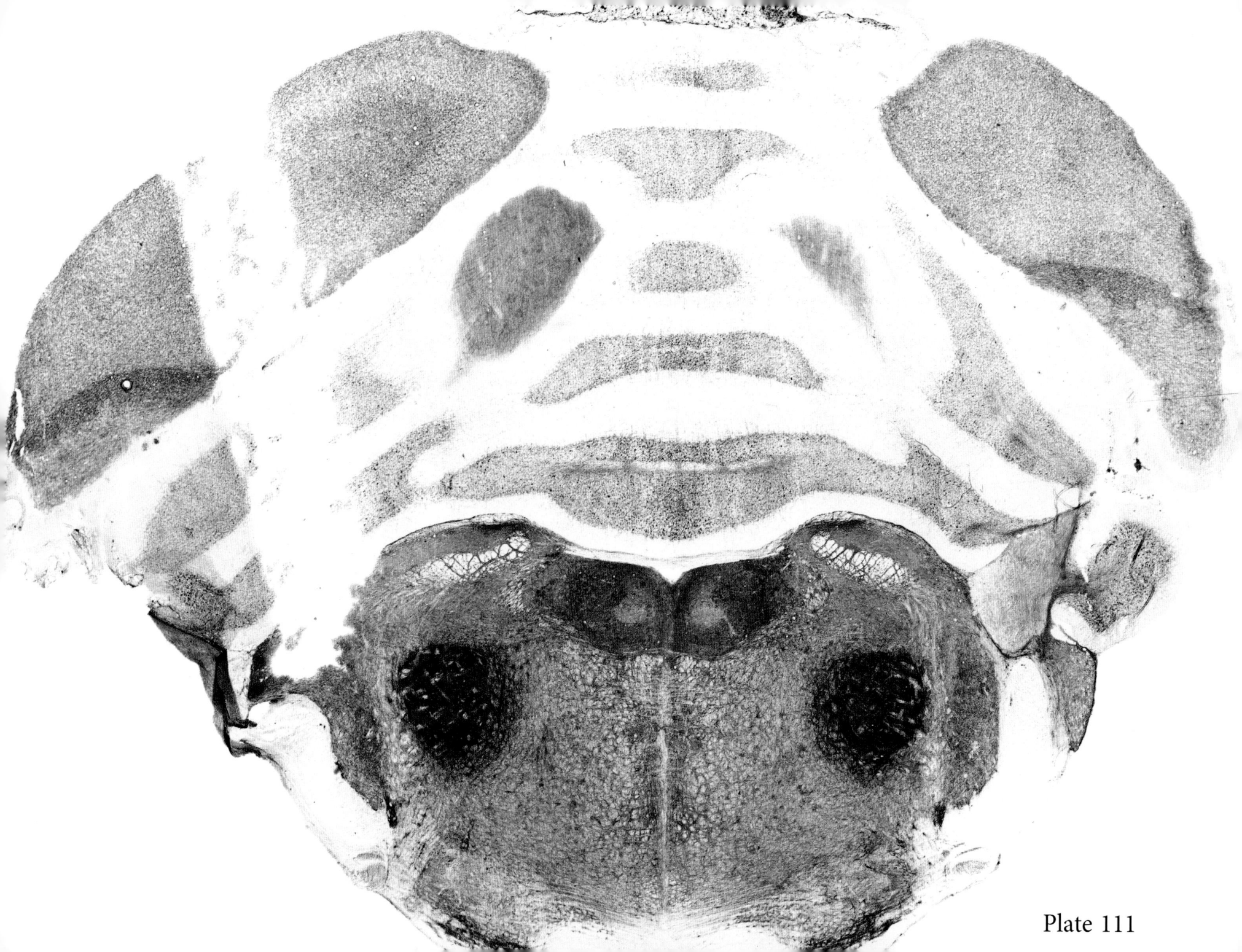

Plate 111

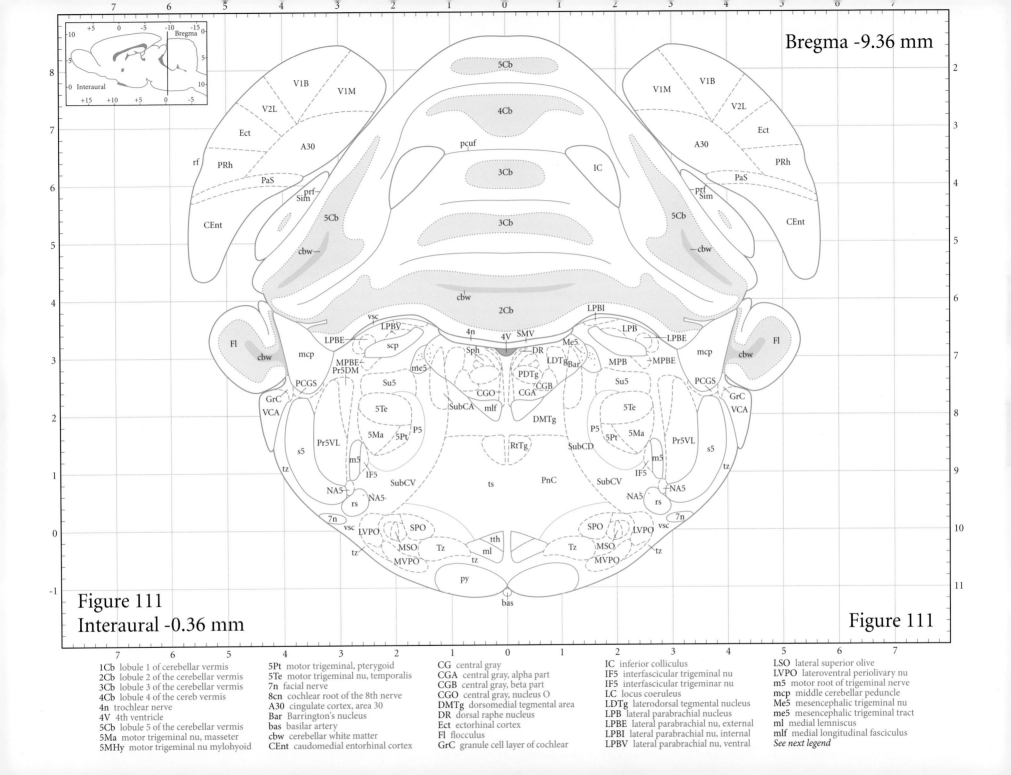

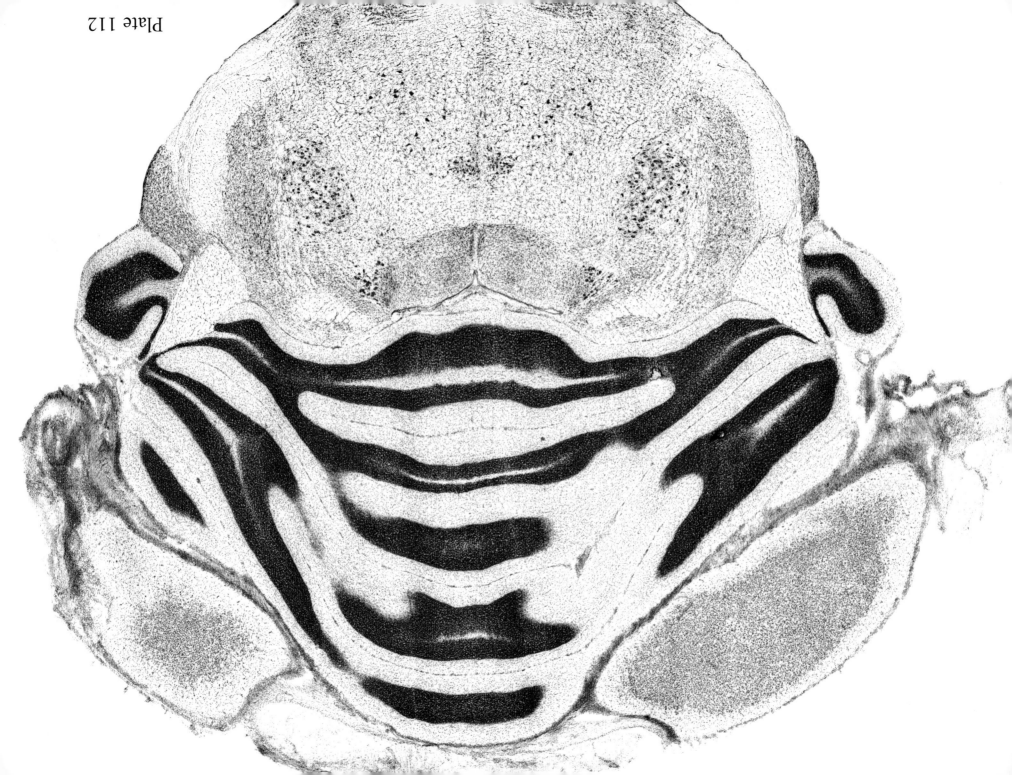

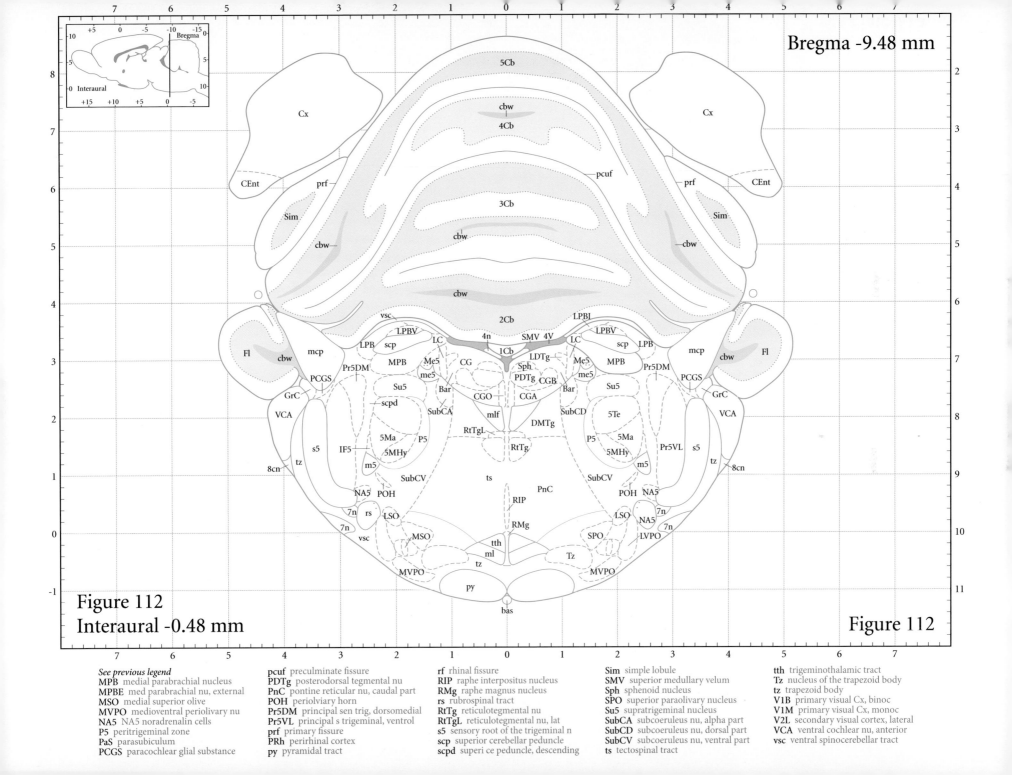

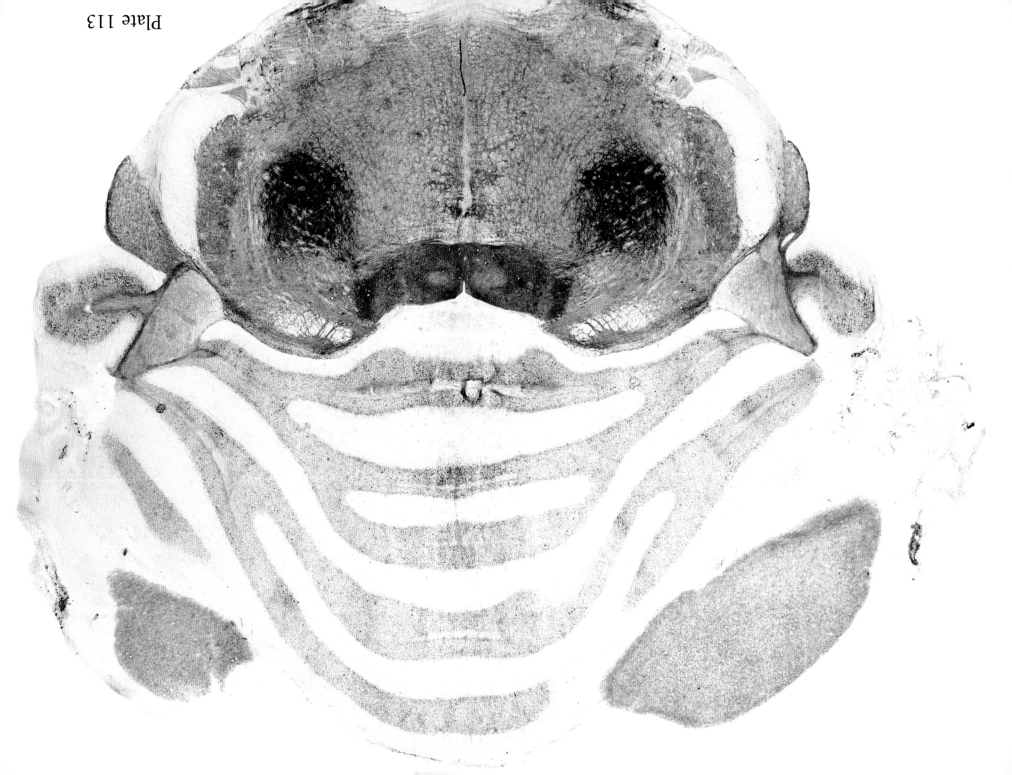

Plate 113

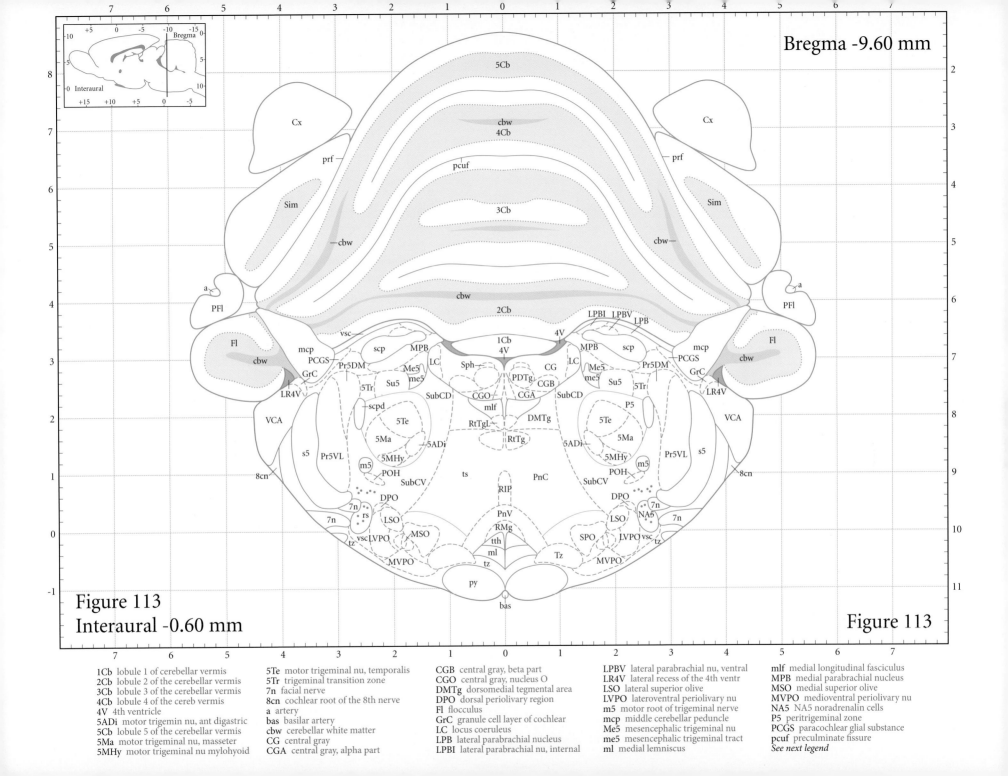

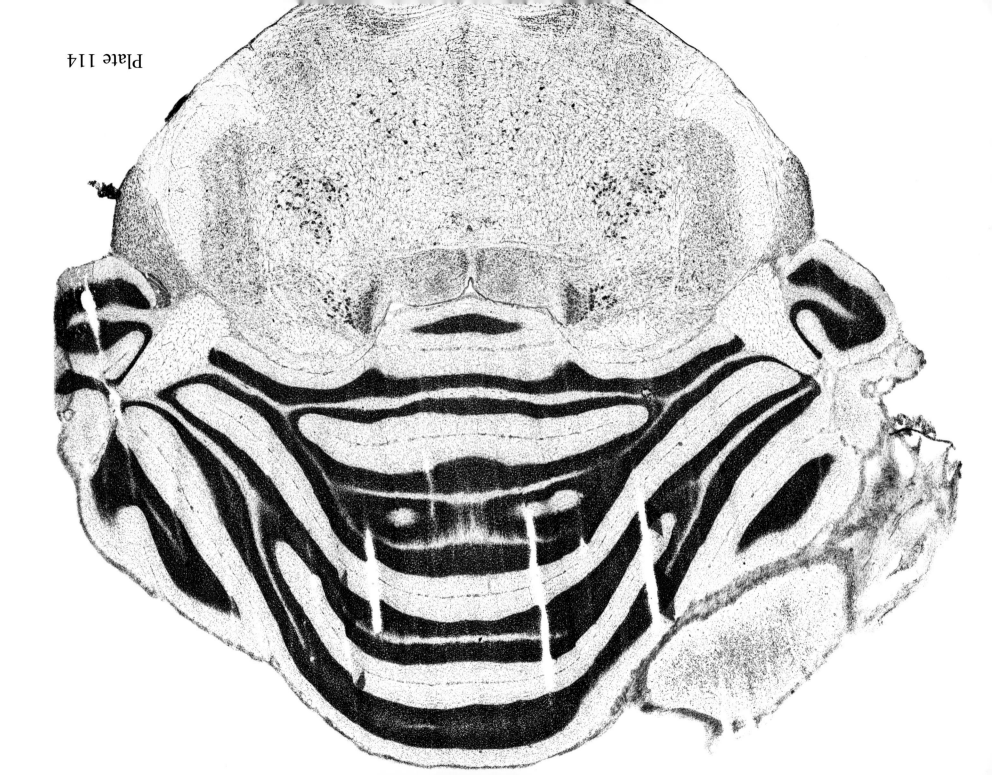

Plate 114

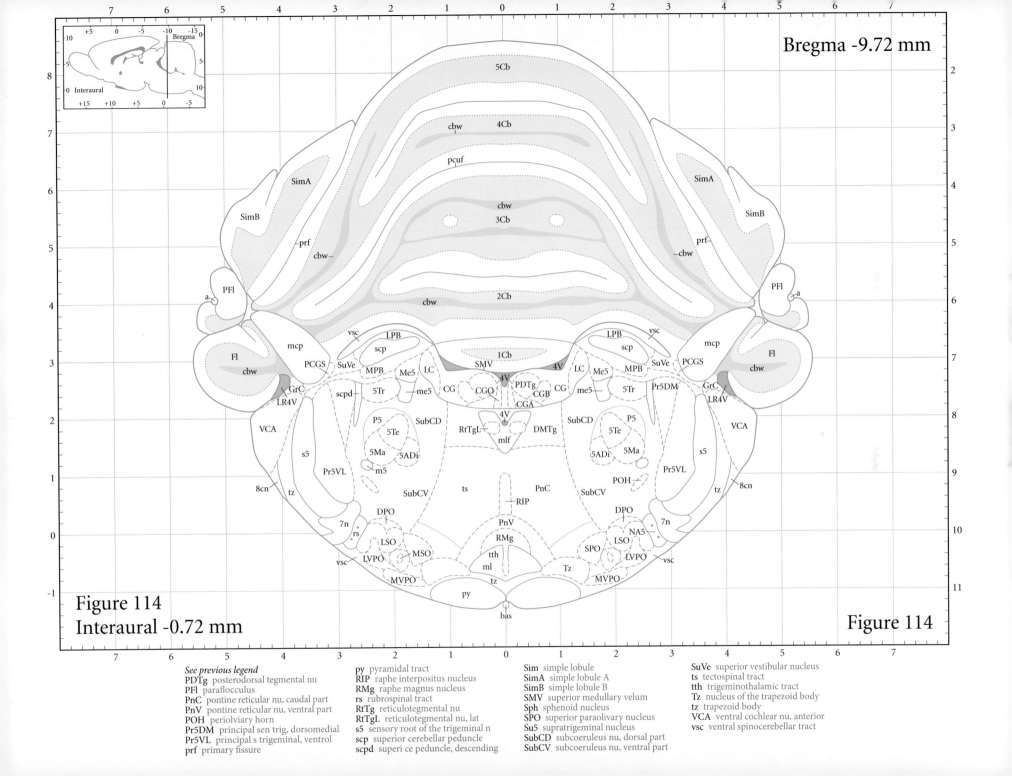

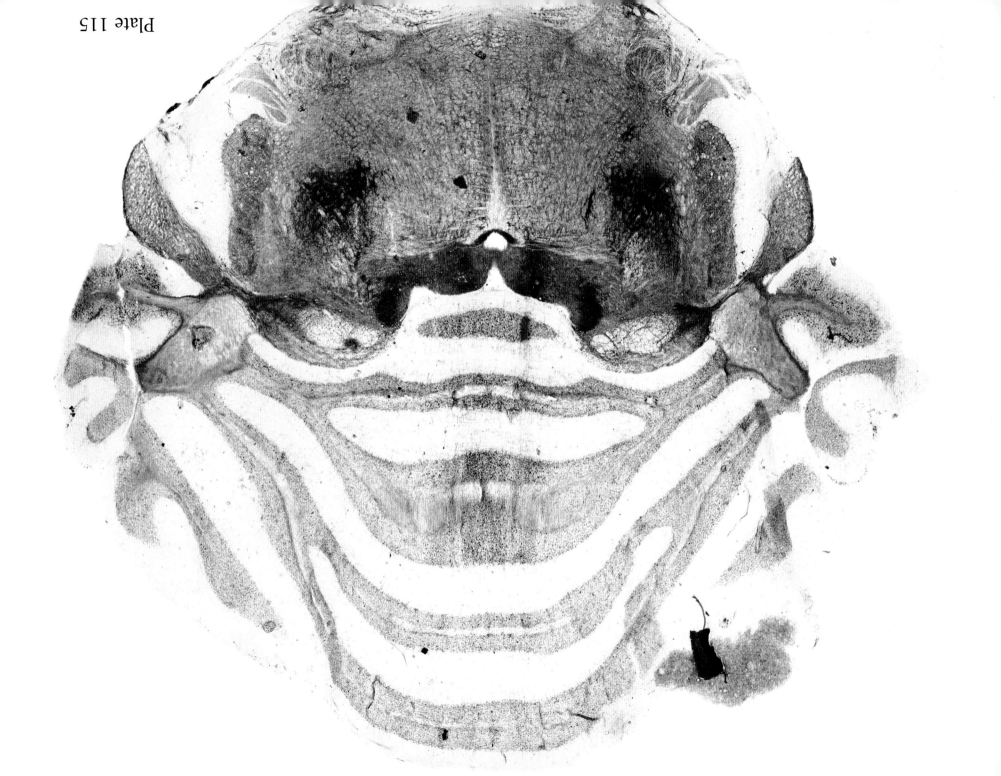

Plate 115

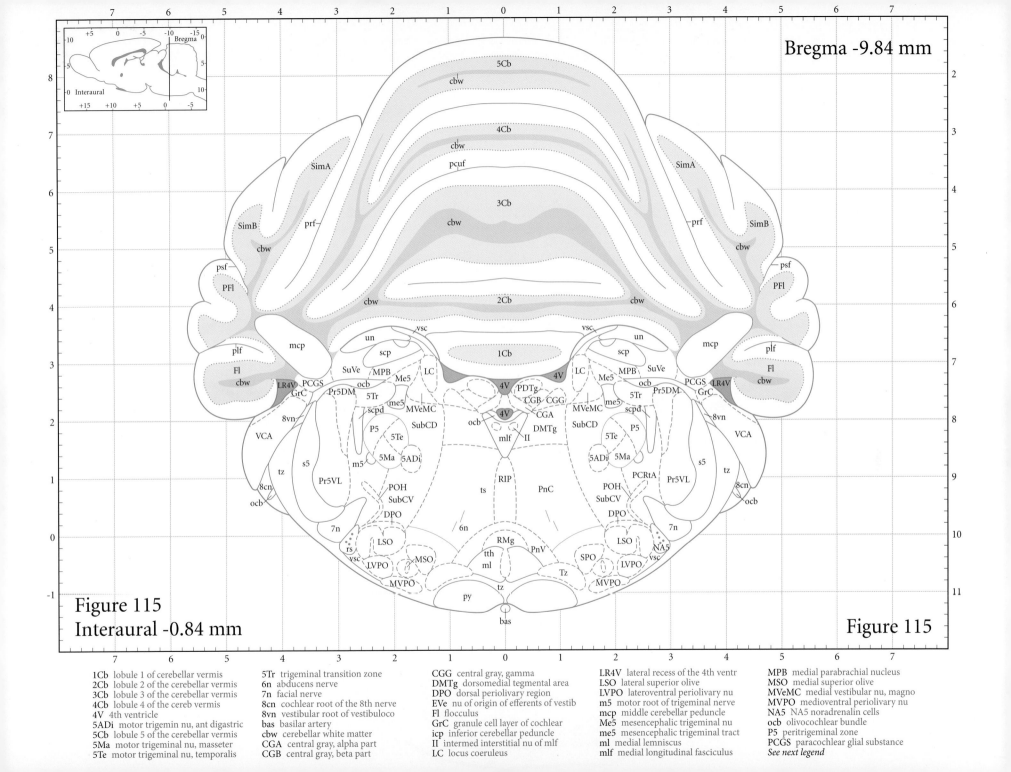

Figure 115
Interaural -0.84 mm
Bregma -9.84 mm

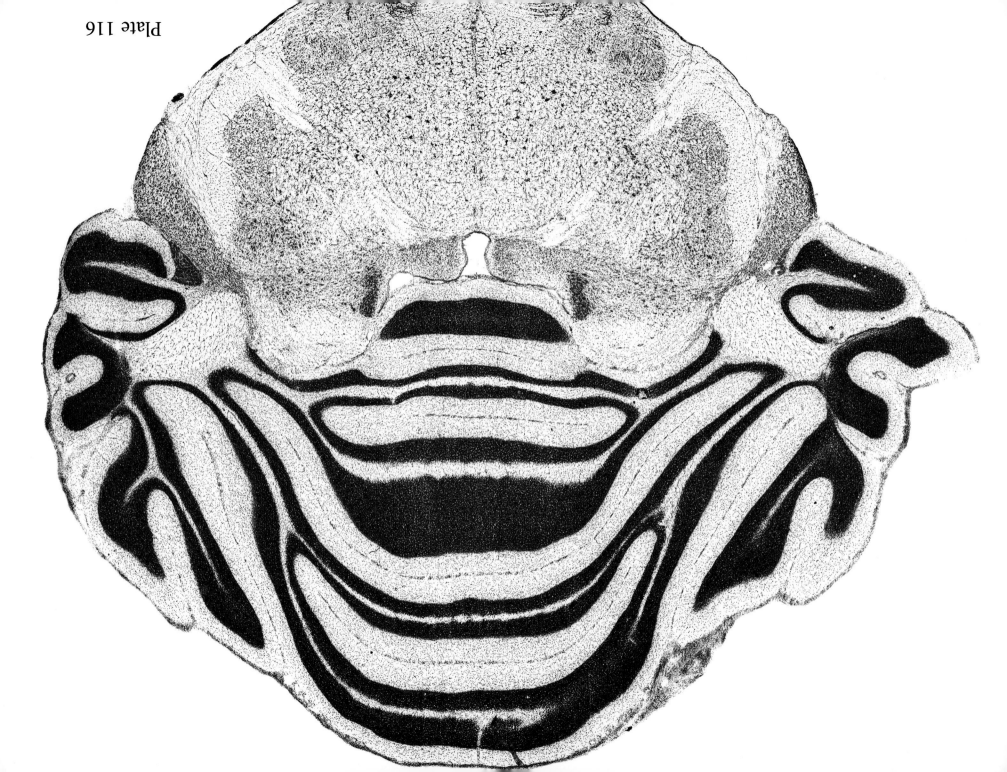

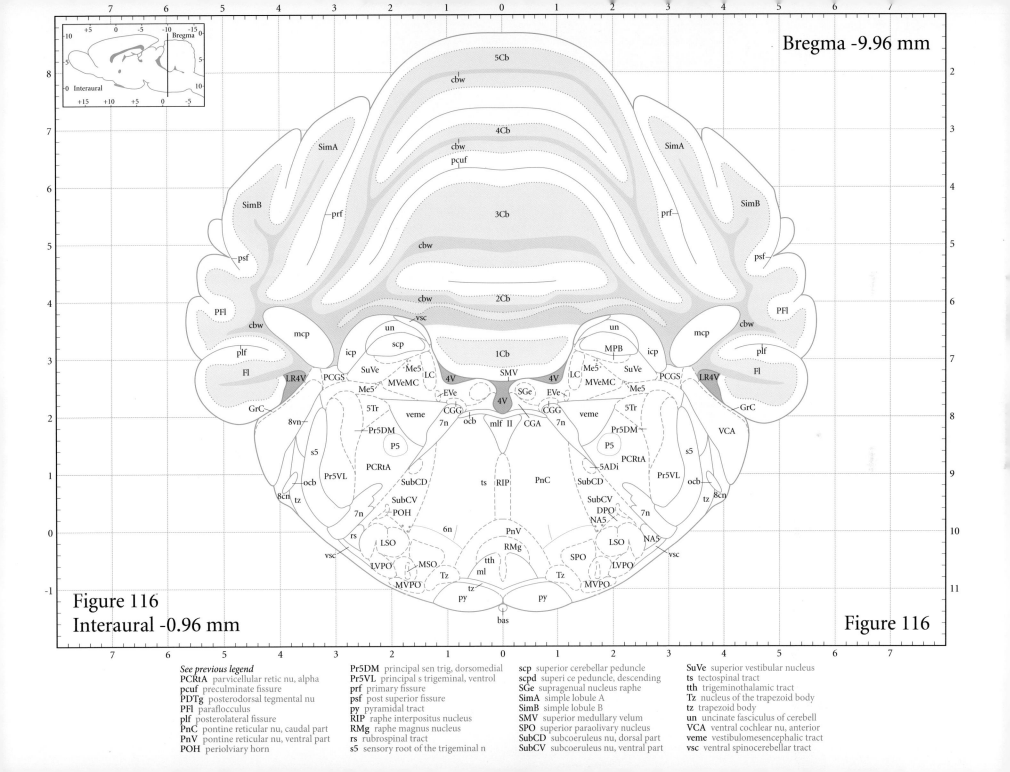

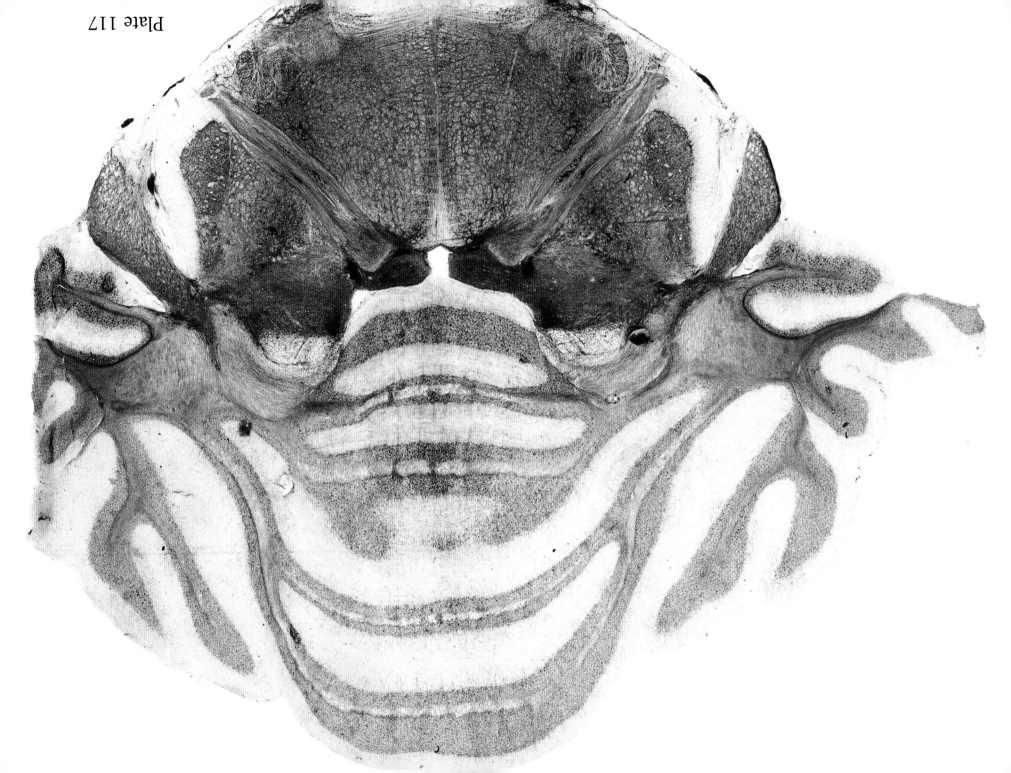

Plate 117

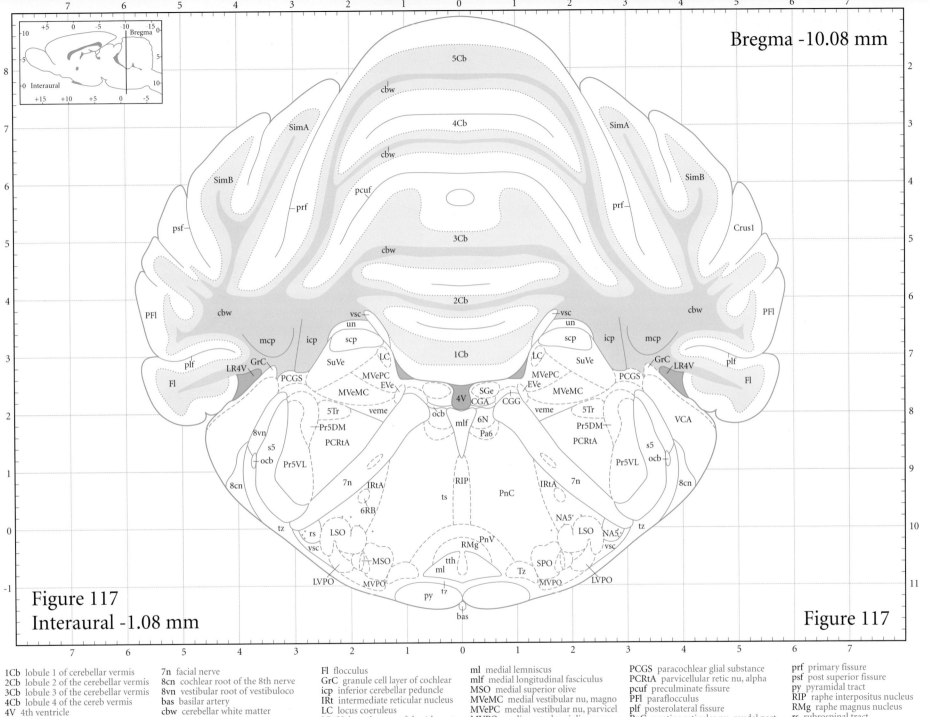

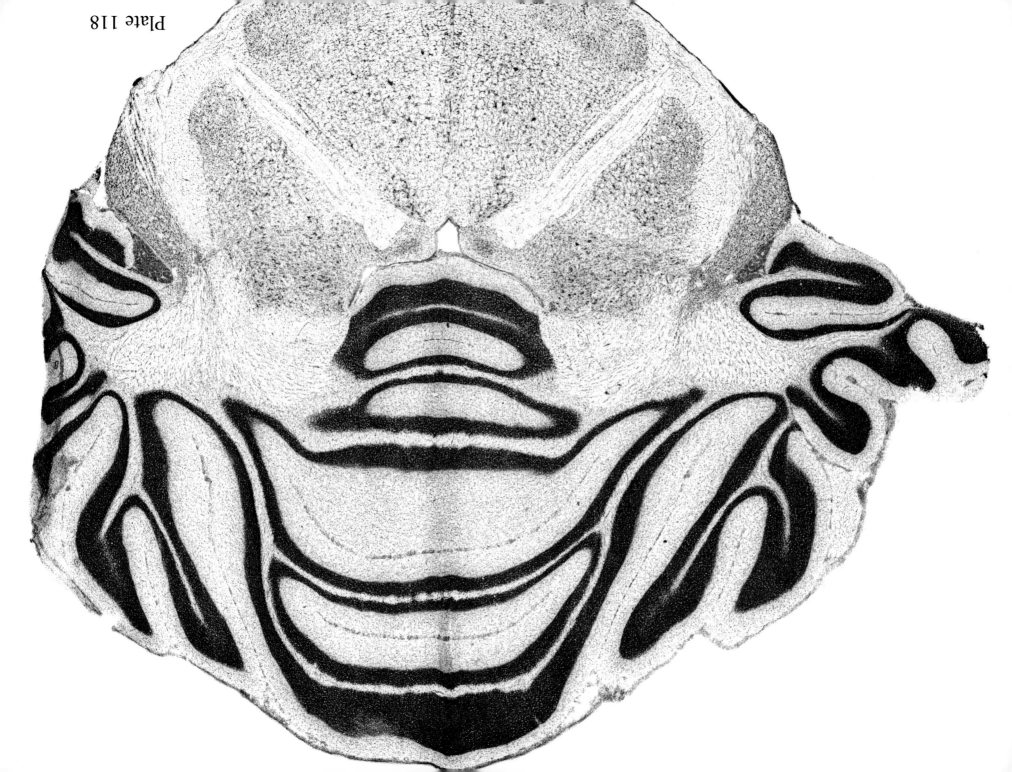

Plate 118

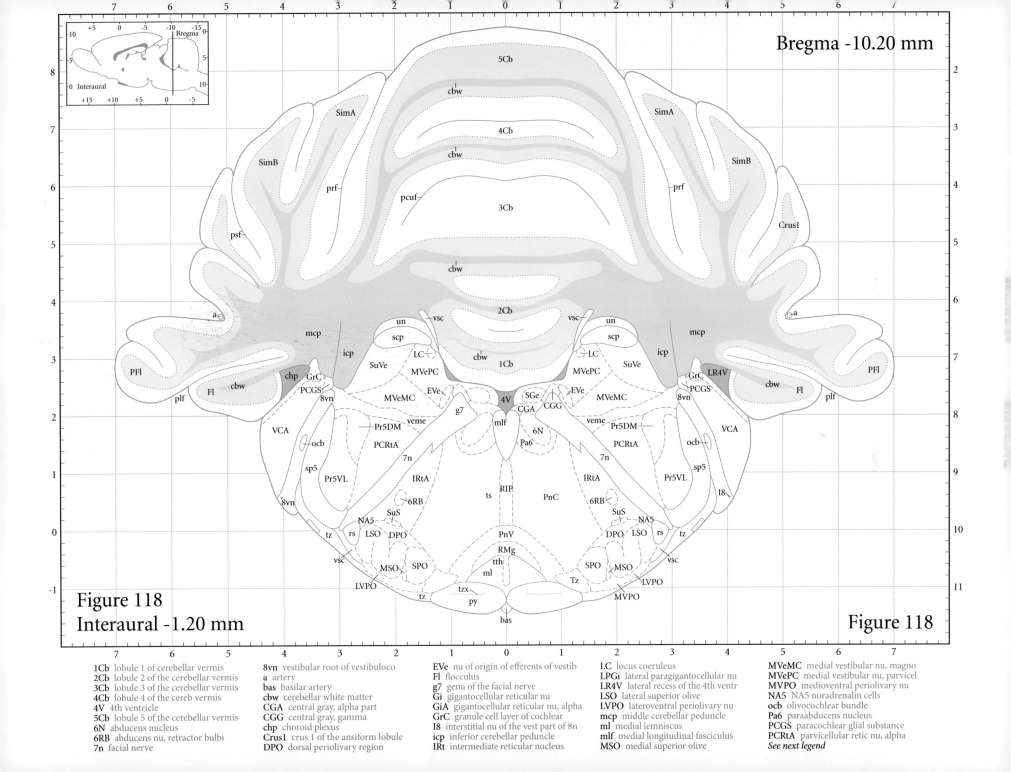

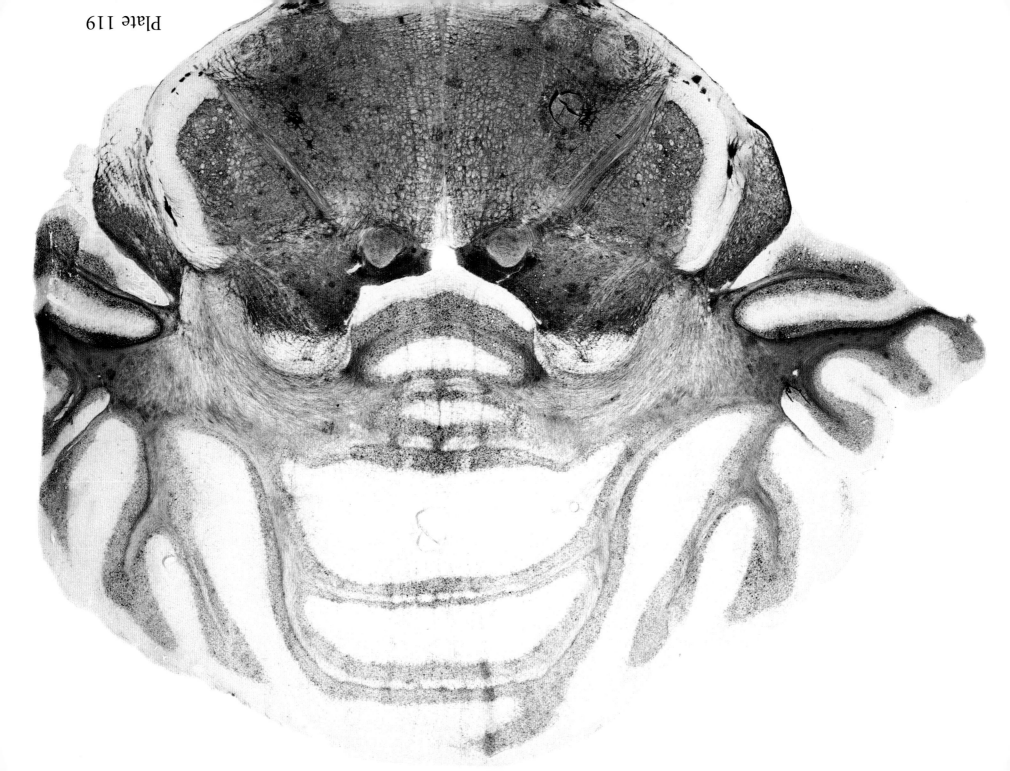

Plate 119

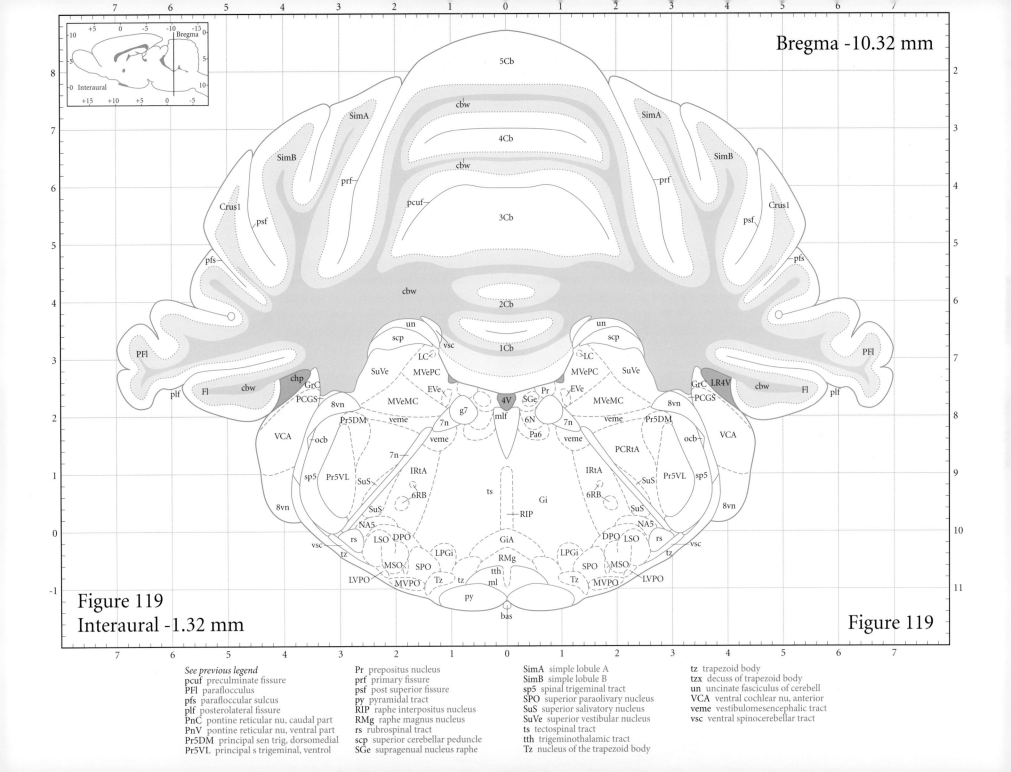

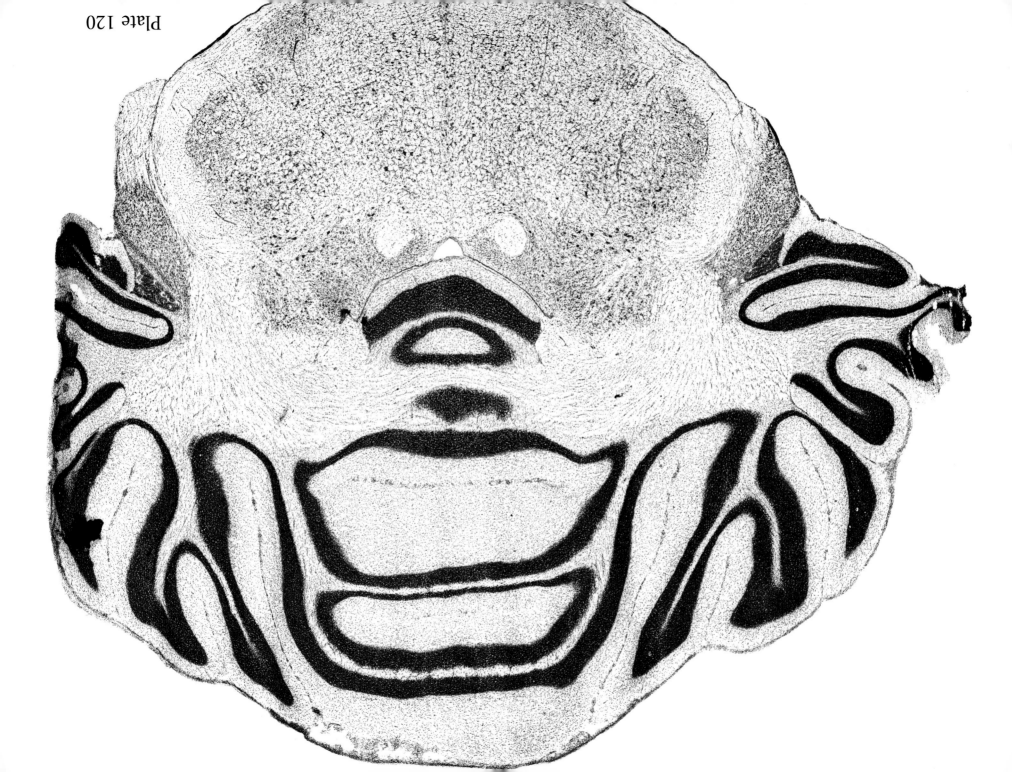

Plate 120

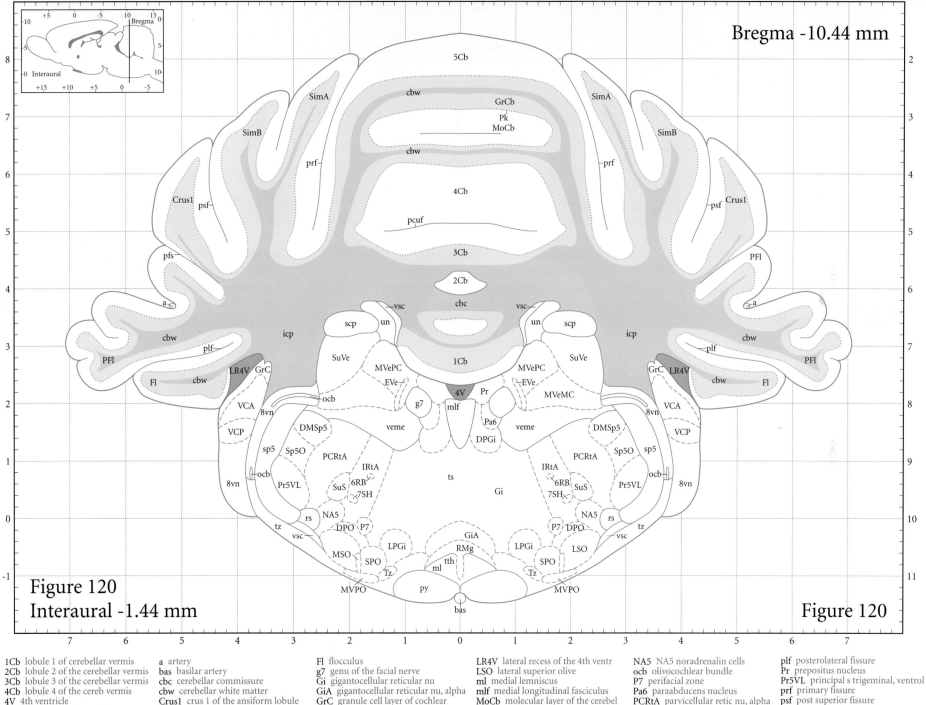

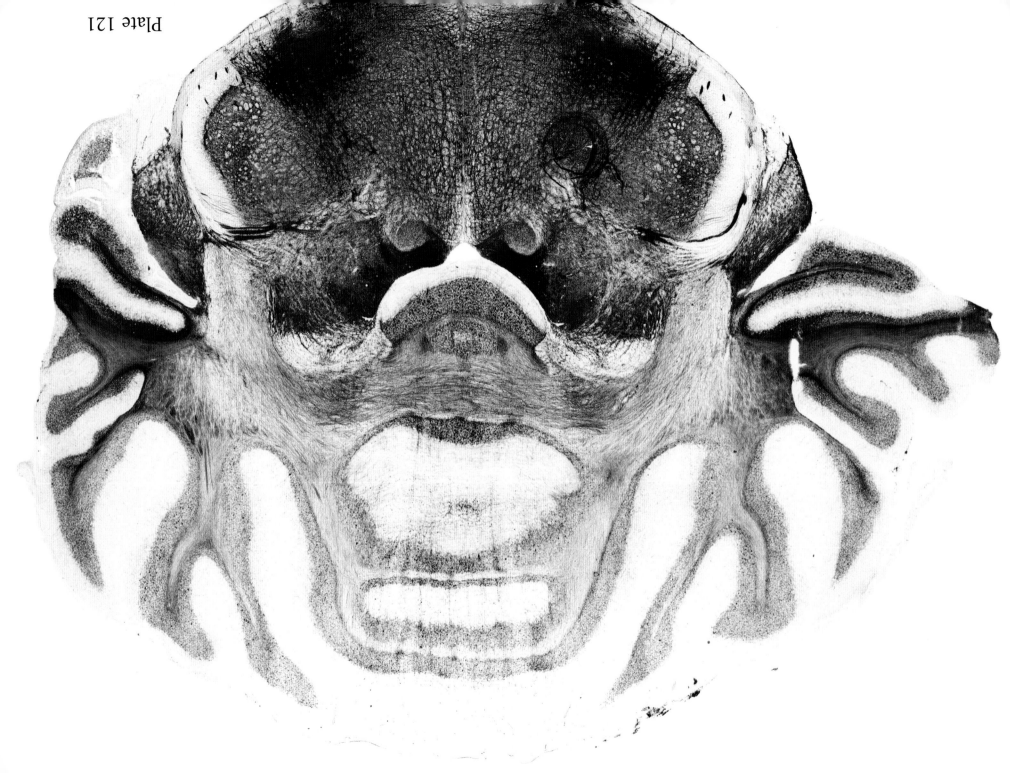

Plate 121

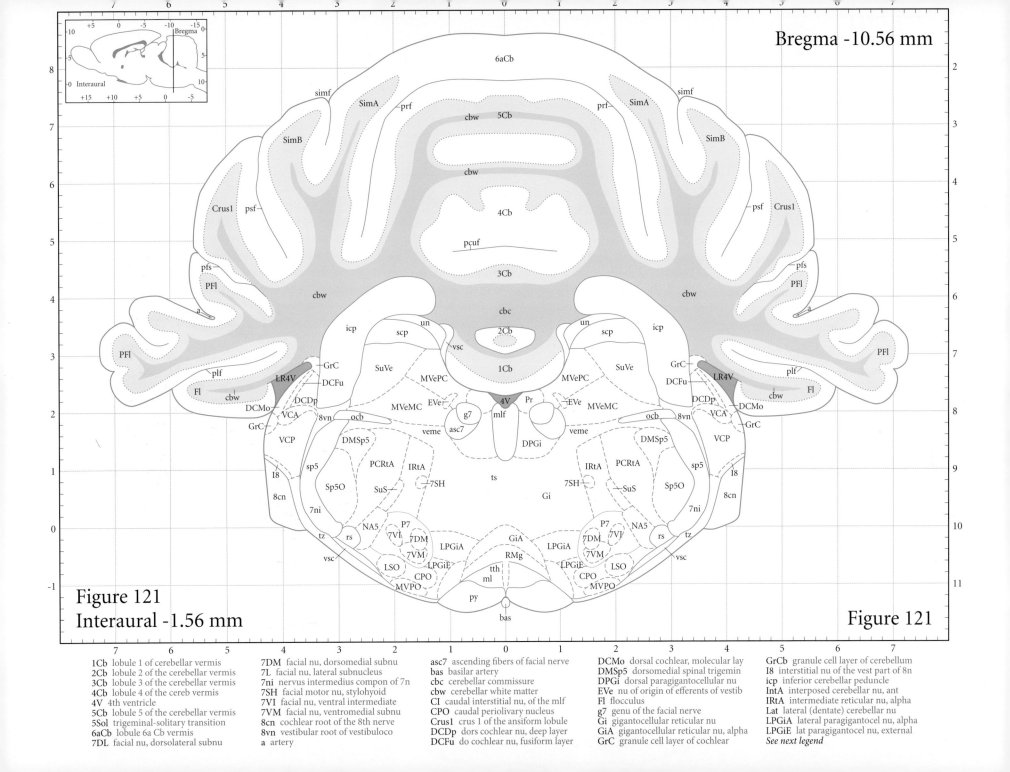

Figure 121
Interaural -1.56 mm
Bregma -10.56 mm

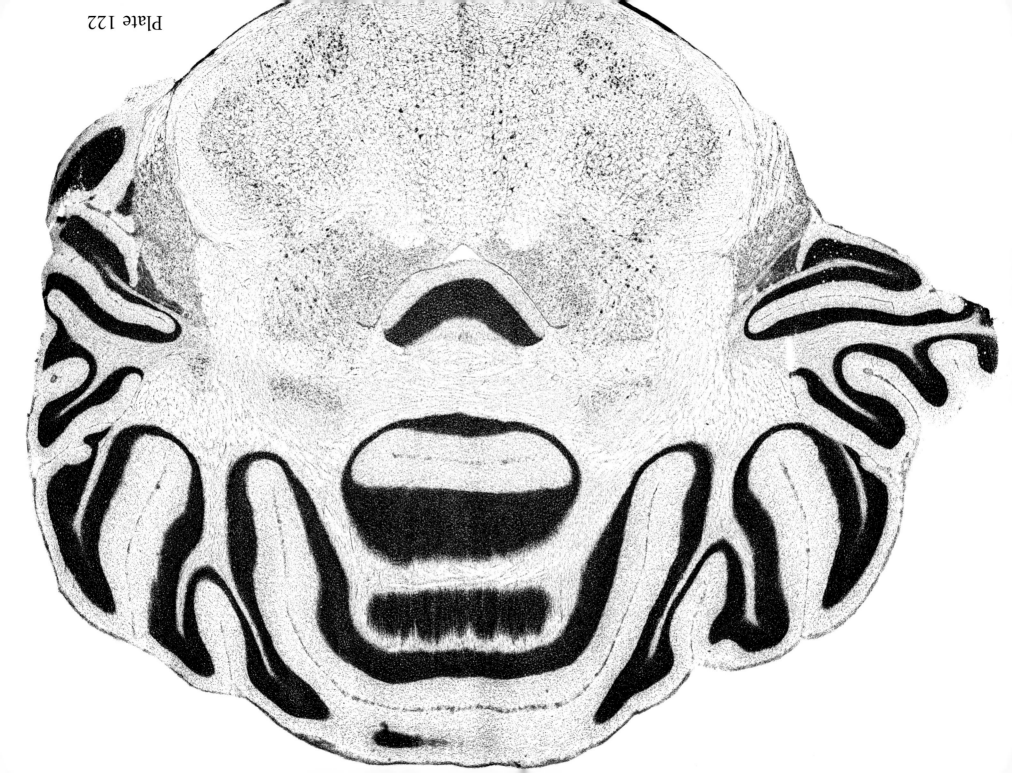

Plate 122

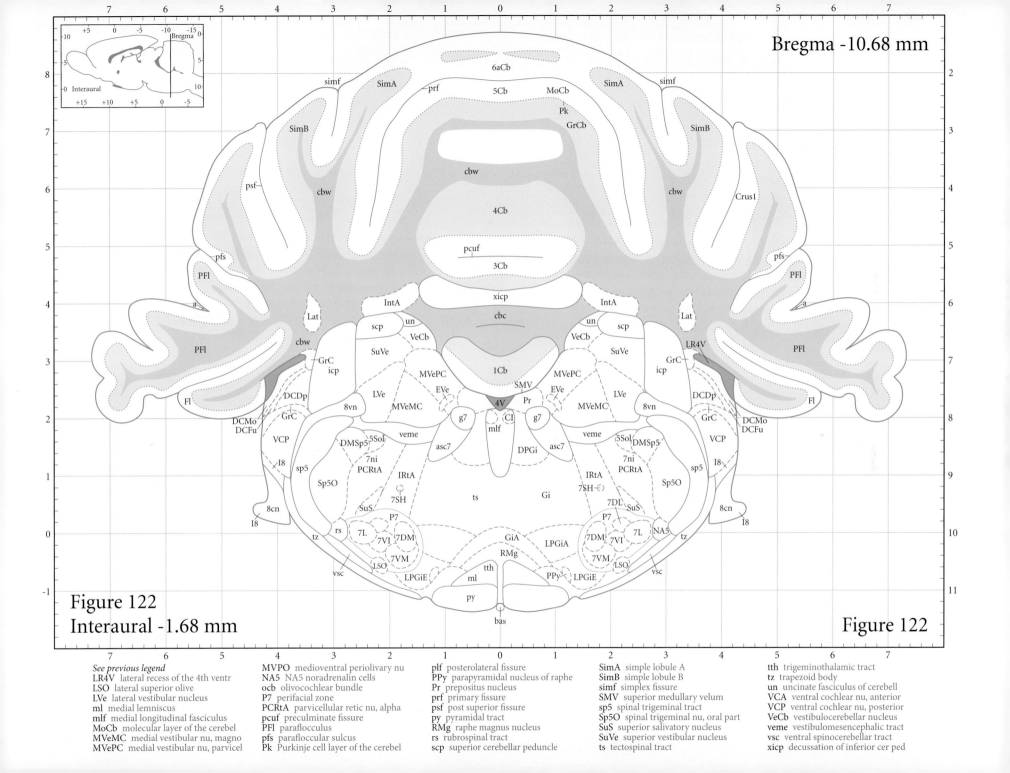

Figure 122
Interaural -1.68 mm
Bregma -10.68 mm

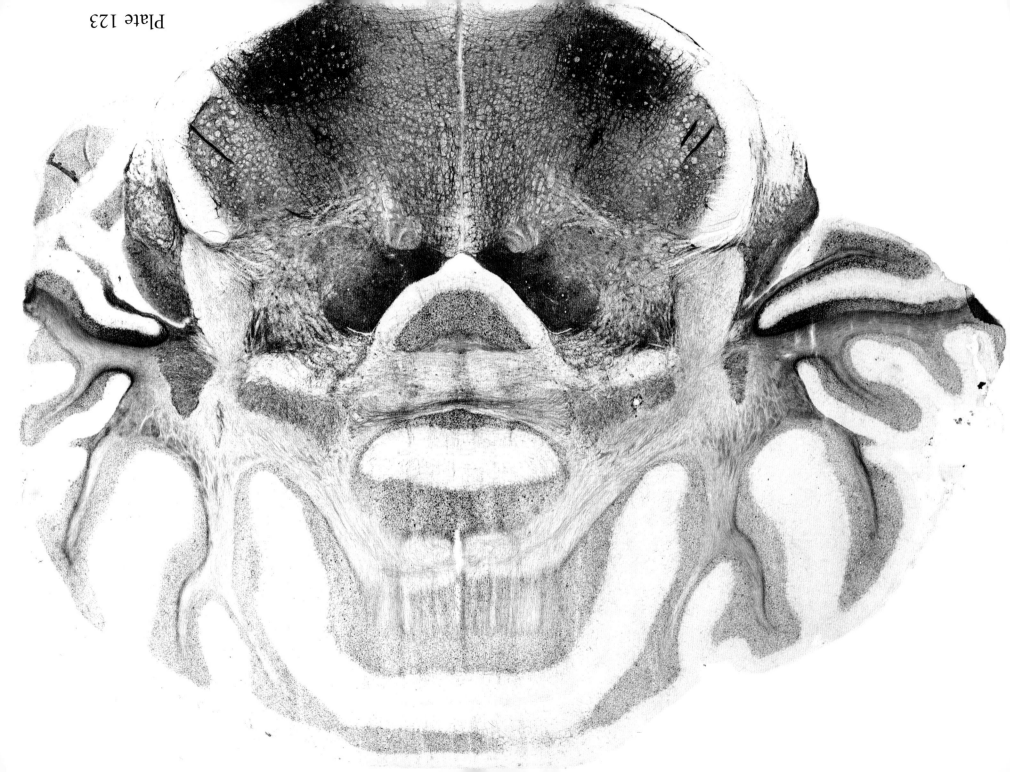

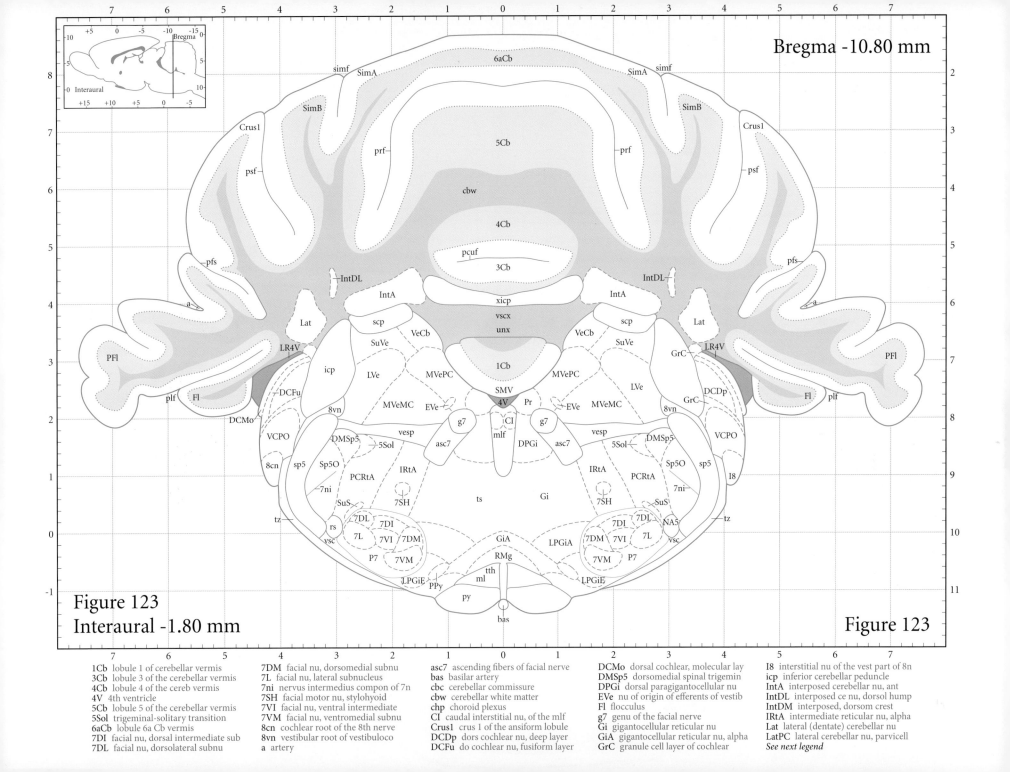

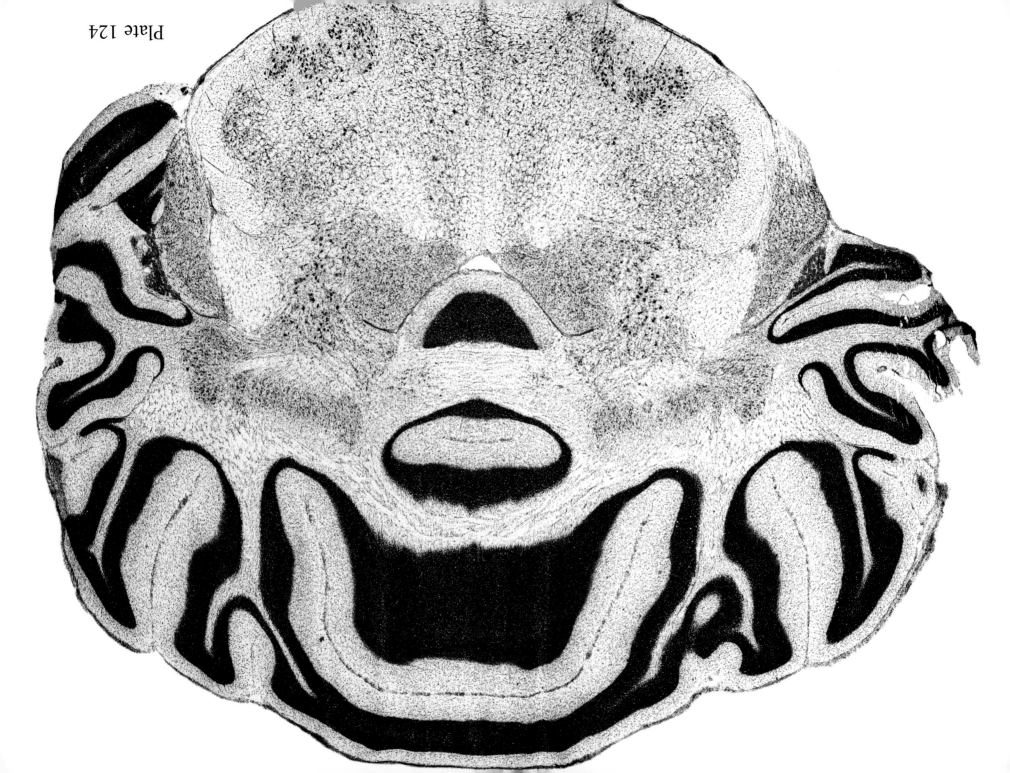

Plate 124

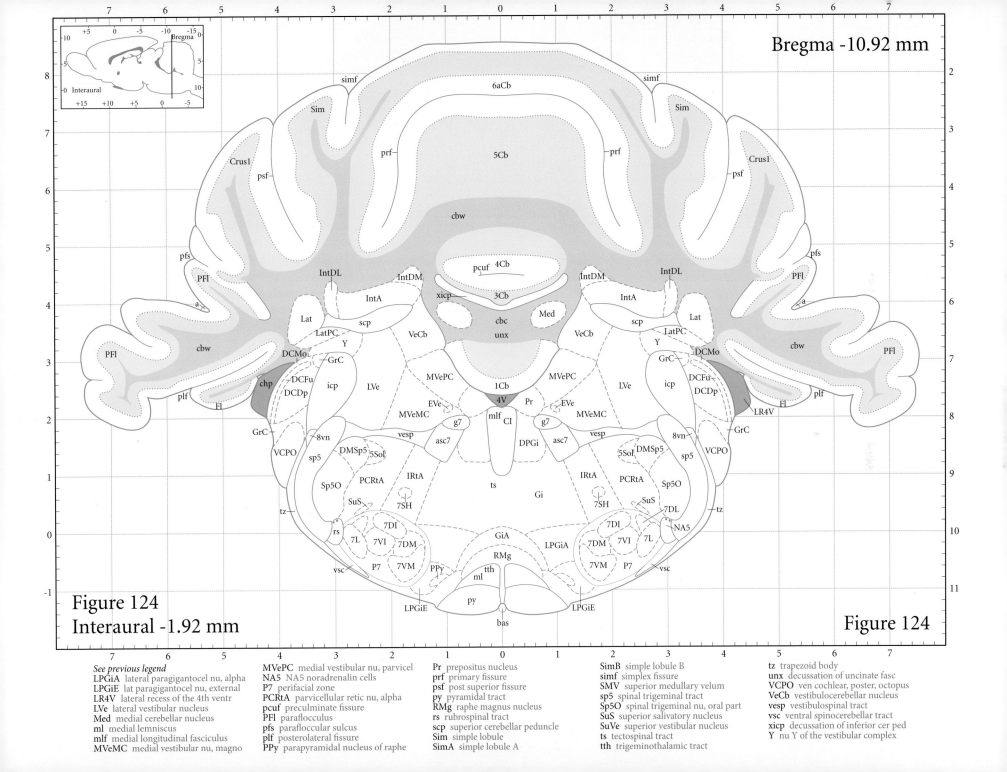

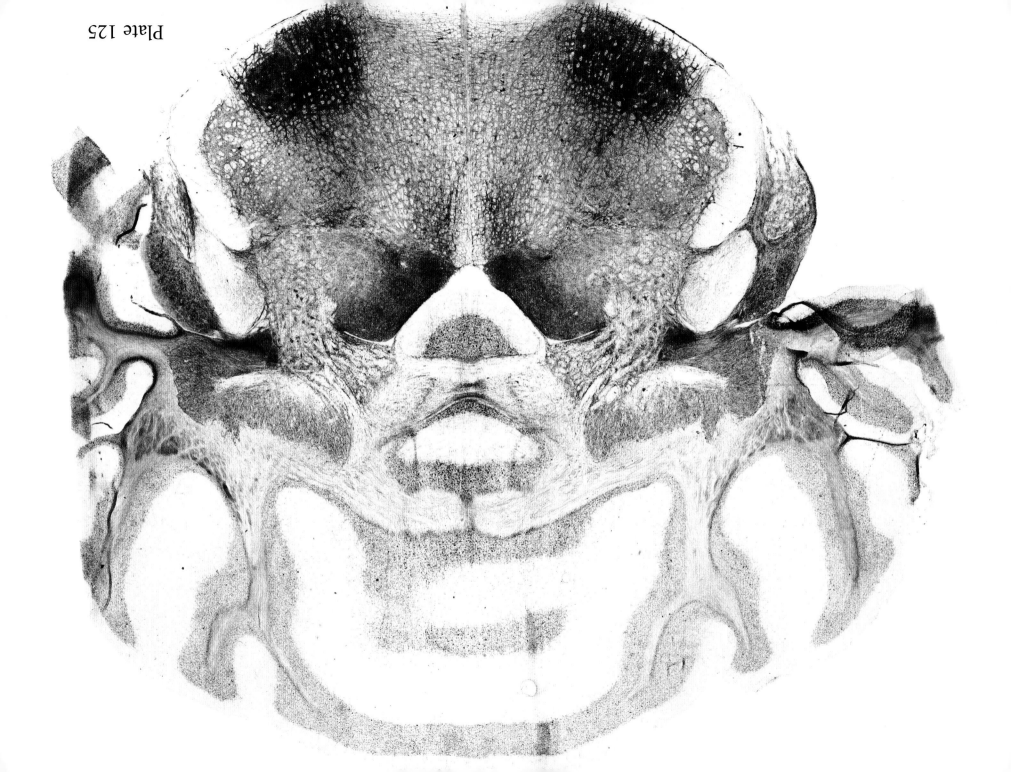

Plate 125

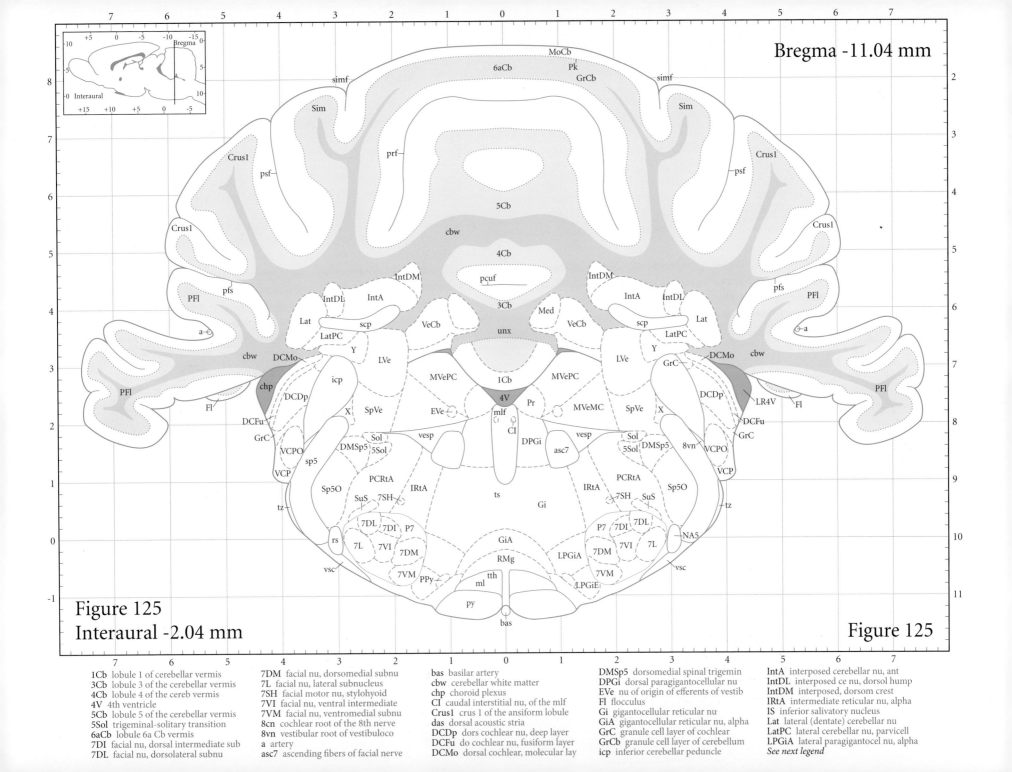

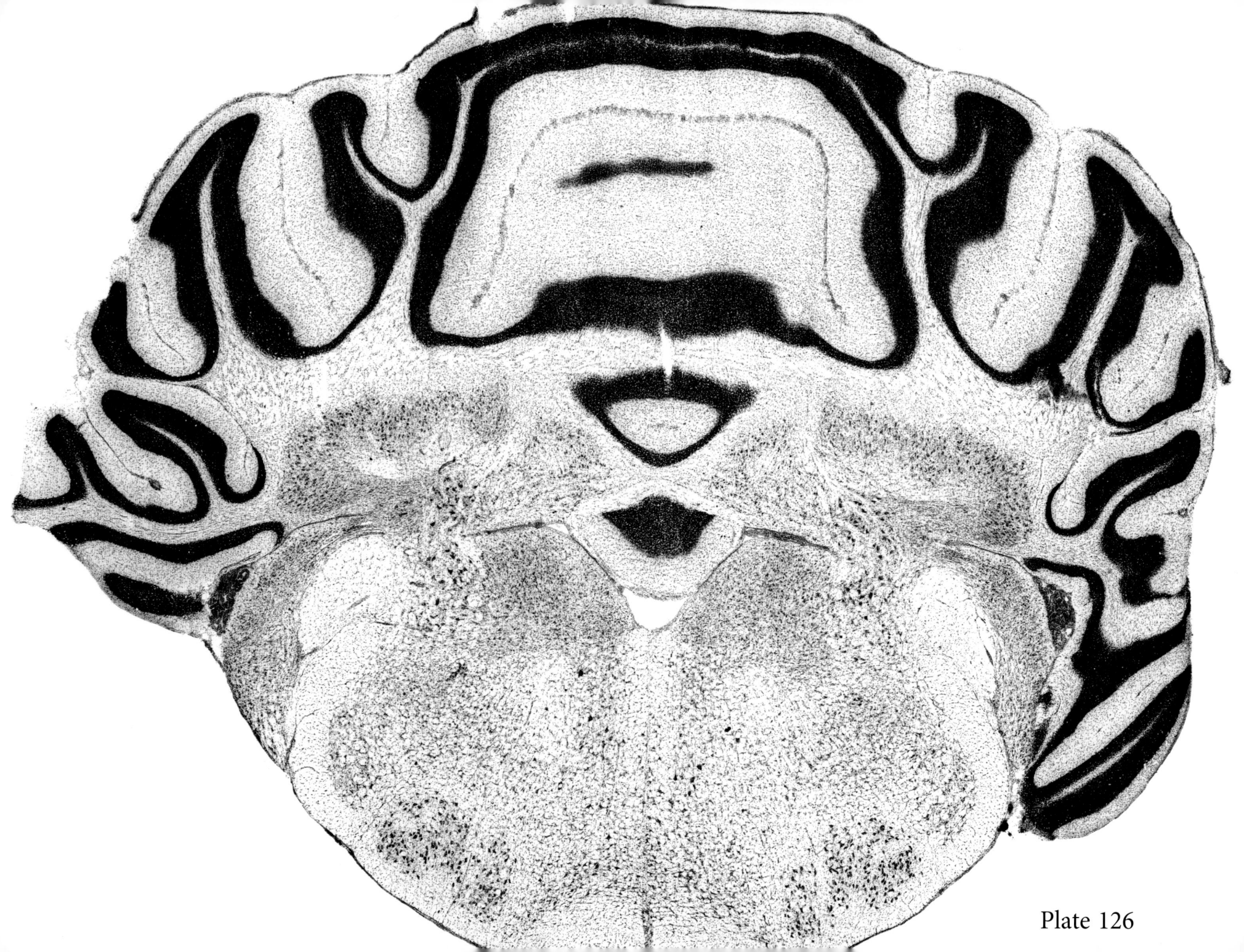

Plate 126

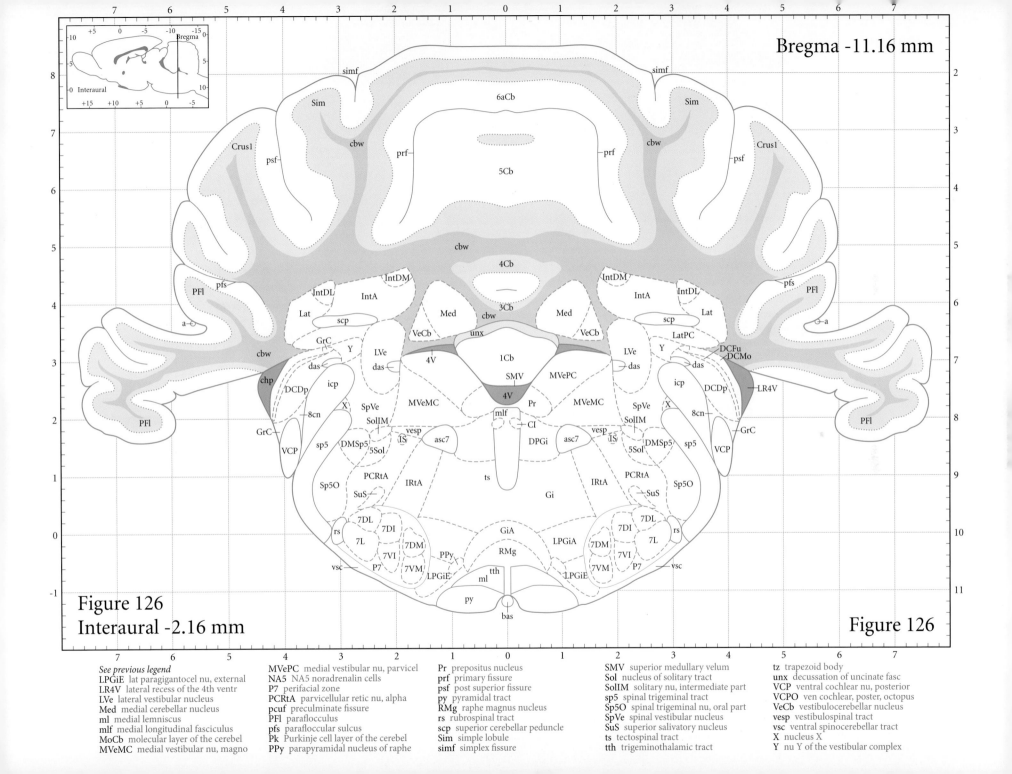

Figure 126
Interaural -2.16 mm
Bregma -11.16 mm

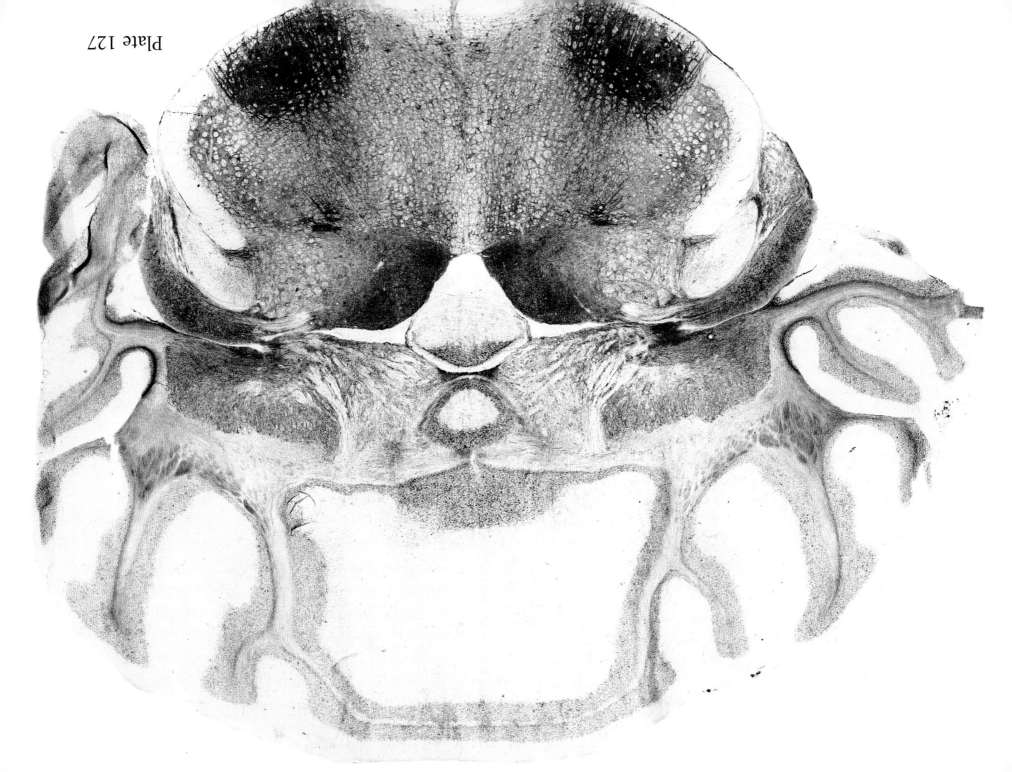

Plate 127

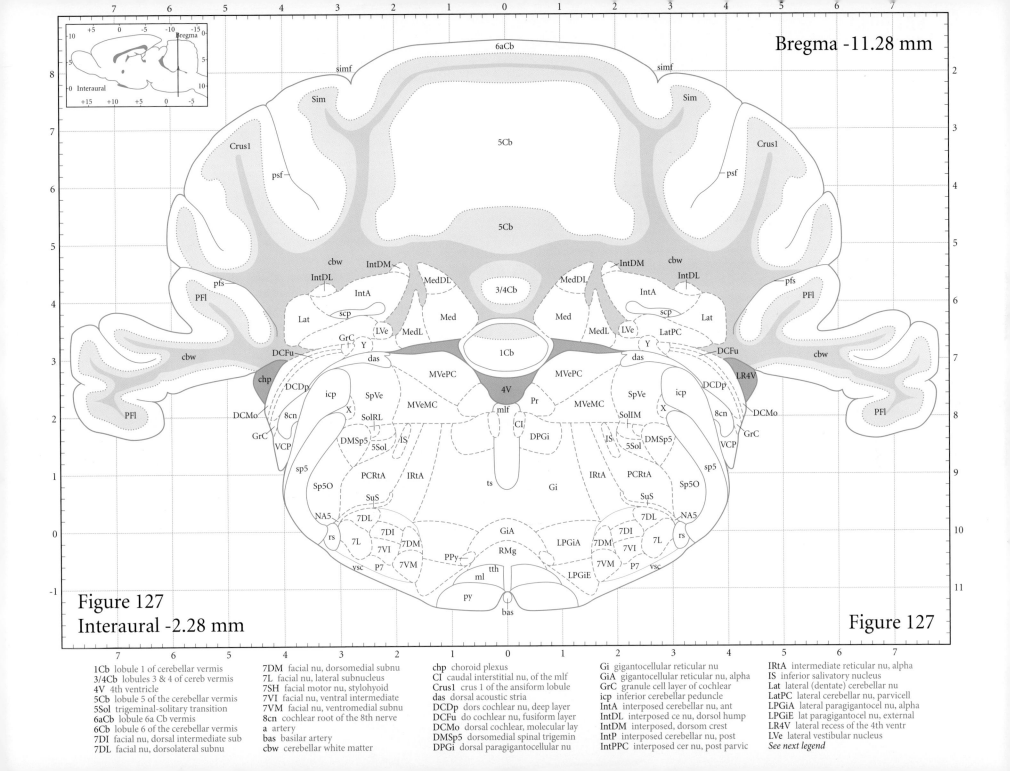

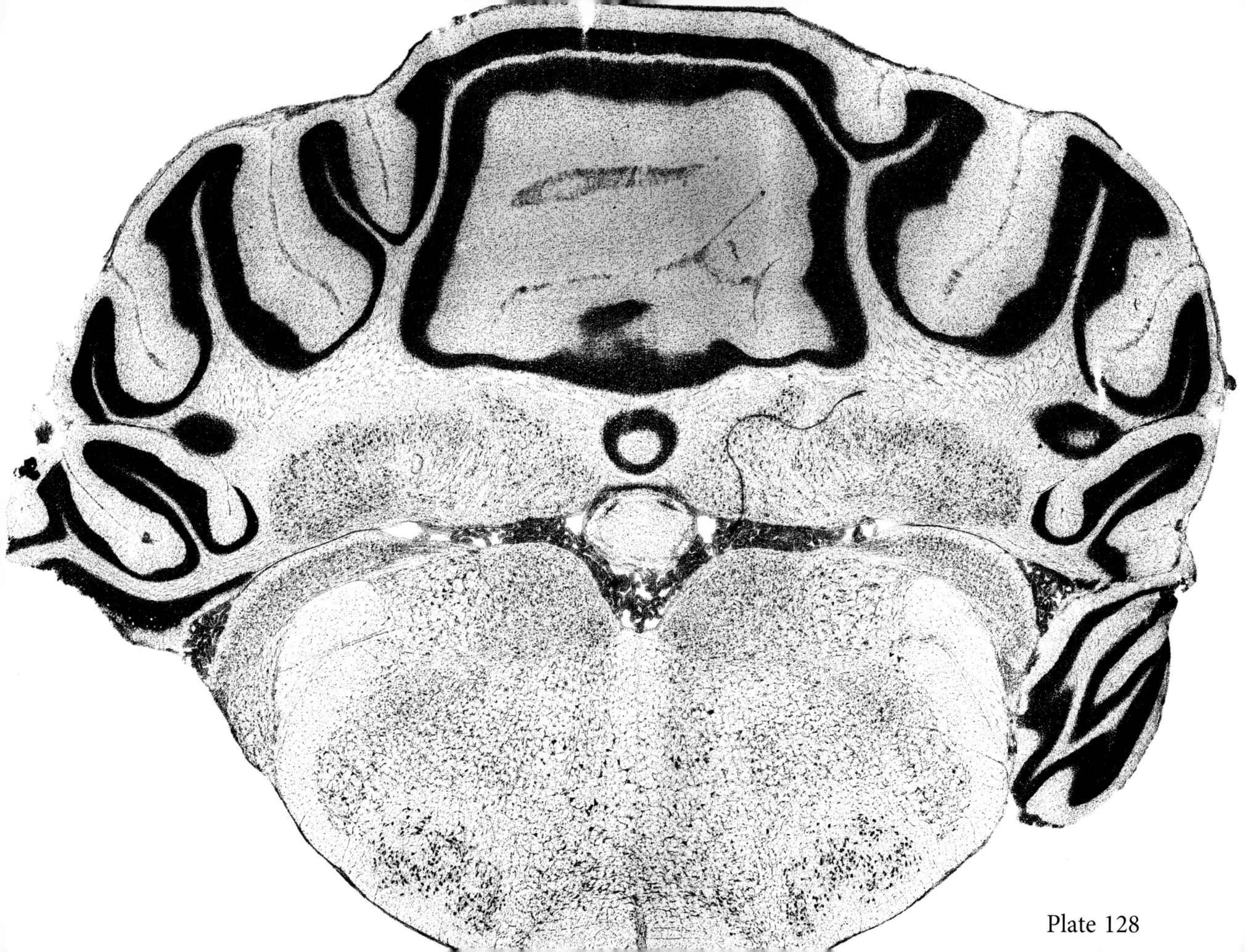

Plate 128

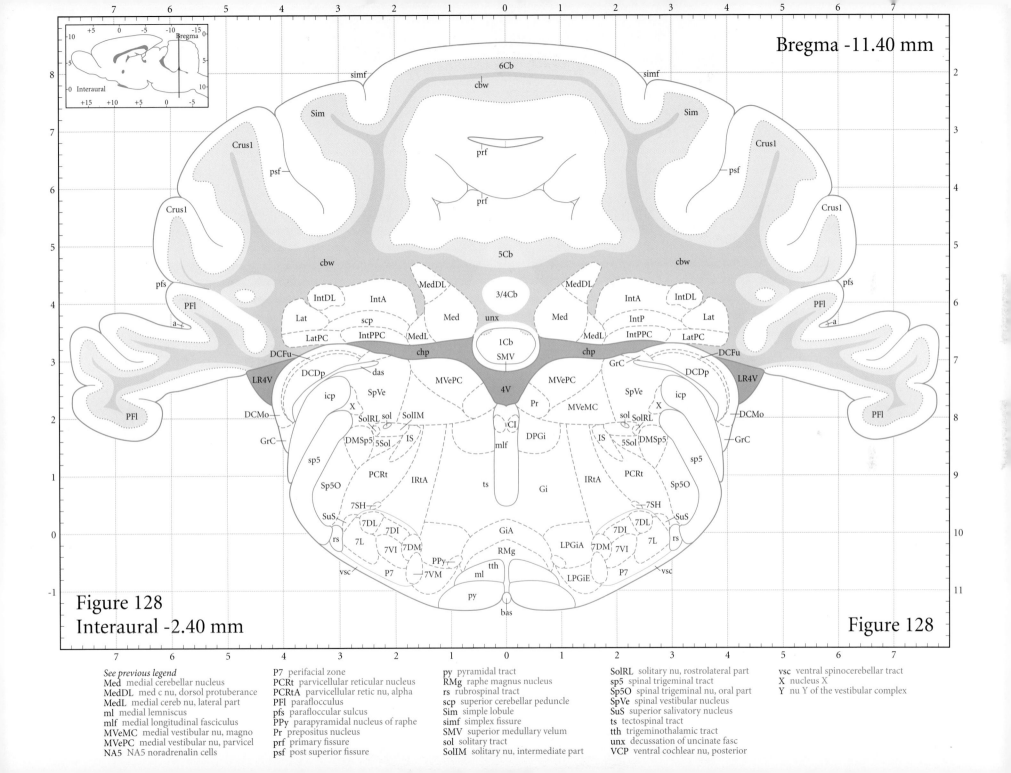

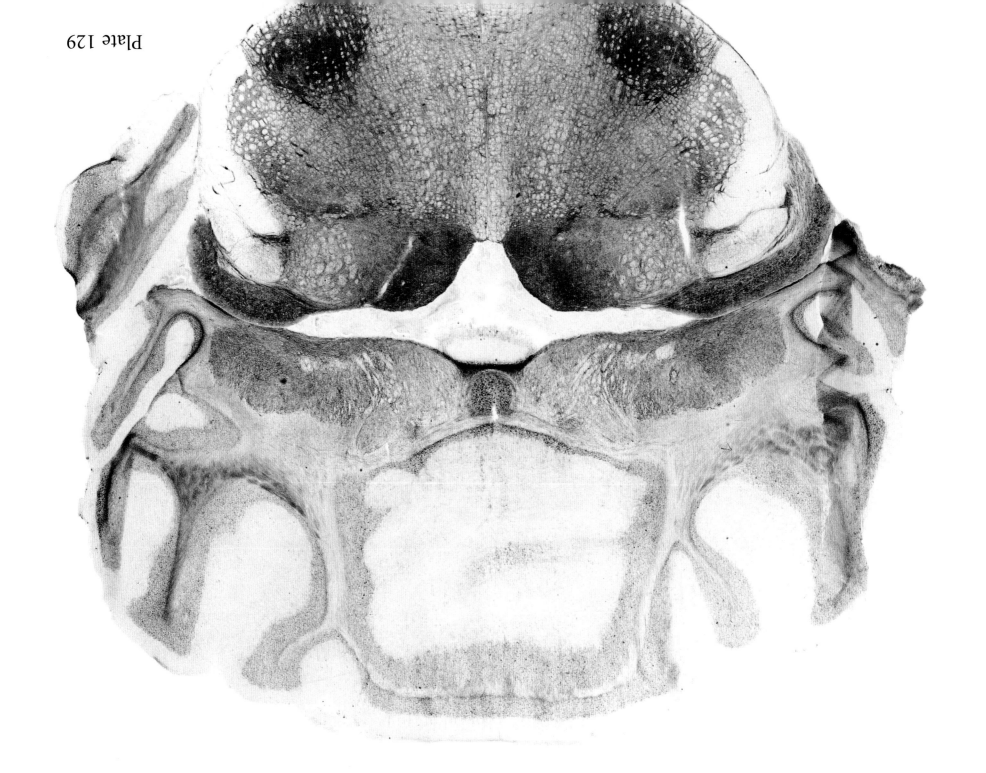

Plate 129

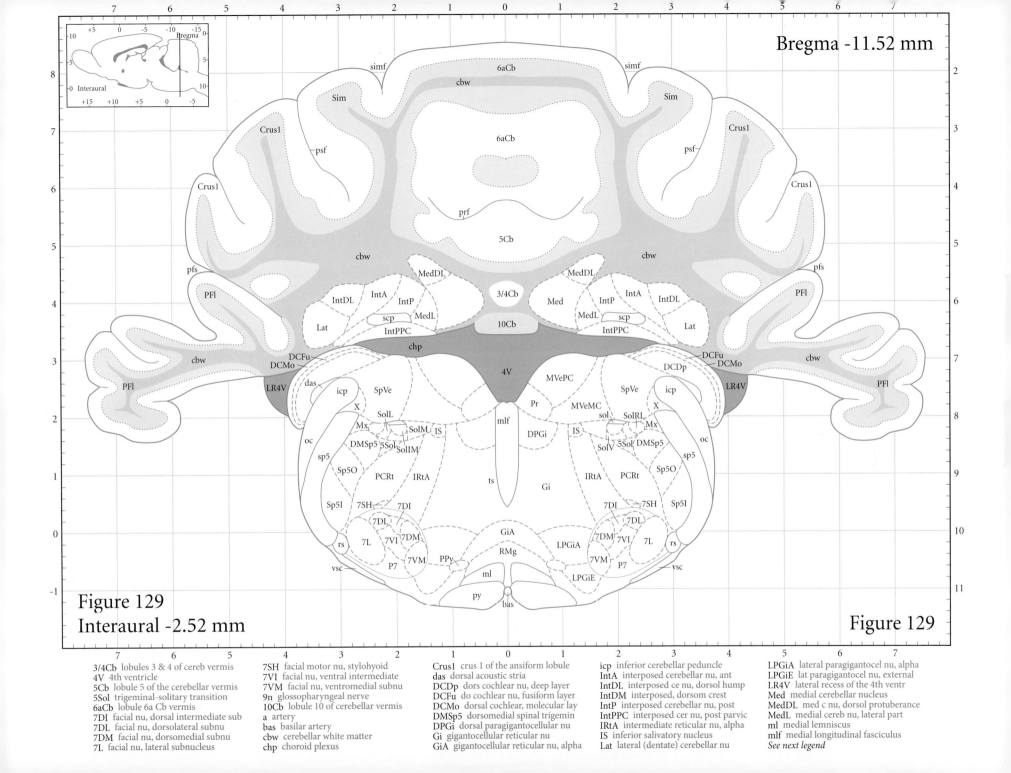

Figure 129
Interaural -2.52 mm
Bregma -11.52 mm

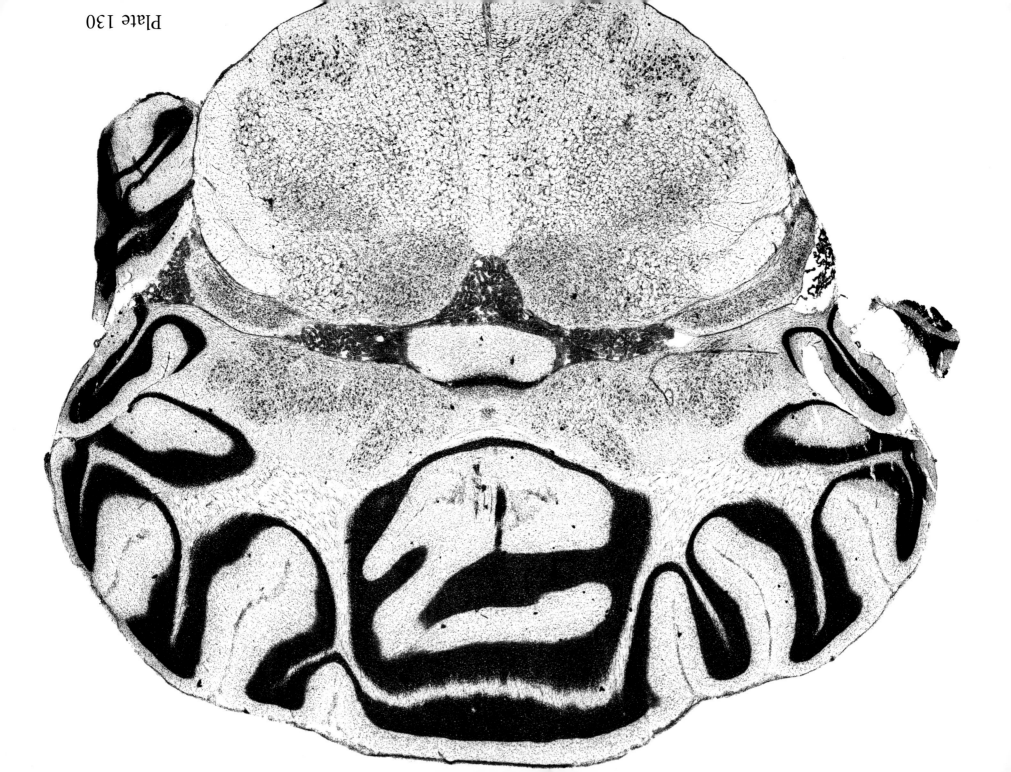
Plate 130

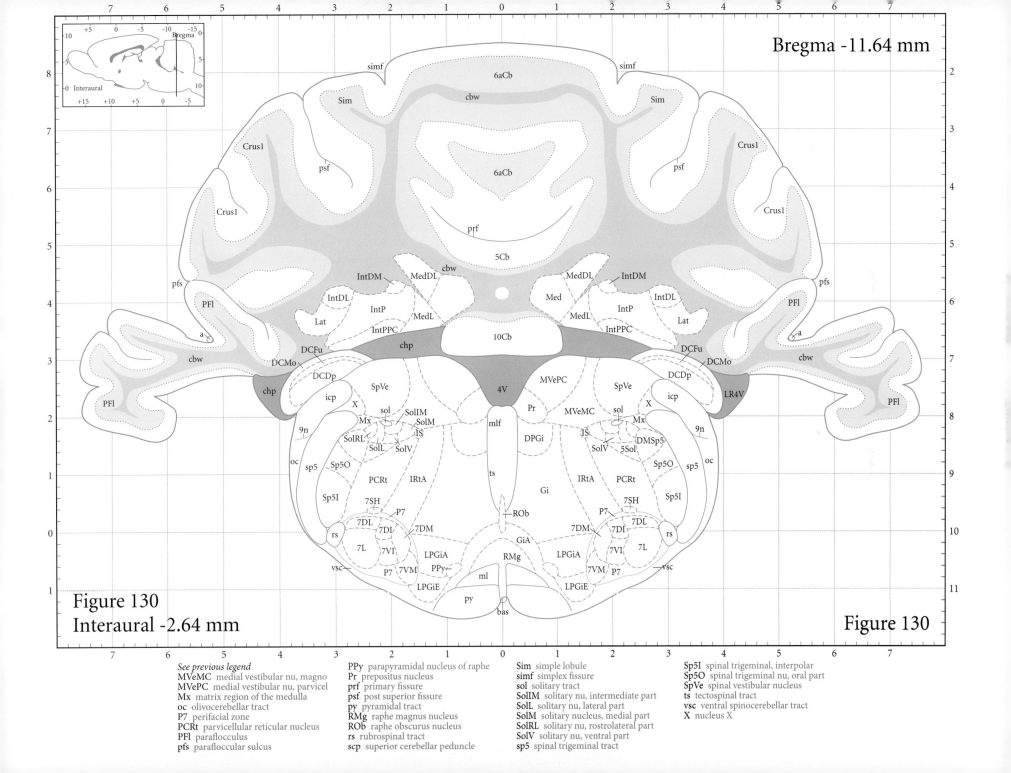

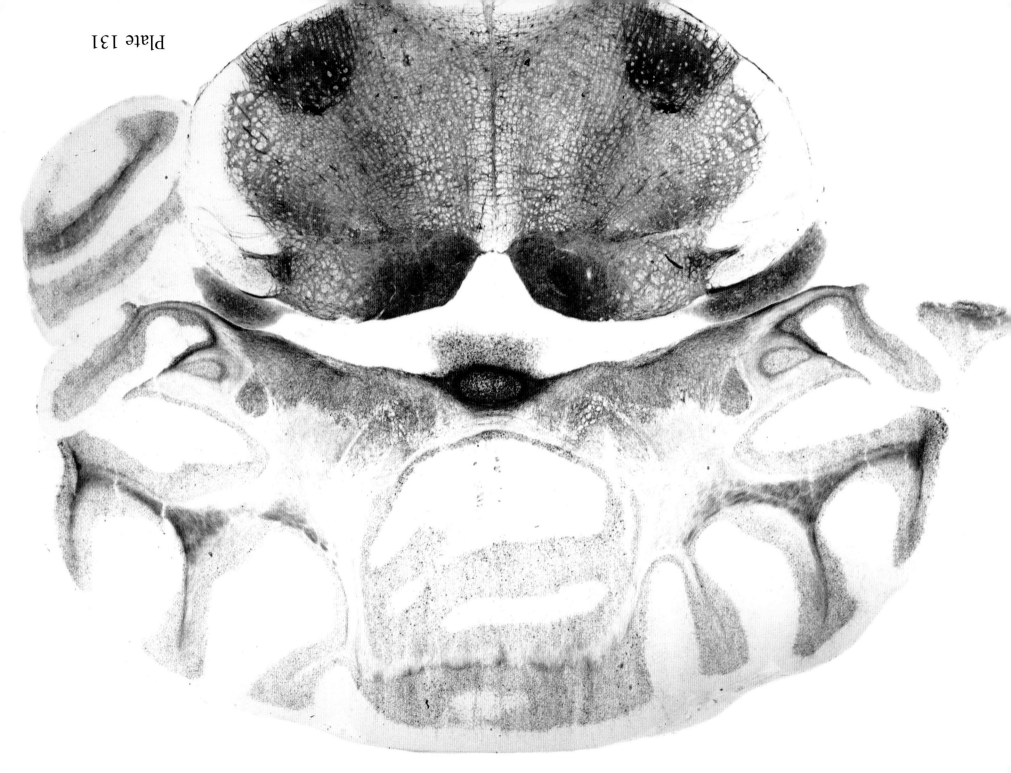

Plate 131

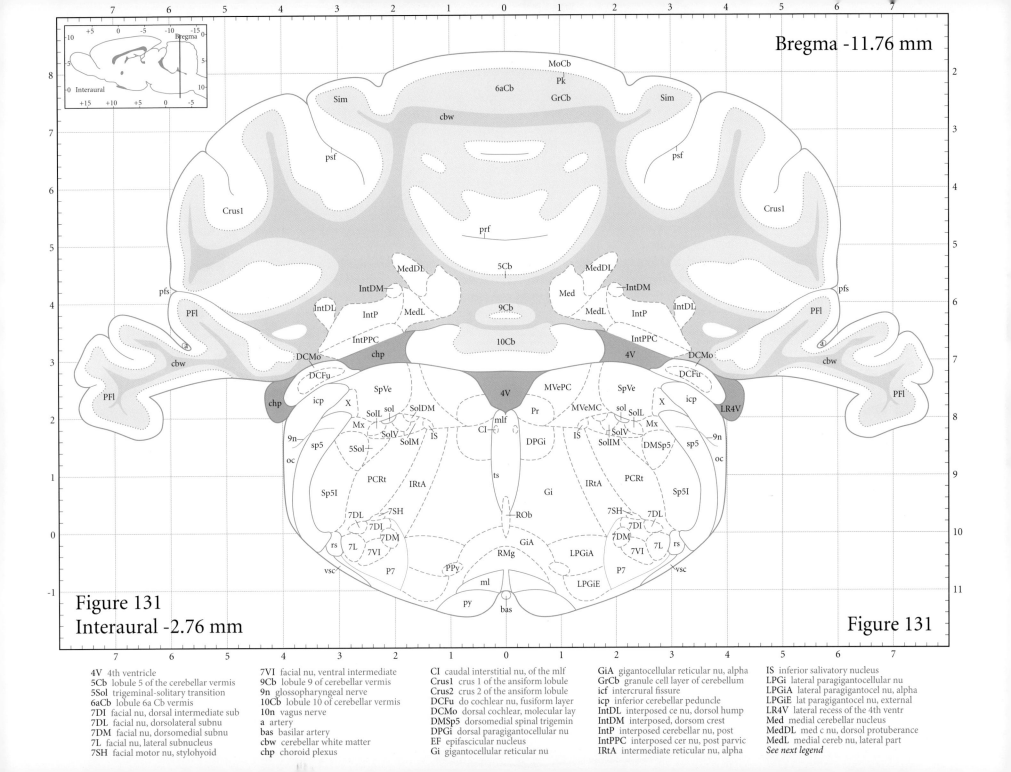

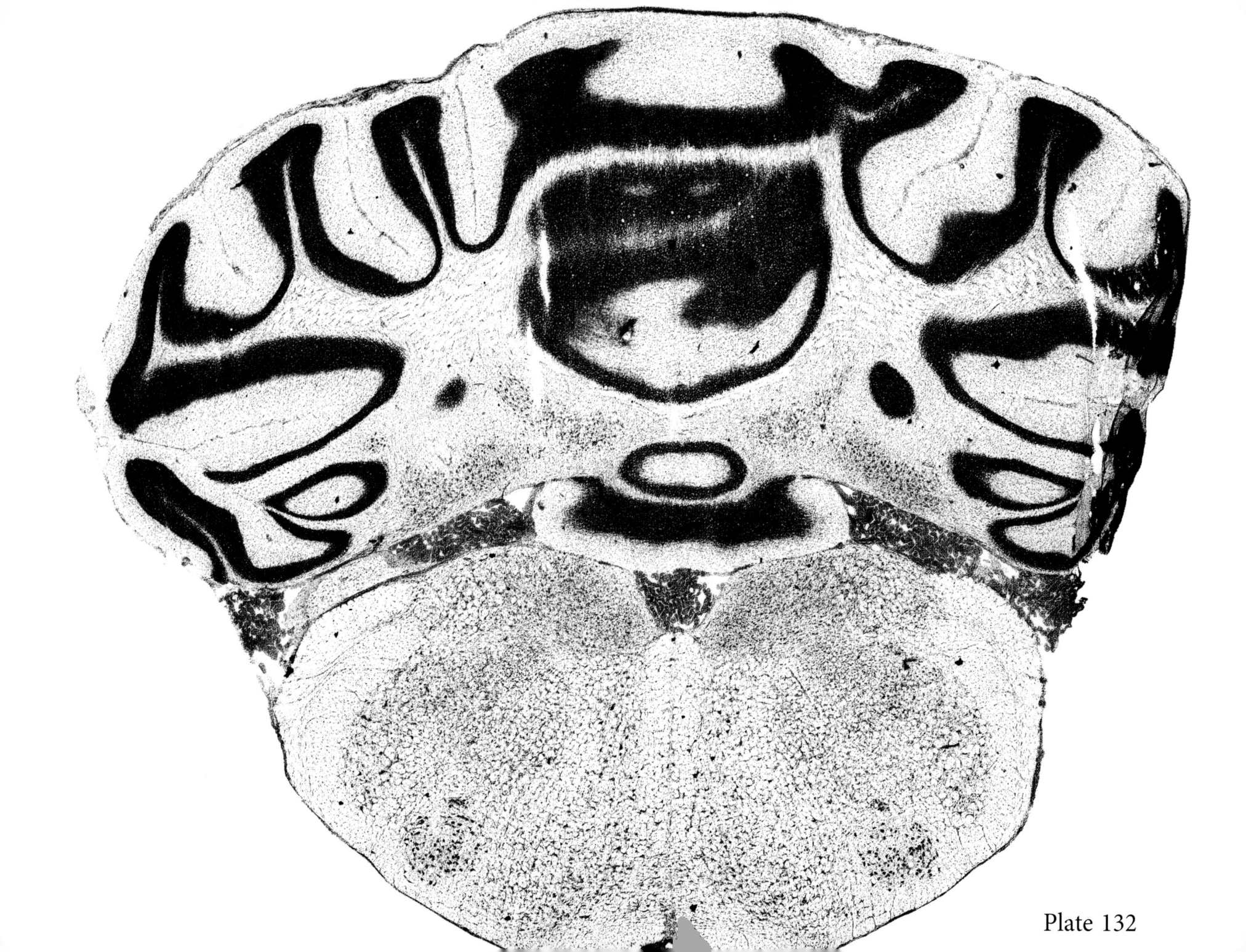

Plate 132

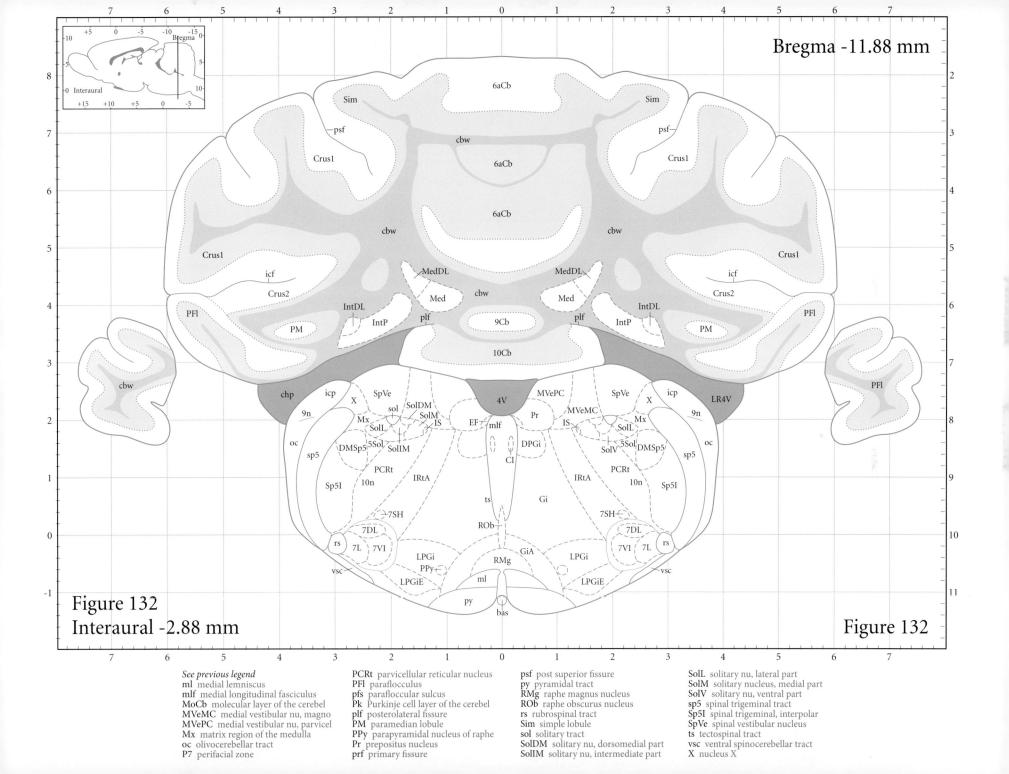

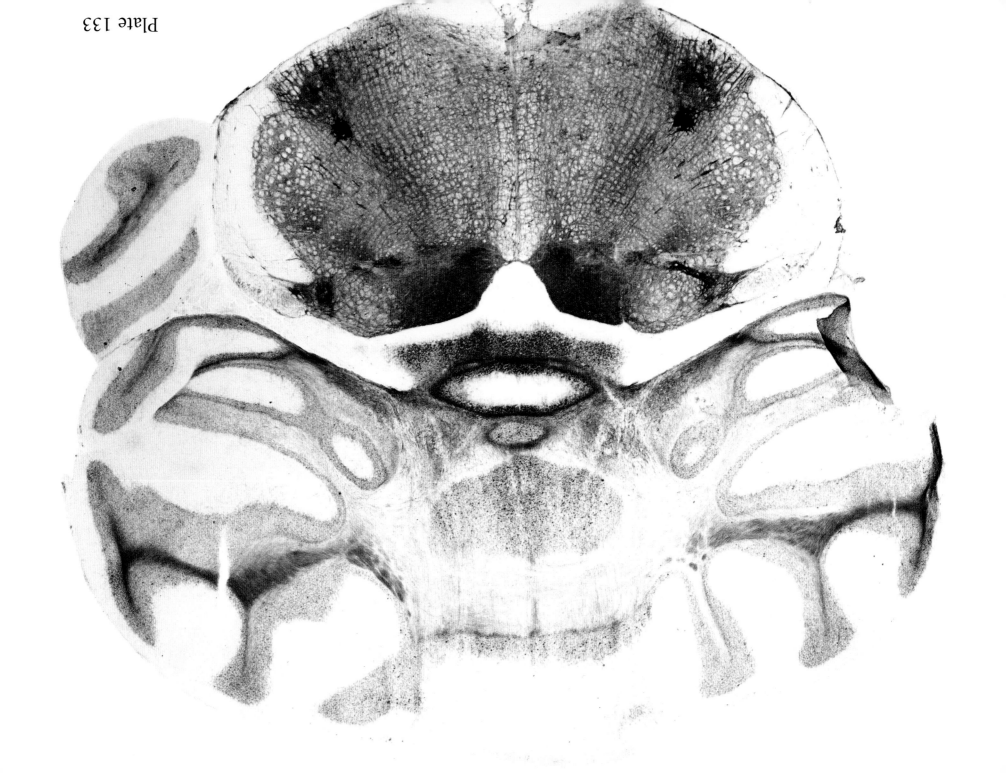

Plate 133

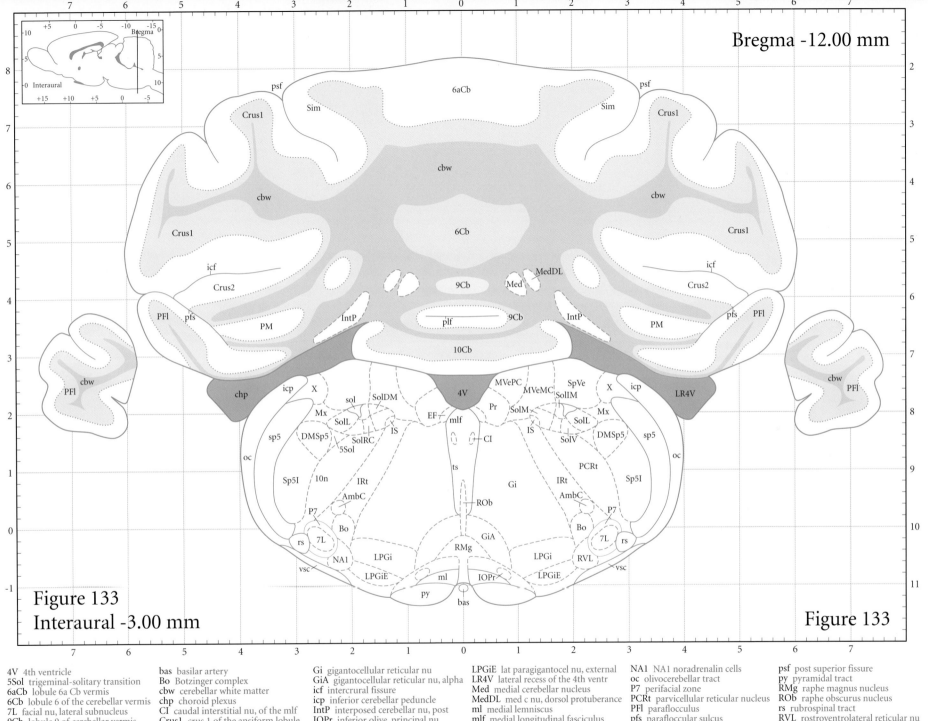

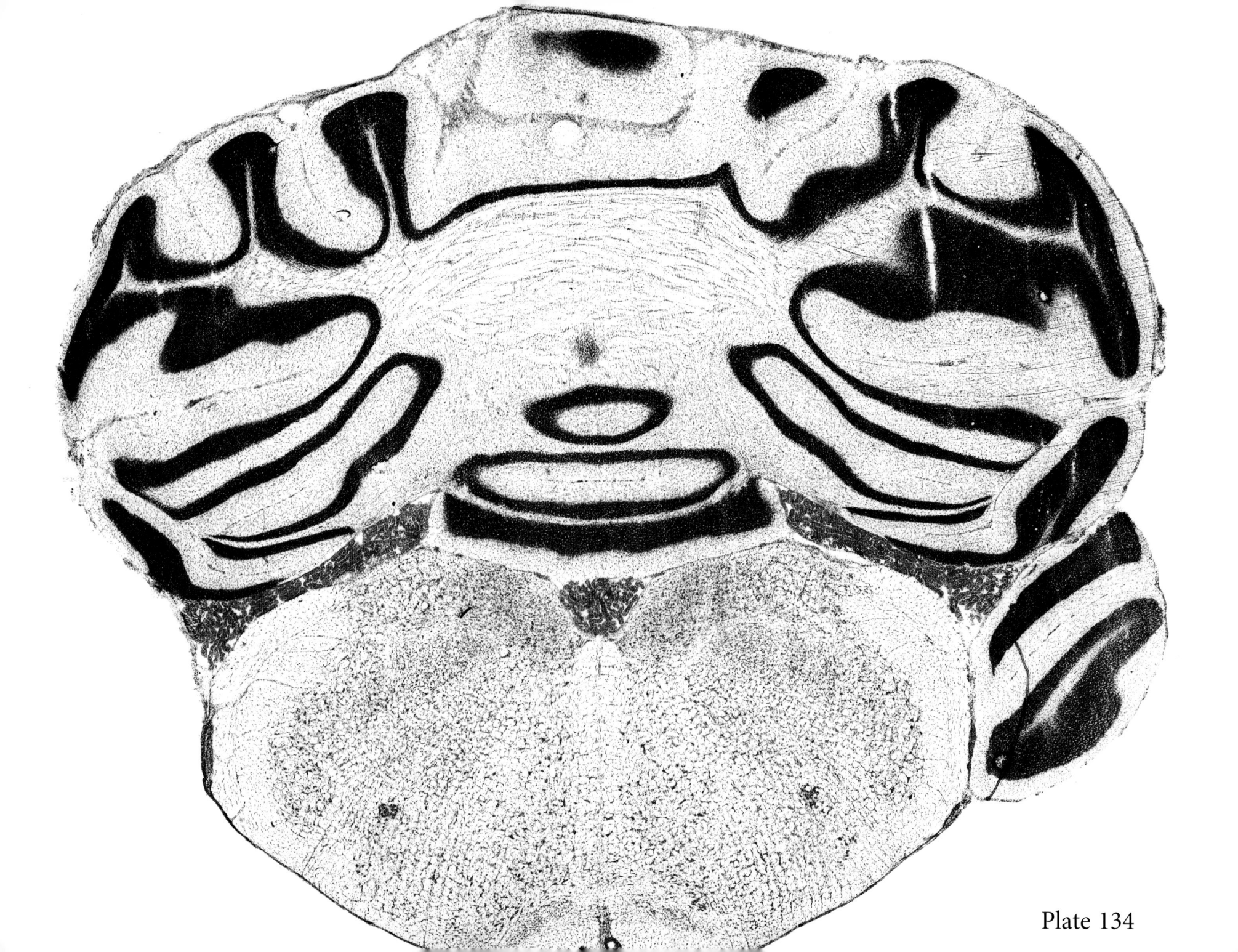

Plate 134

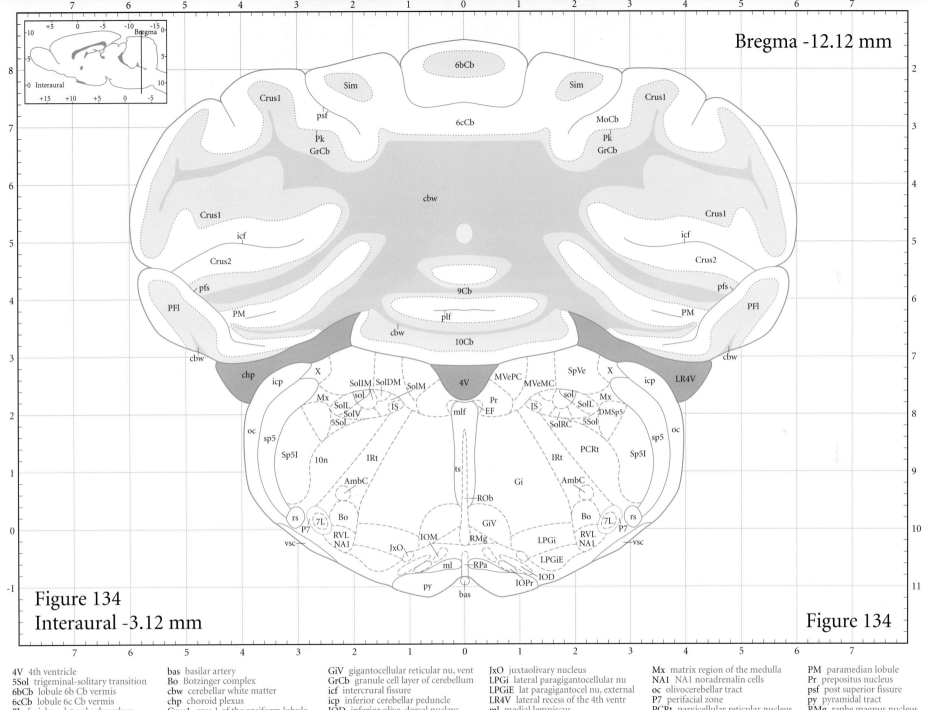

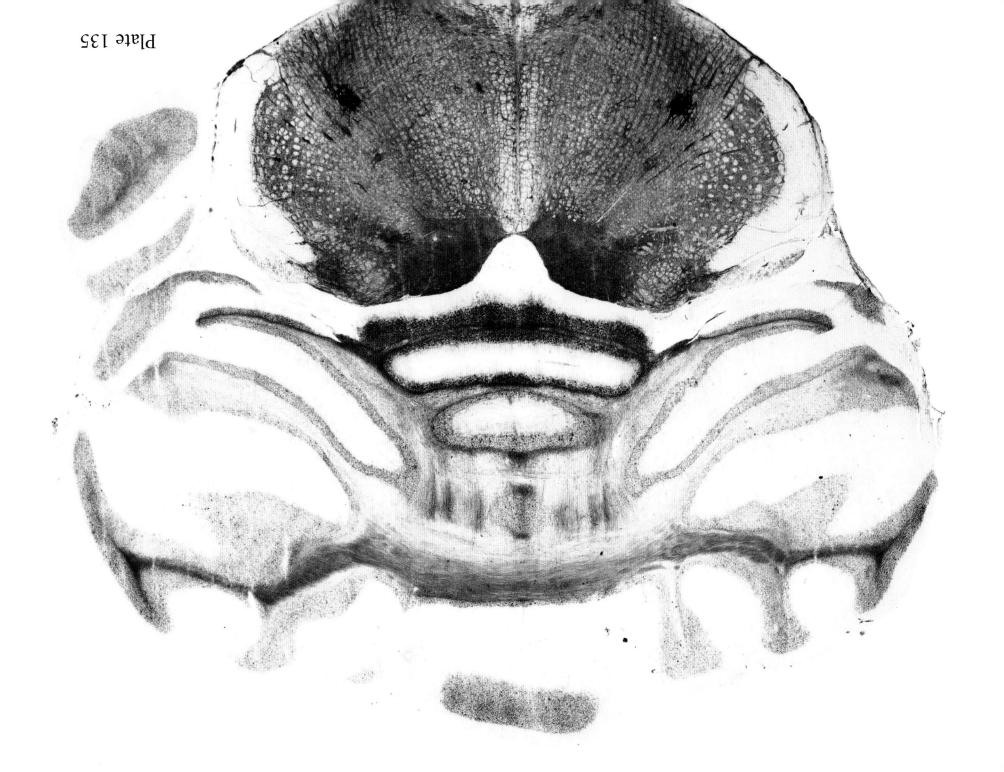

Plate 135

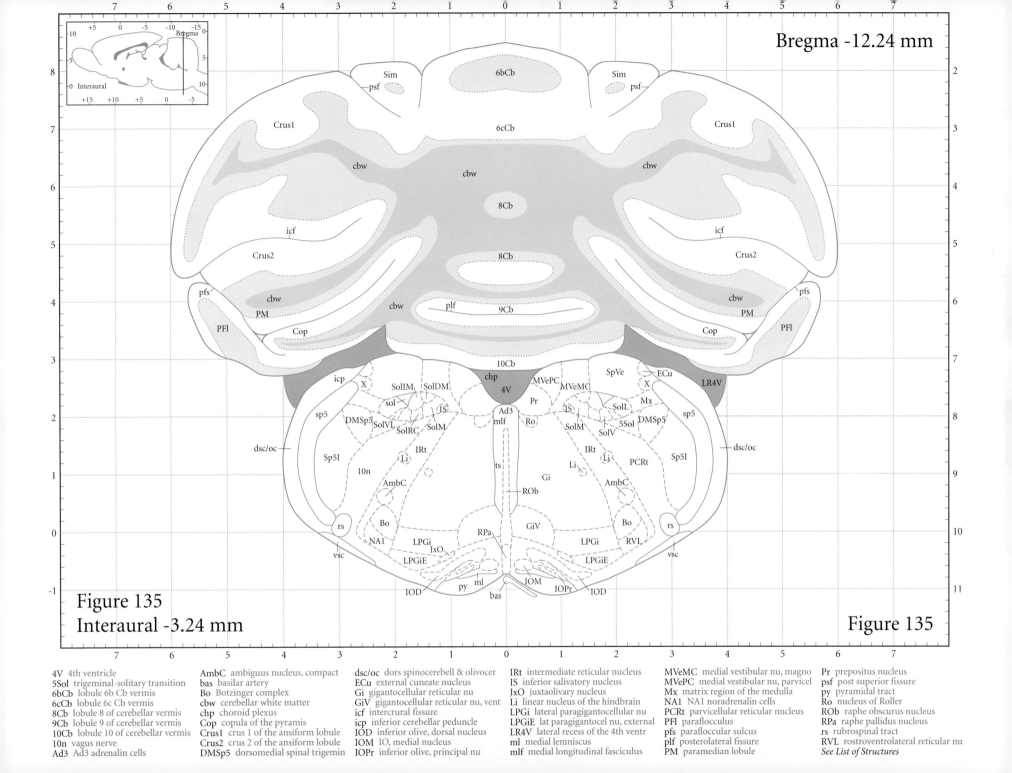

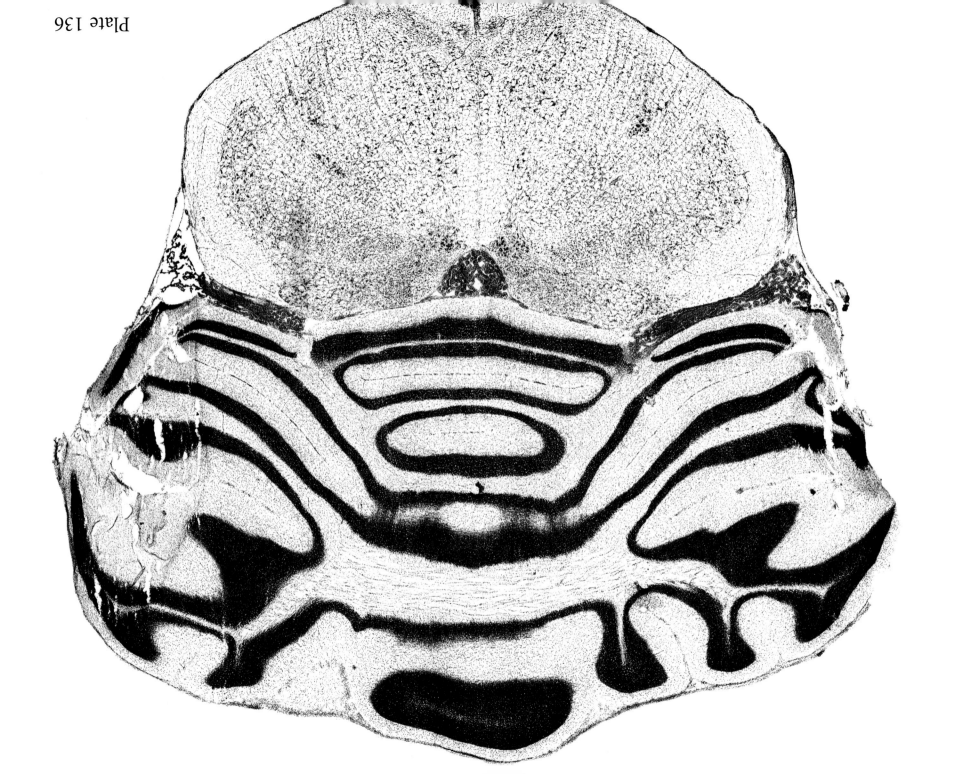

Plate 136

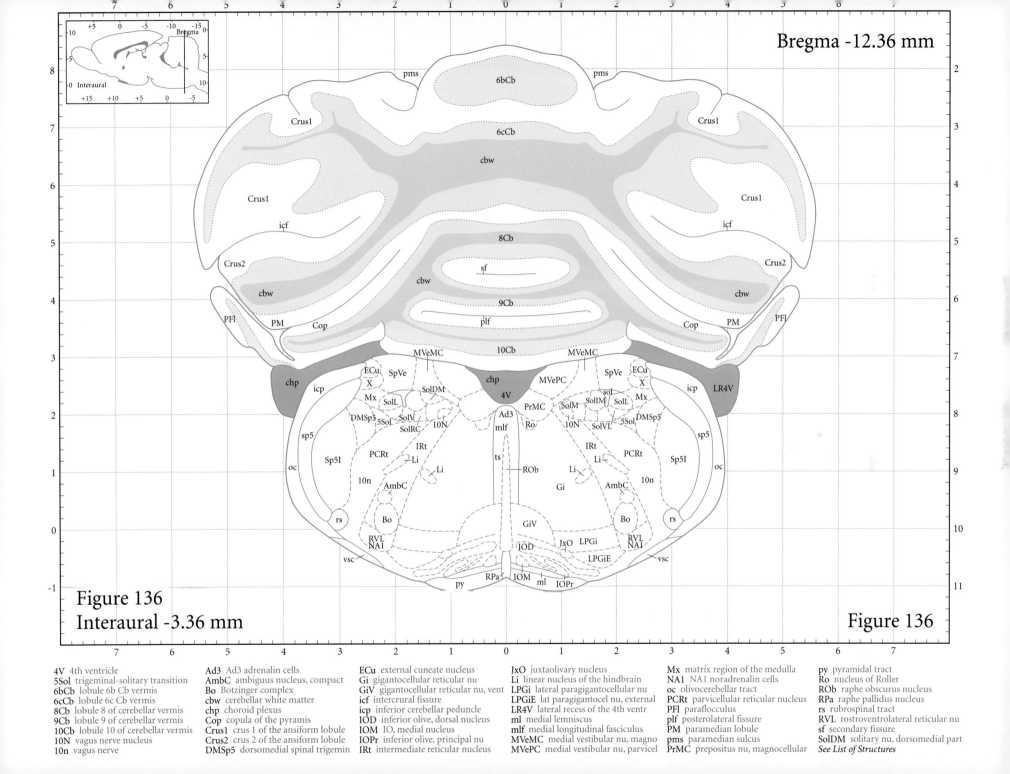

Figure 136
Interaural -3.36 mm
Bregma -12.36 mm

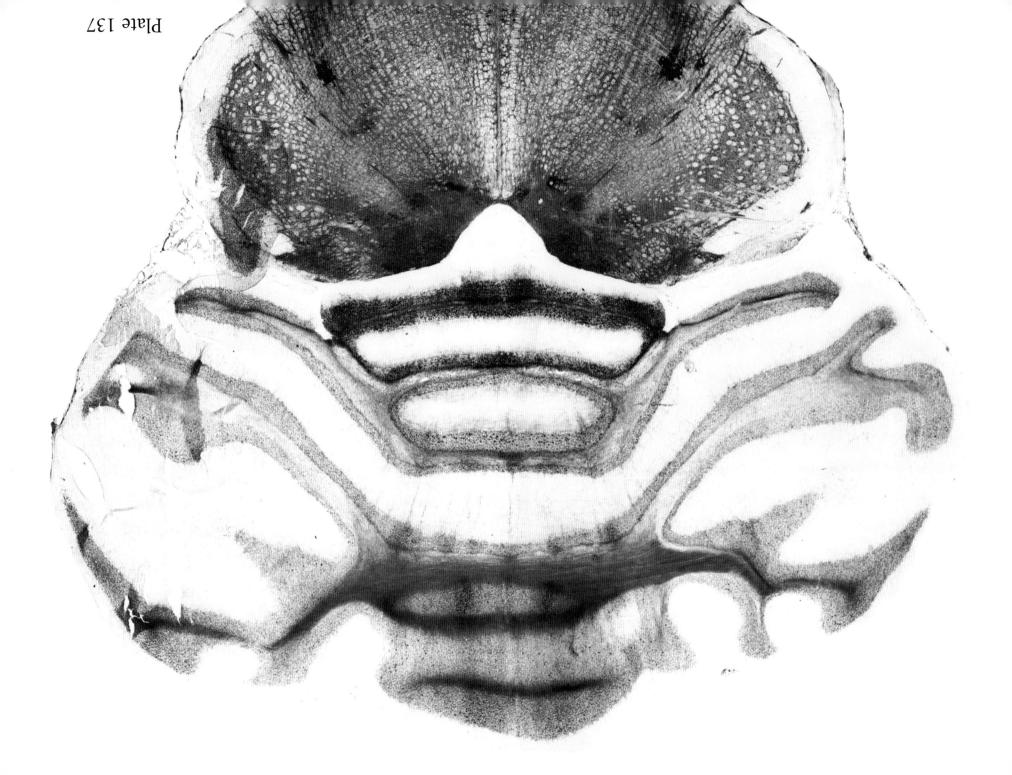

Plate 137

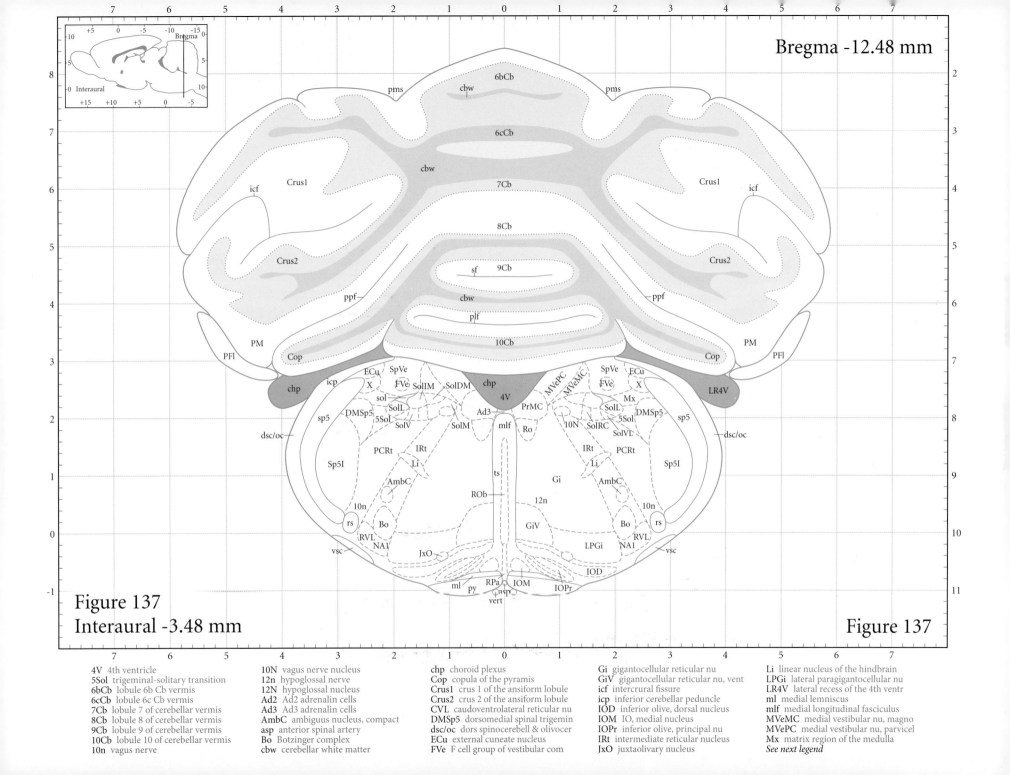

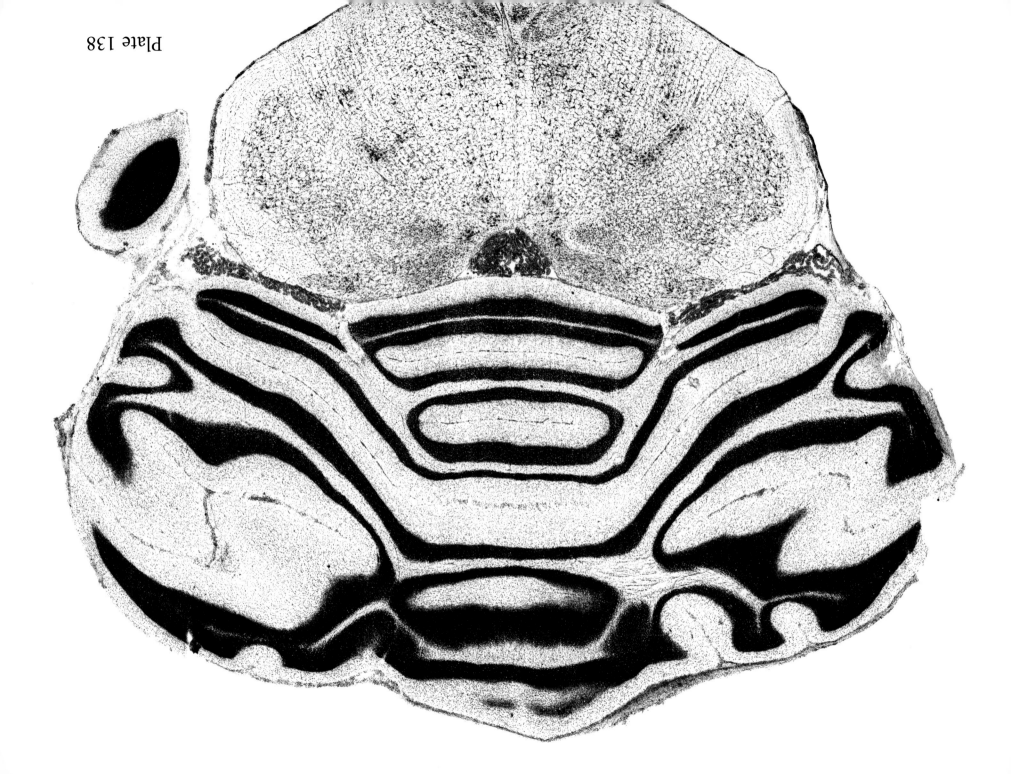

Plate 138

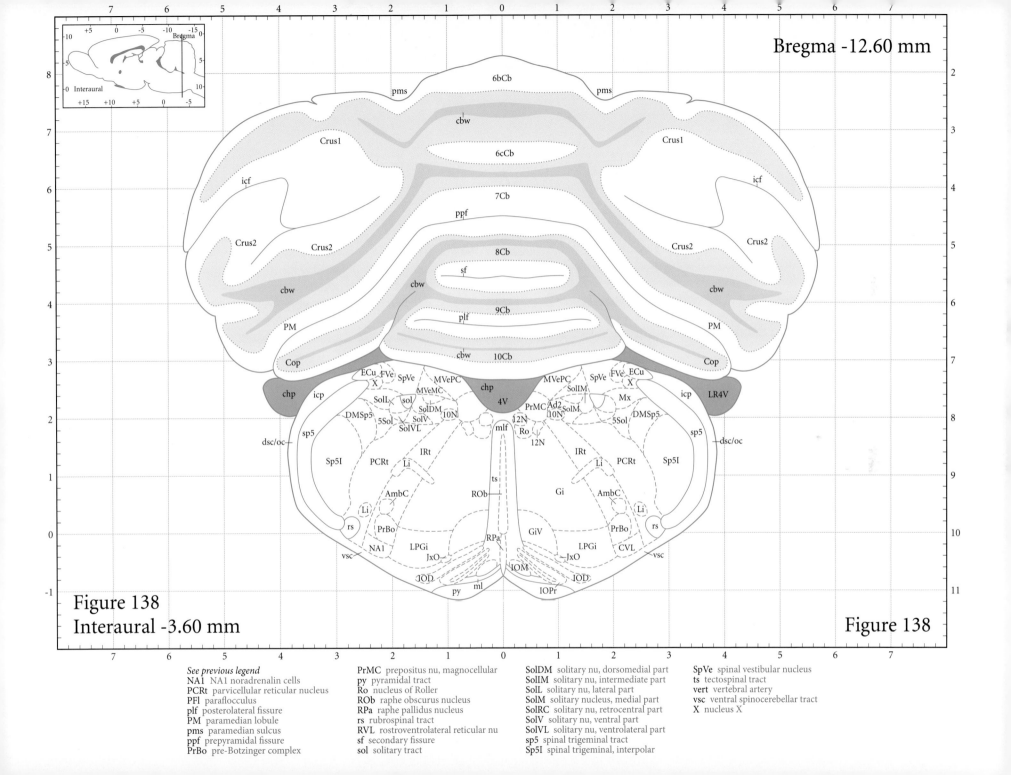

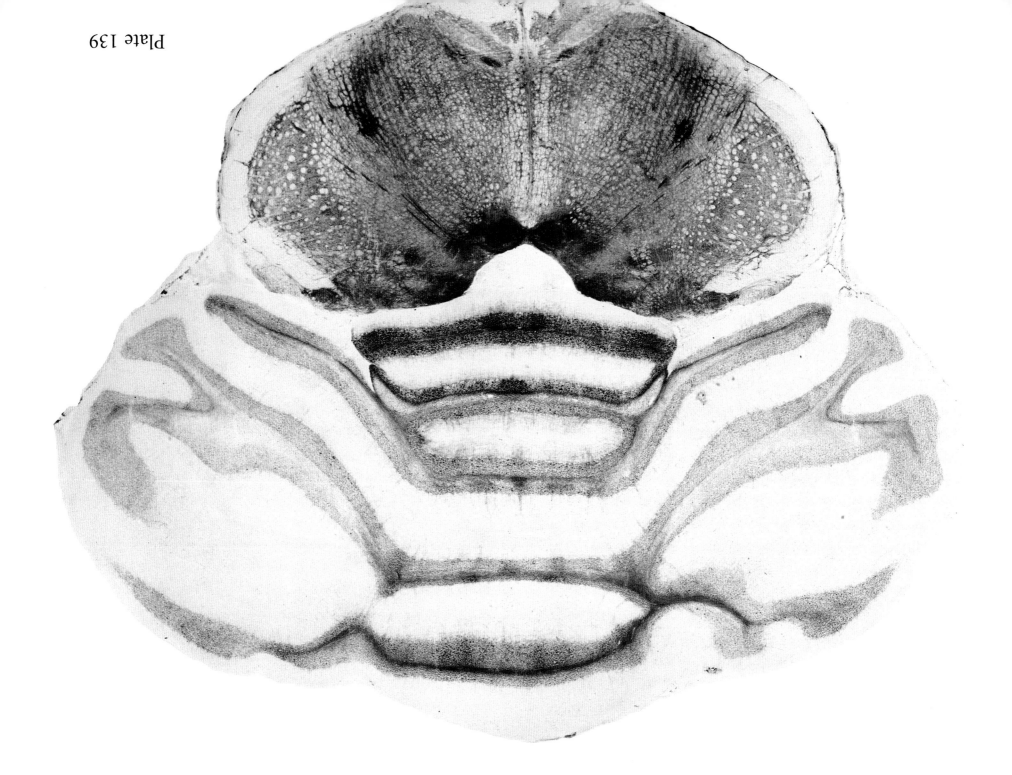

Plate 139

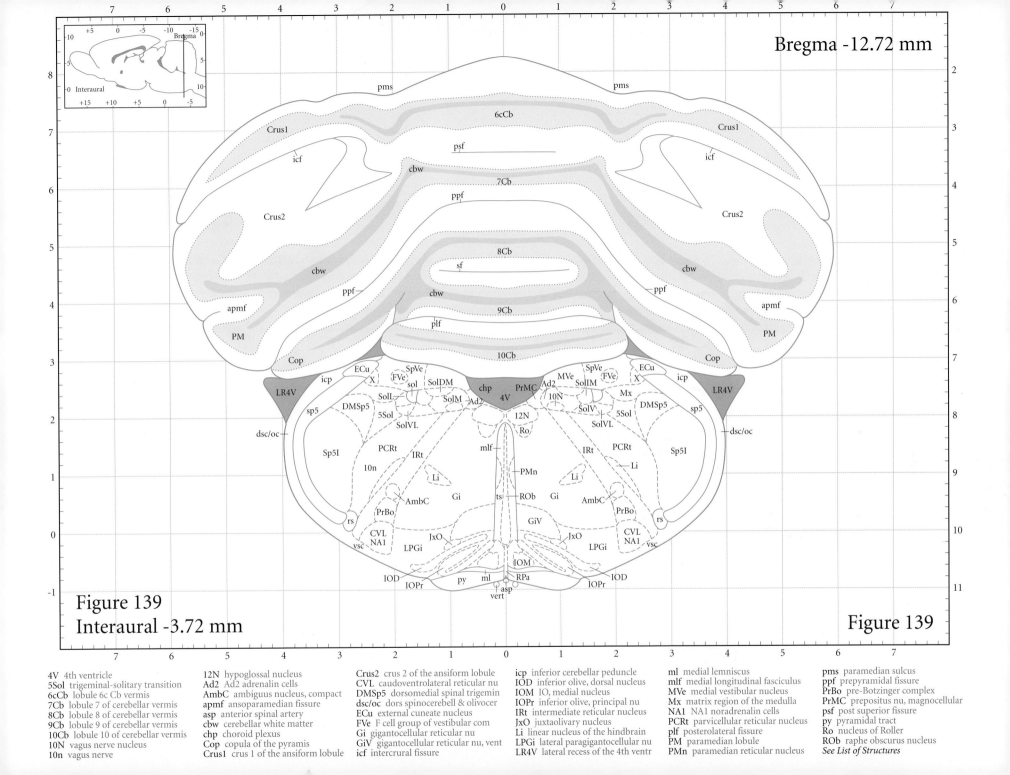

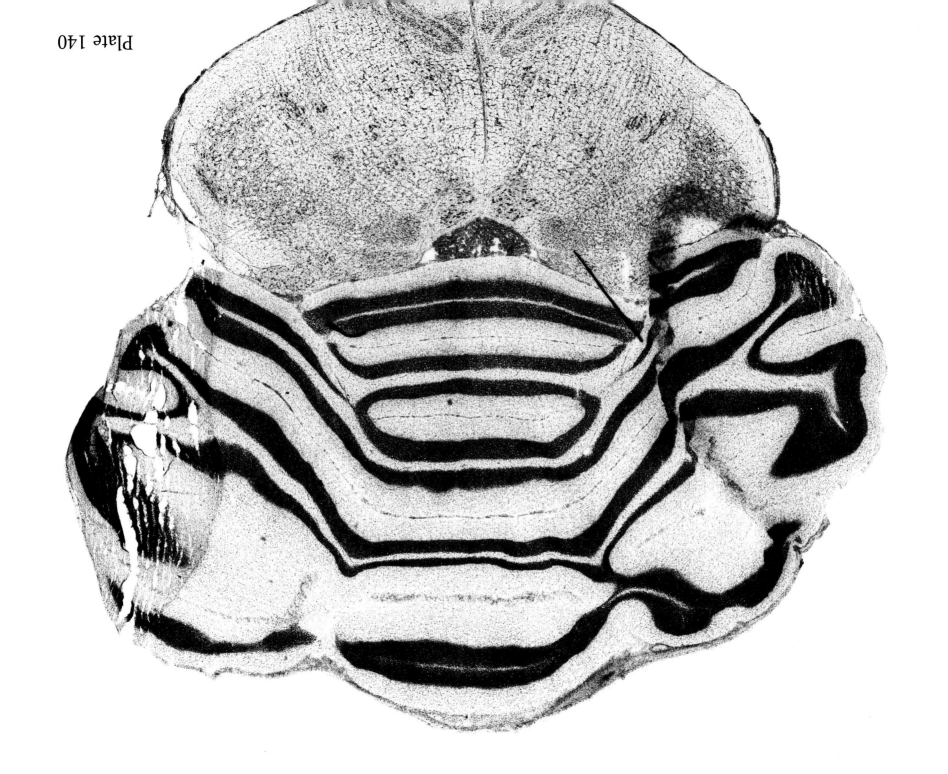

Plate 140

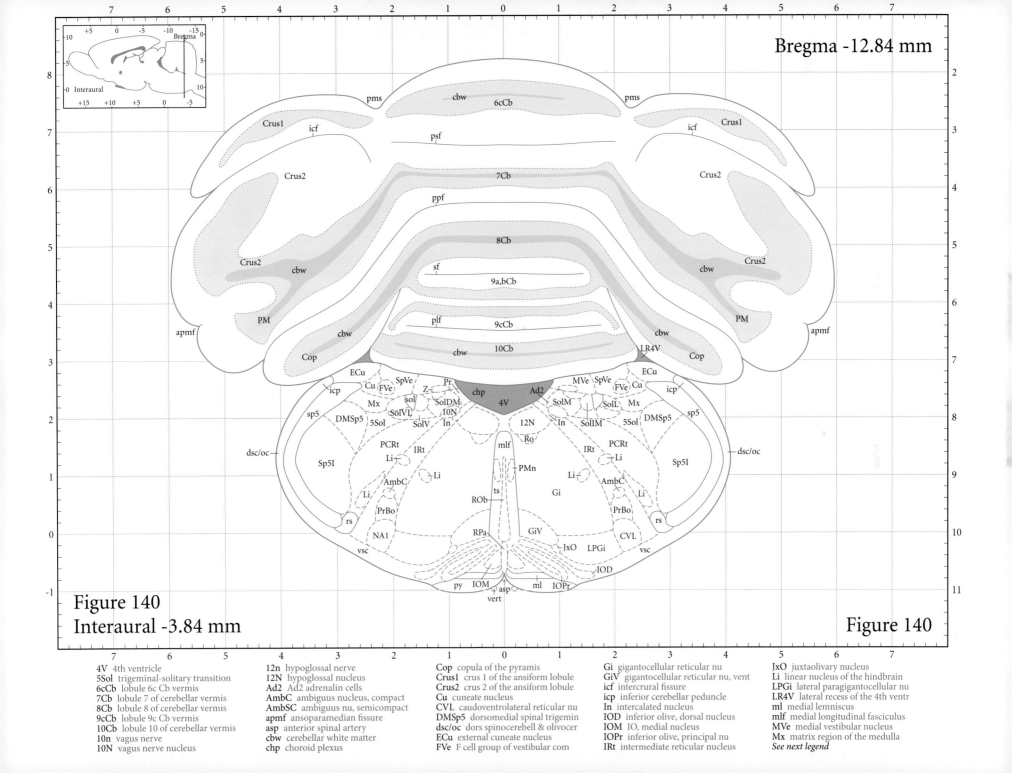

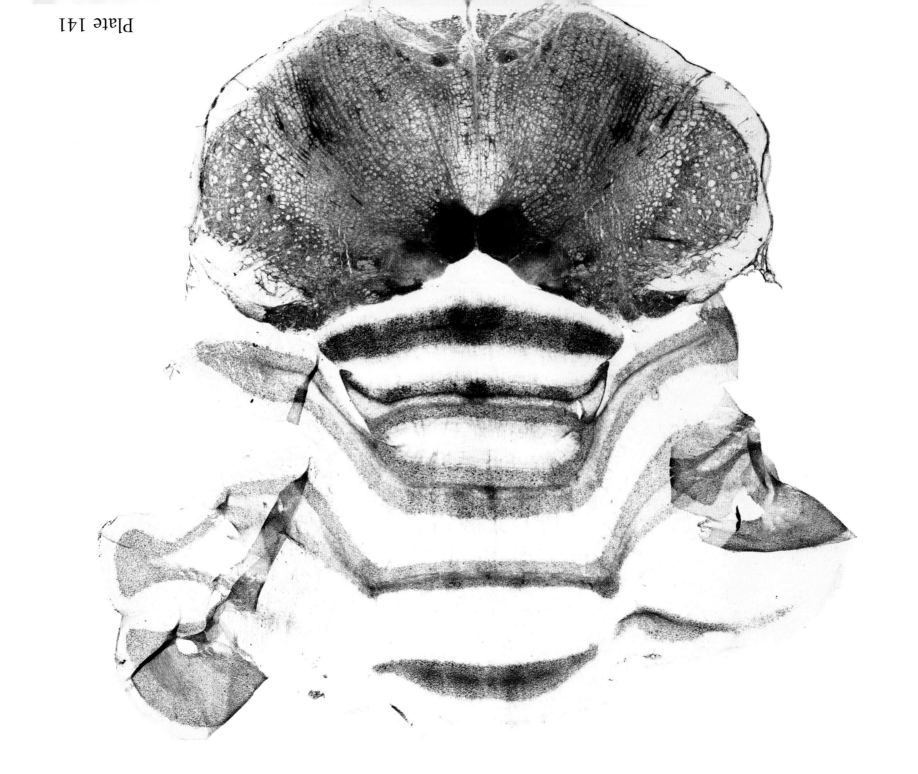

Plate 141

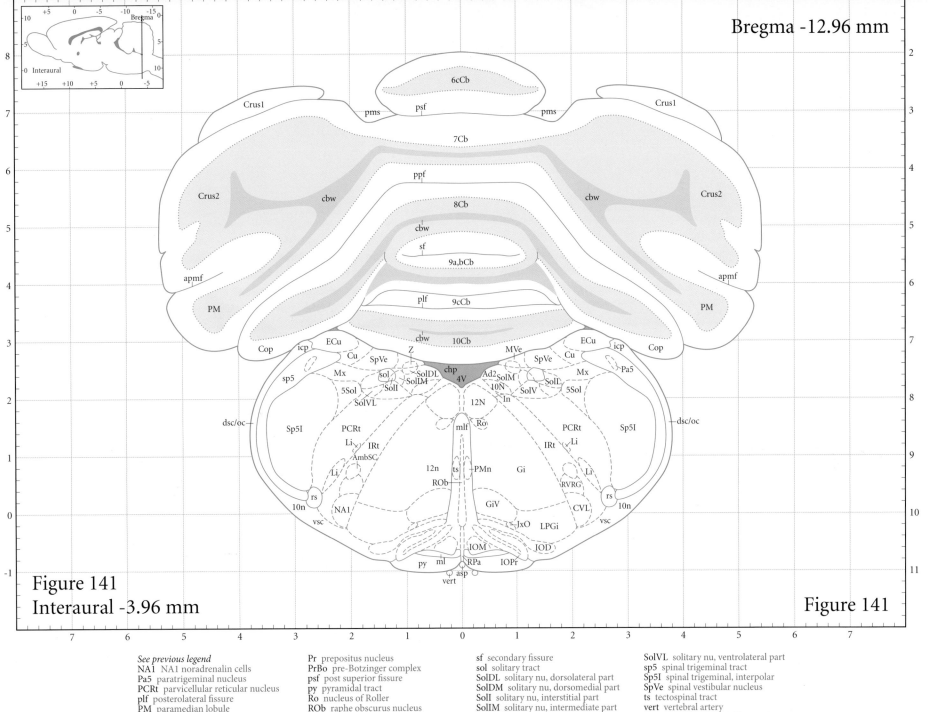

Plate 142
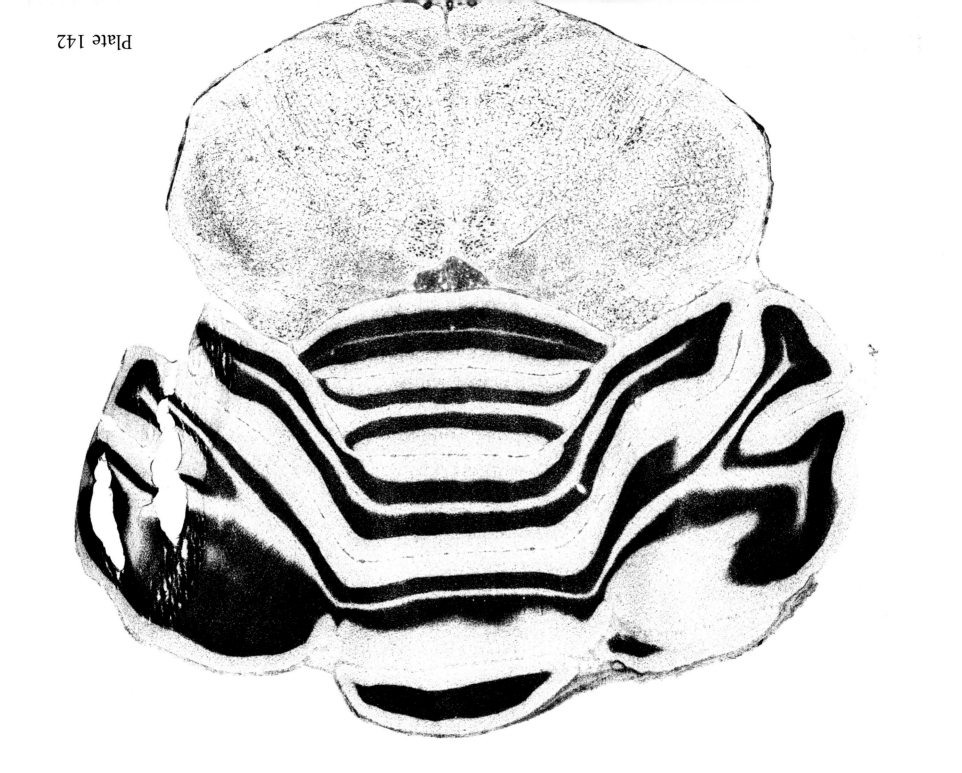

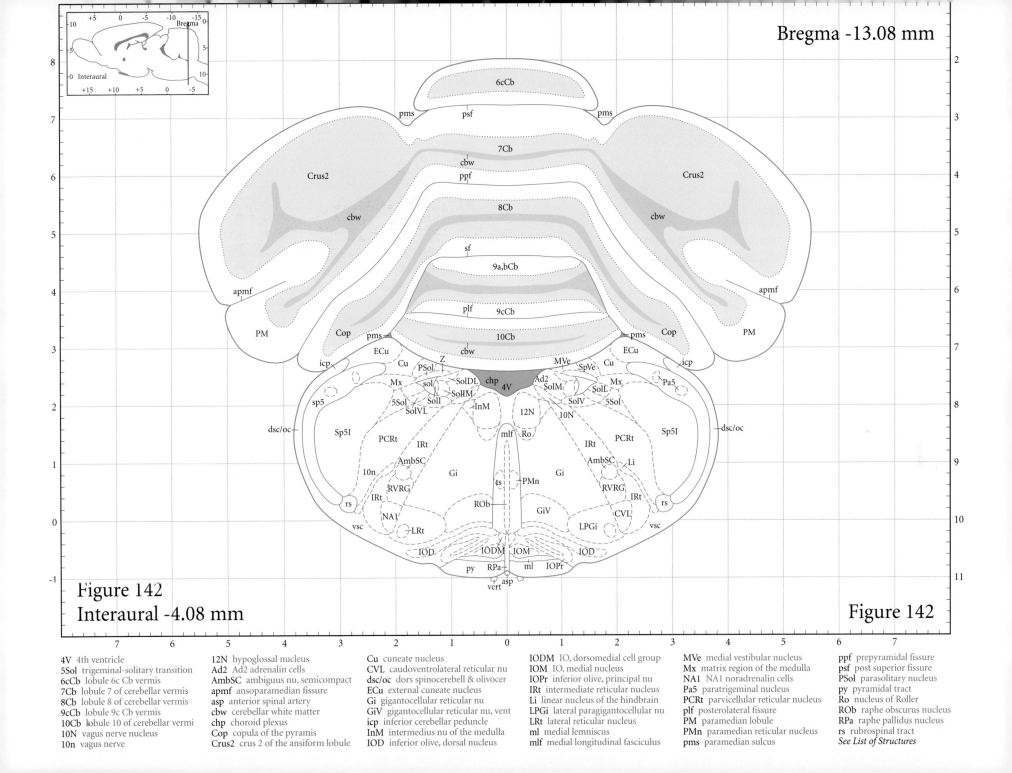

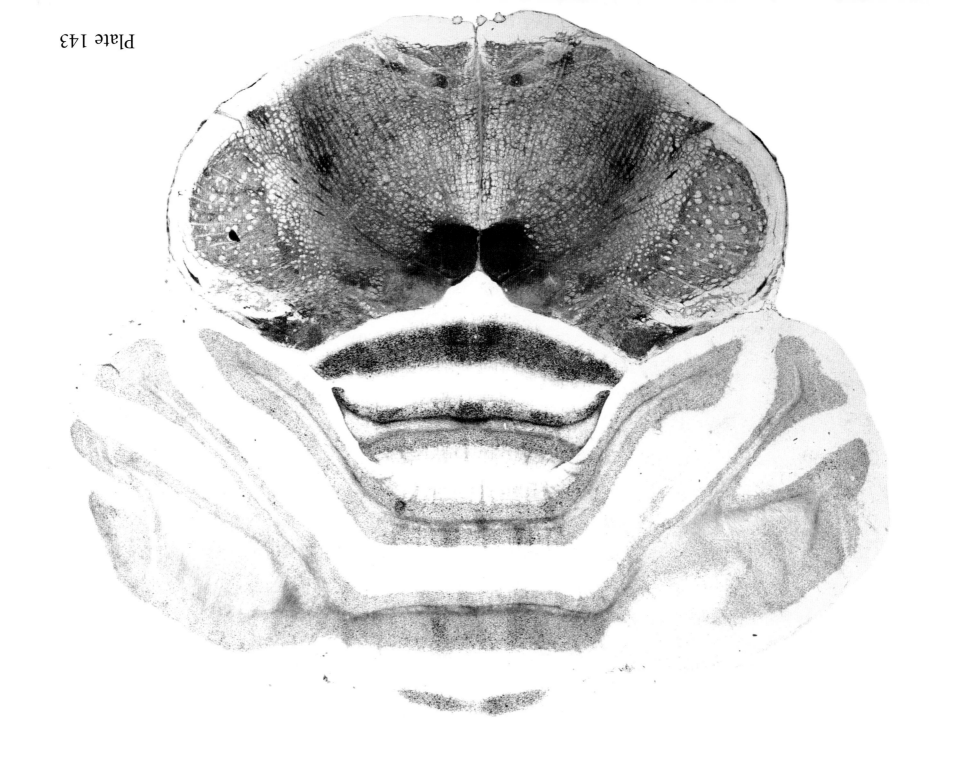

Plate 143

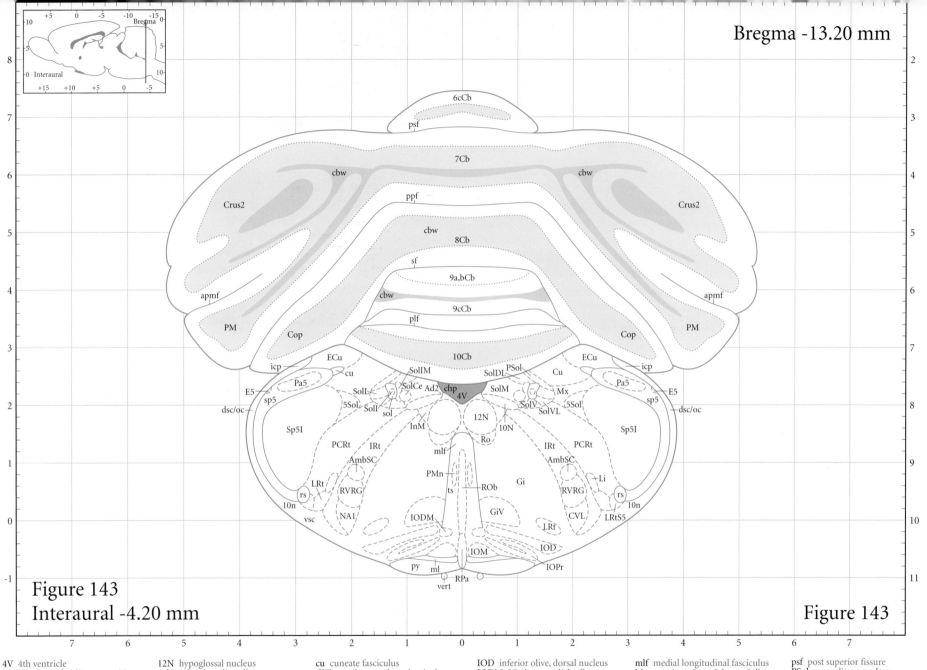

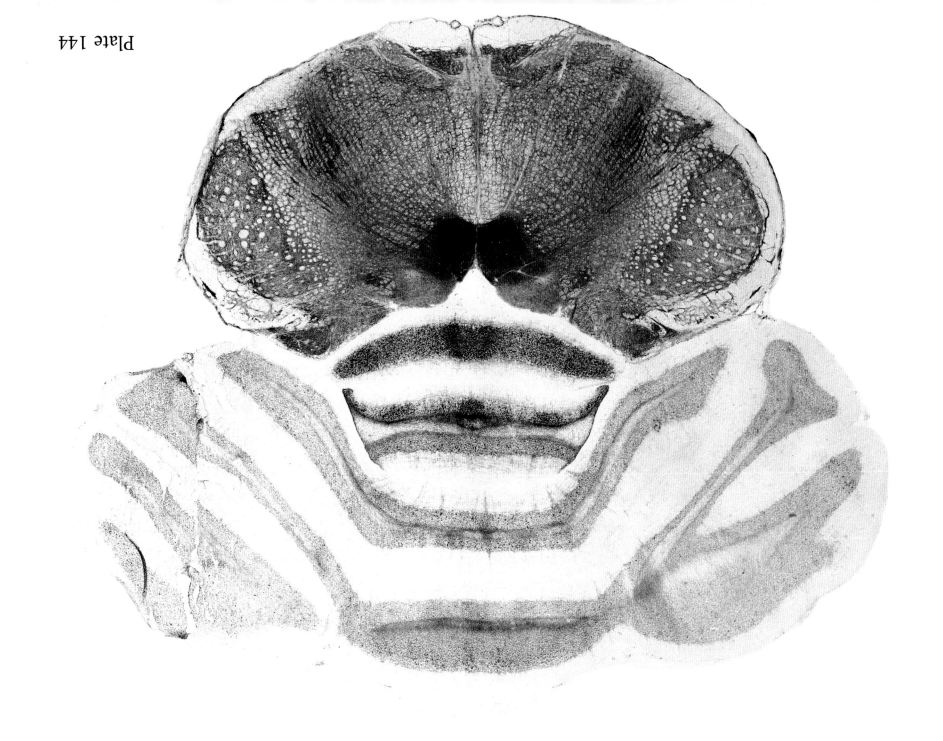

Plate 144

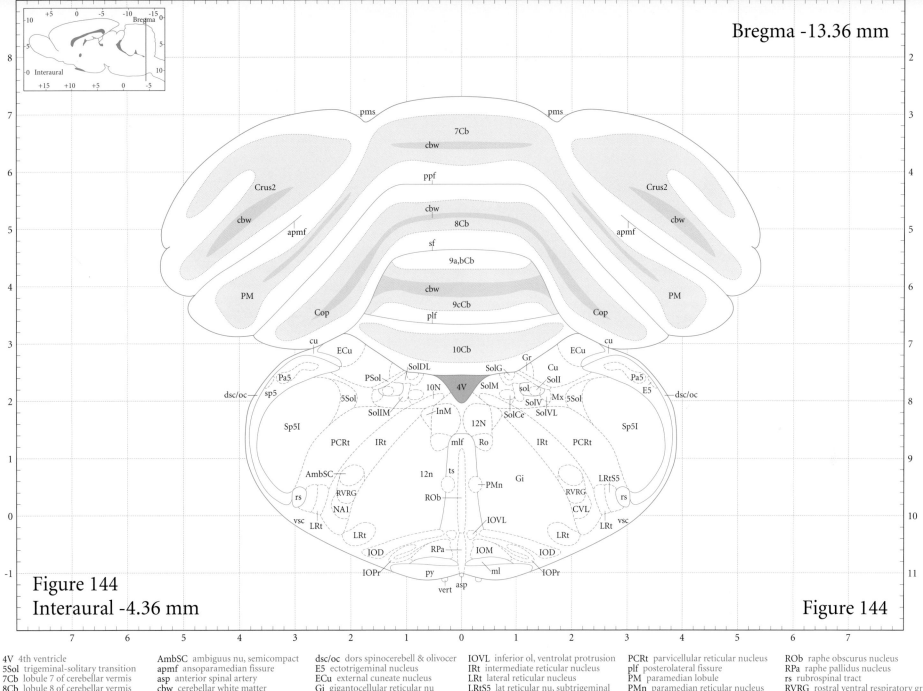

Plate 145

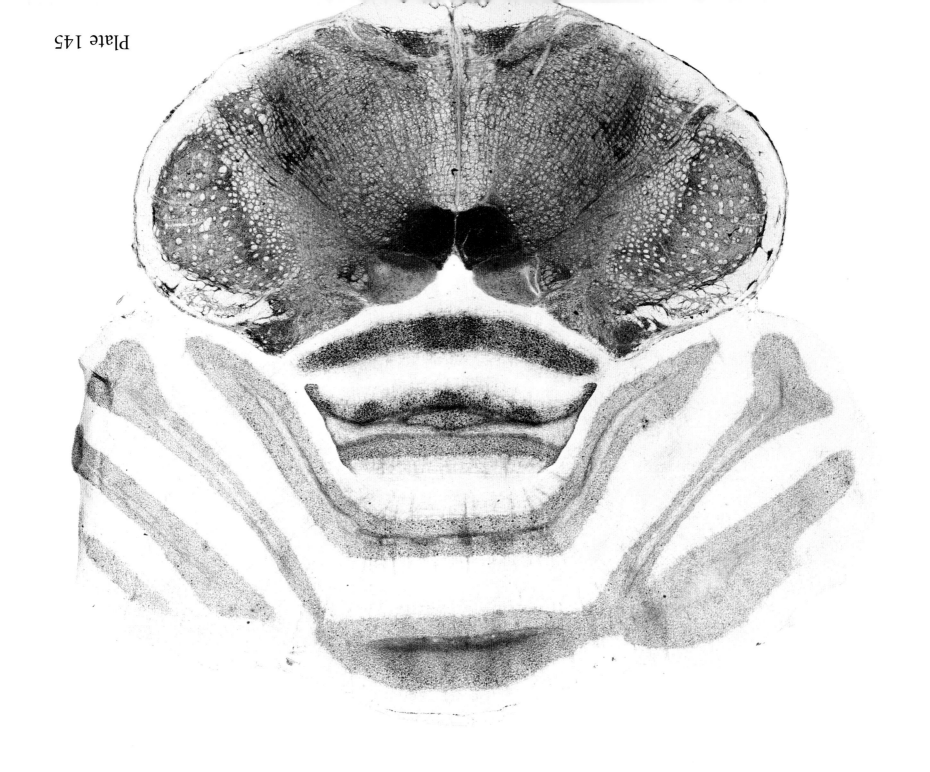

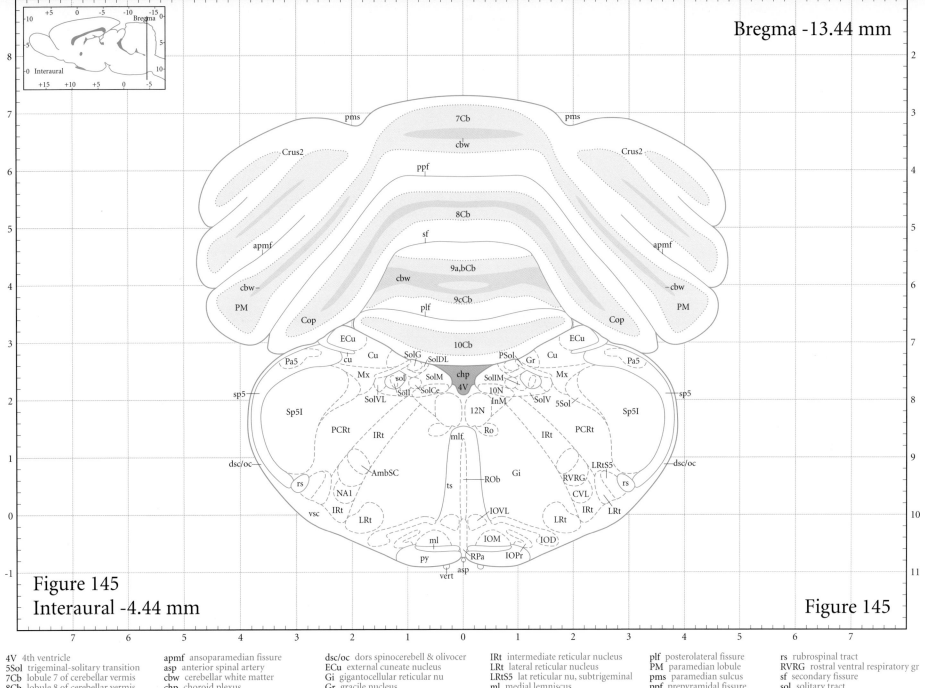

Figure 145
Interaural -4.44 mm
Bregma -13.44 mm

Plate 146

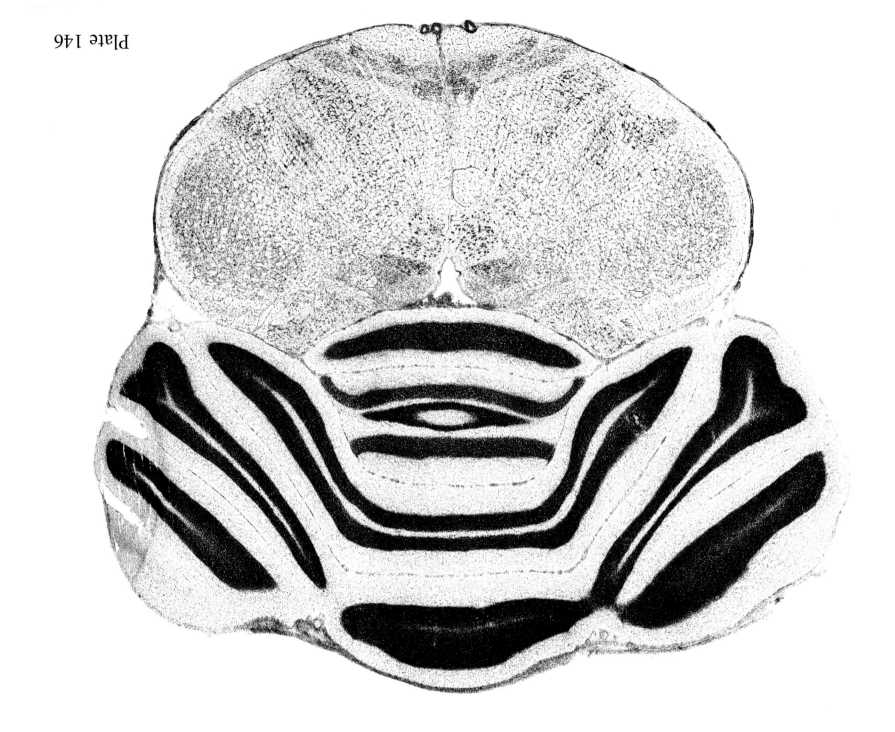

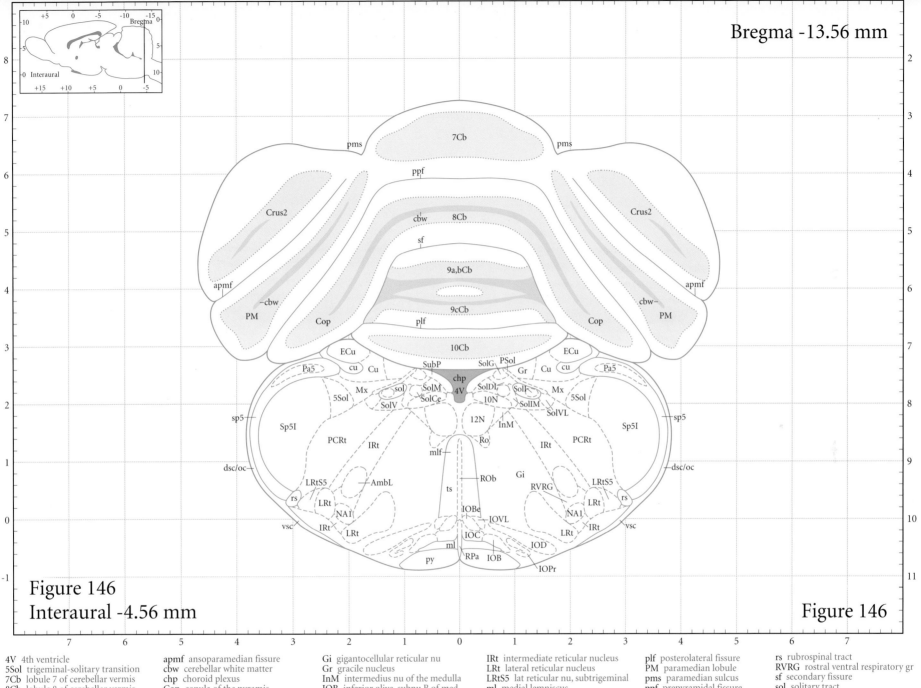

Plate 147

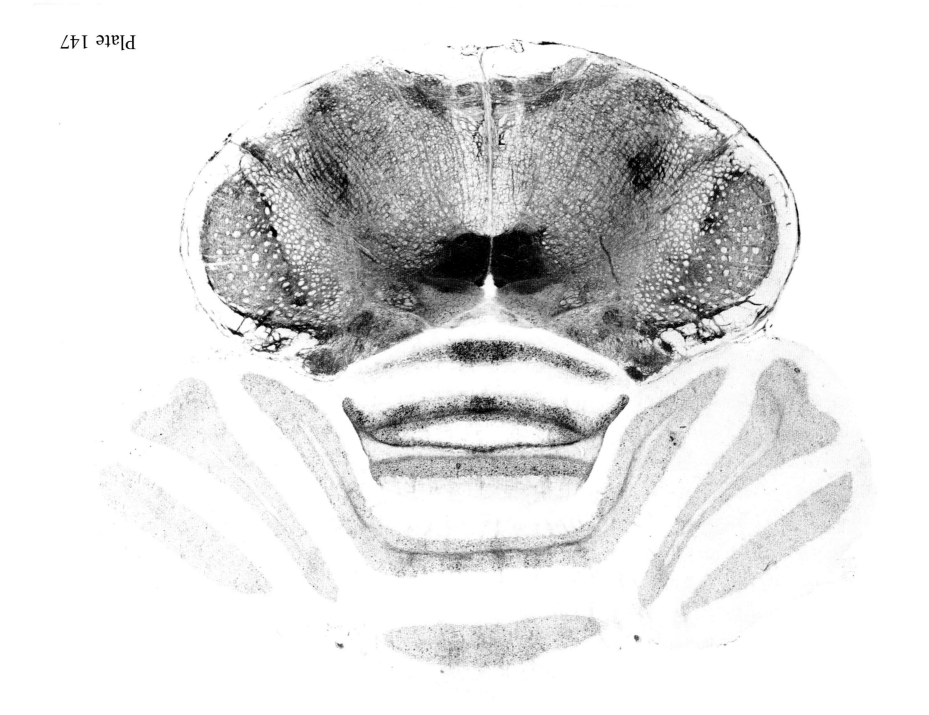

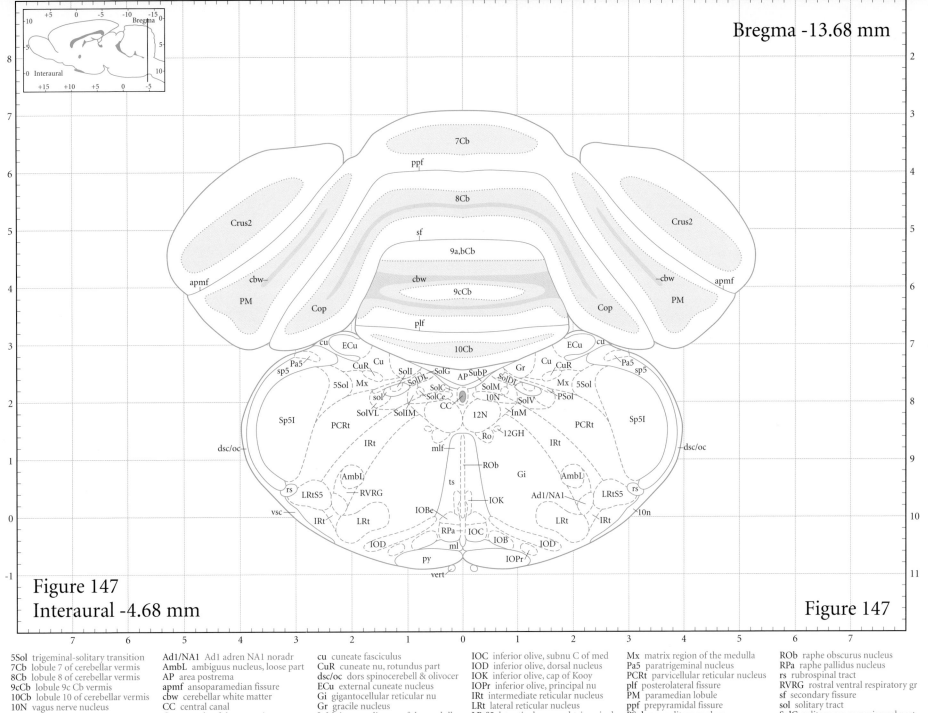

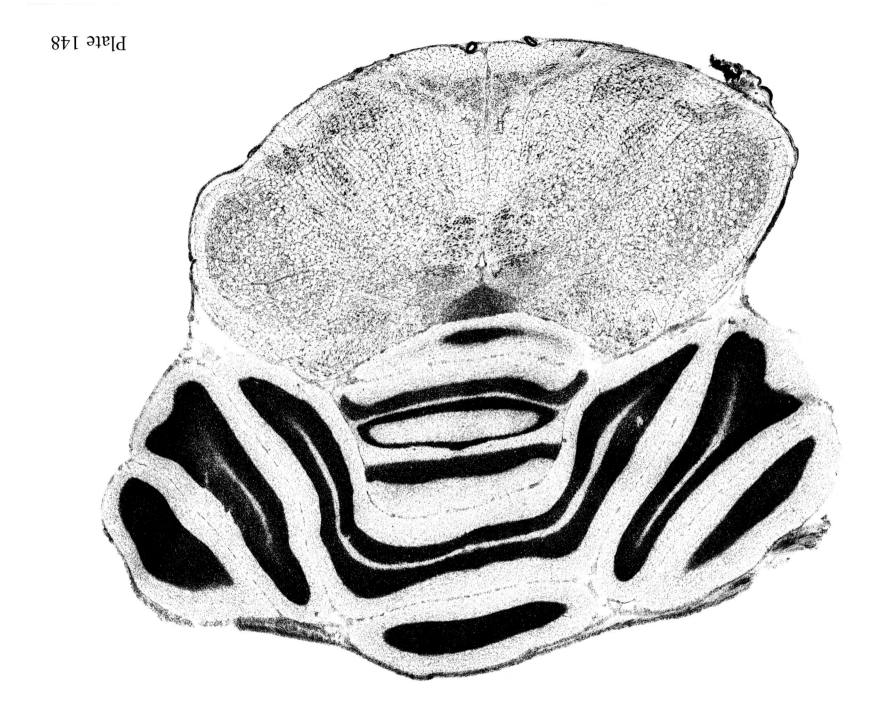

Plate 148

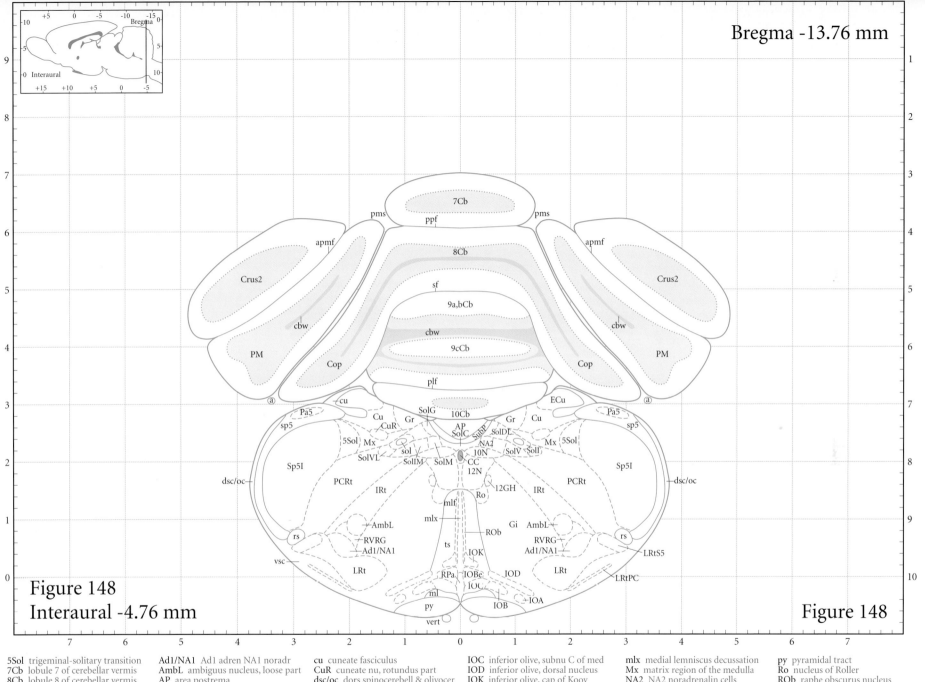

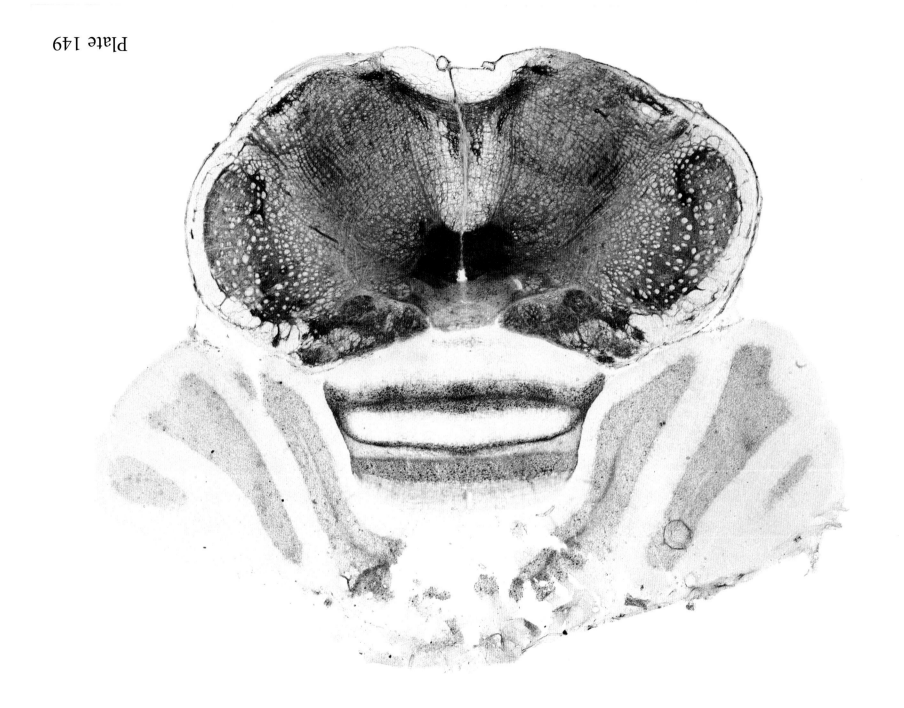

Plate 149

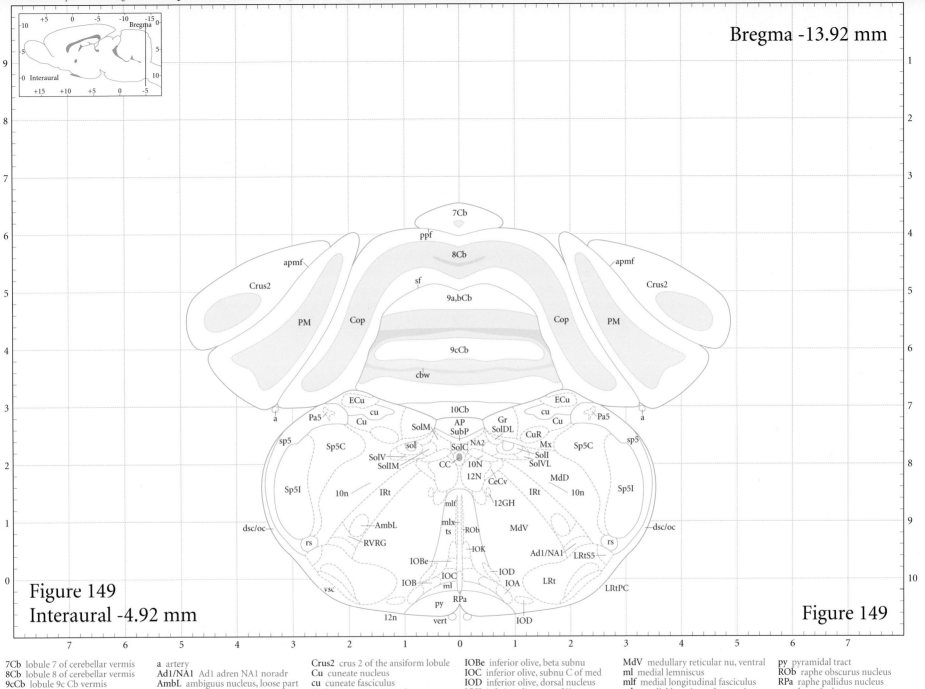

Plate 150

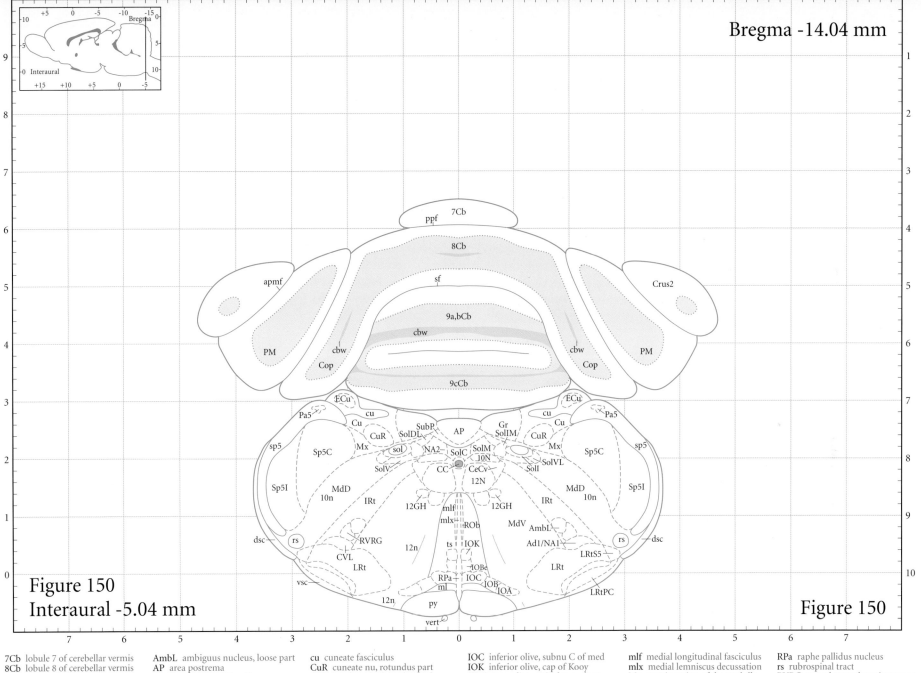

Plate 151

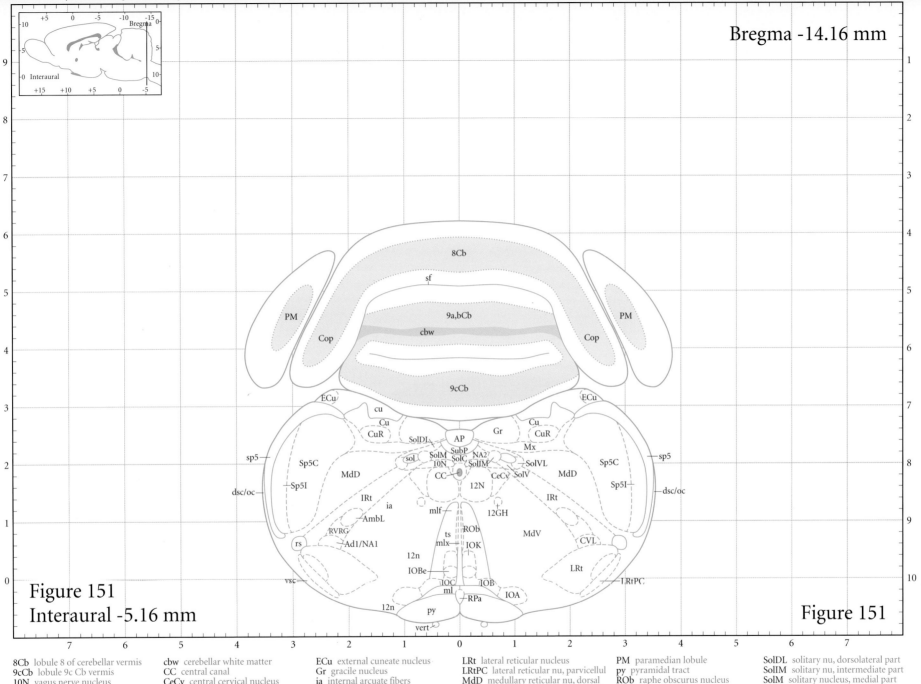

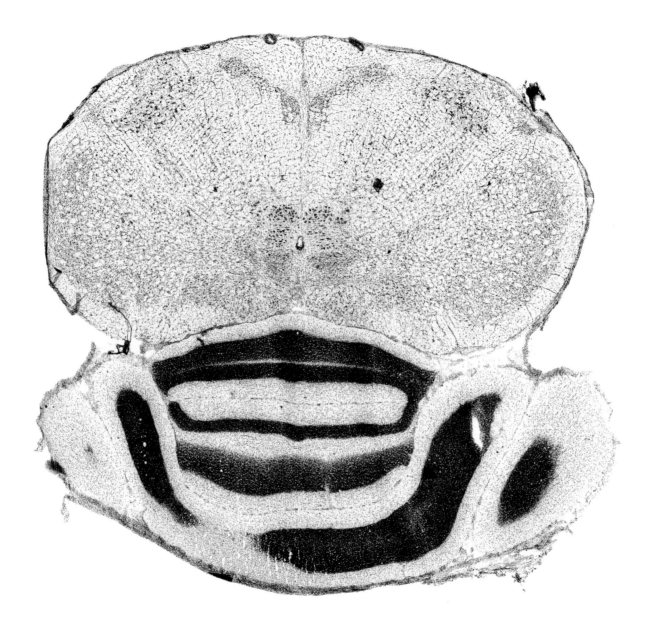

Plate 152

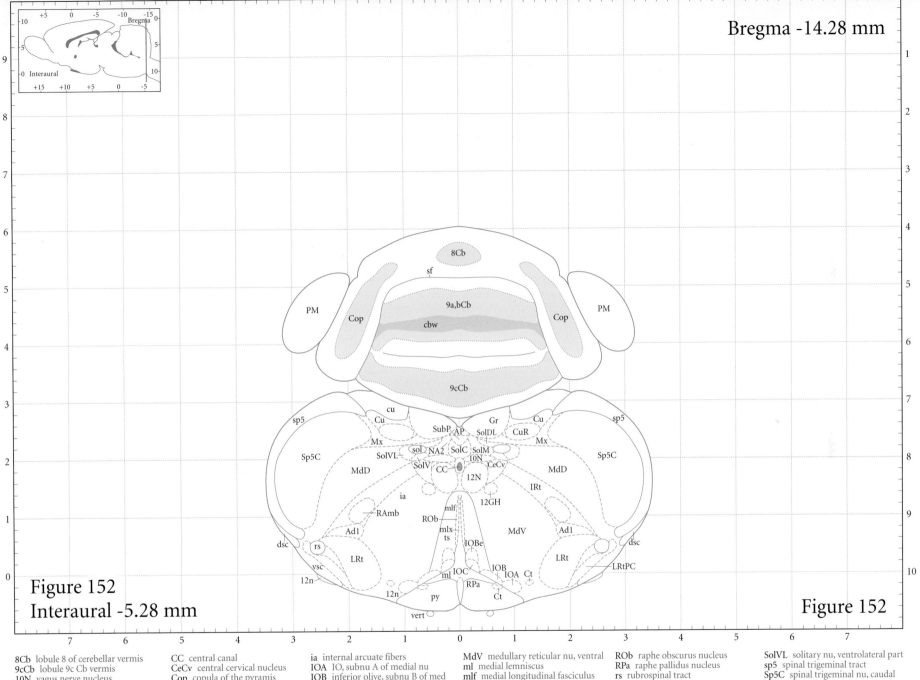

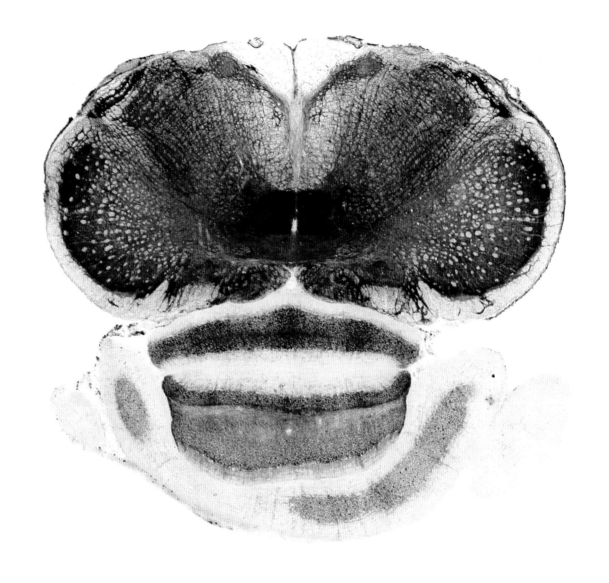

Plate 153

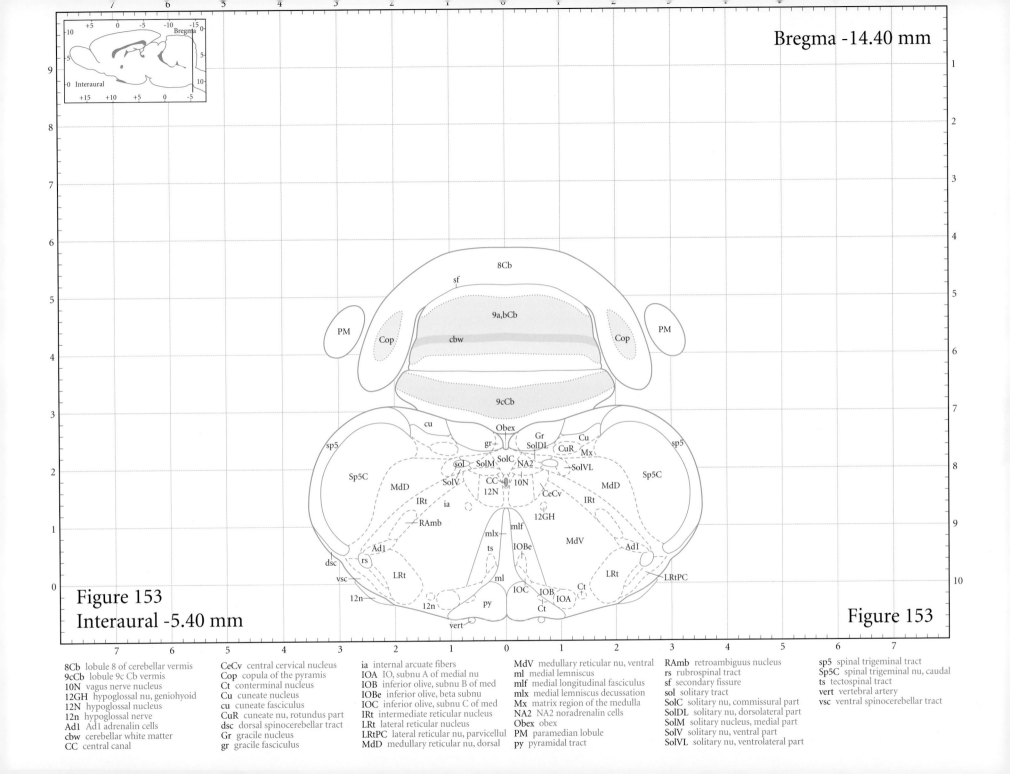

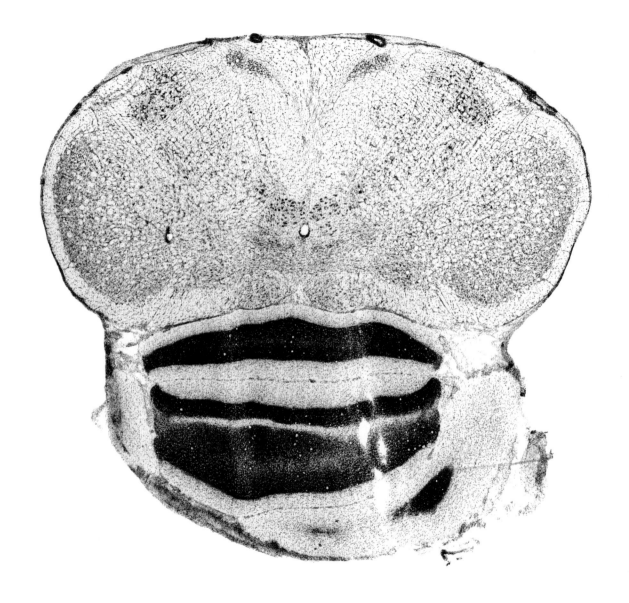

Plate 154

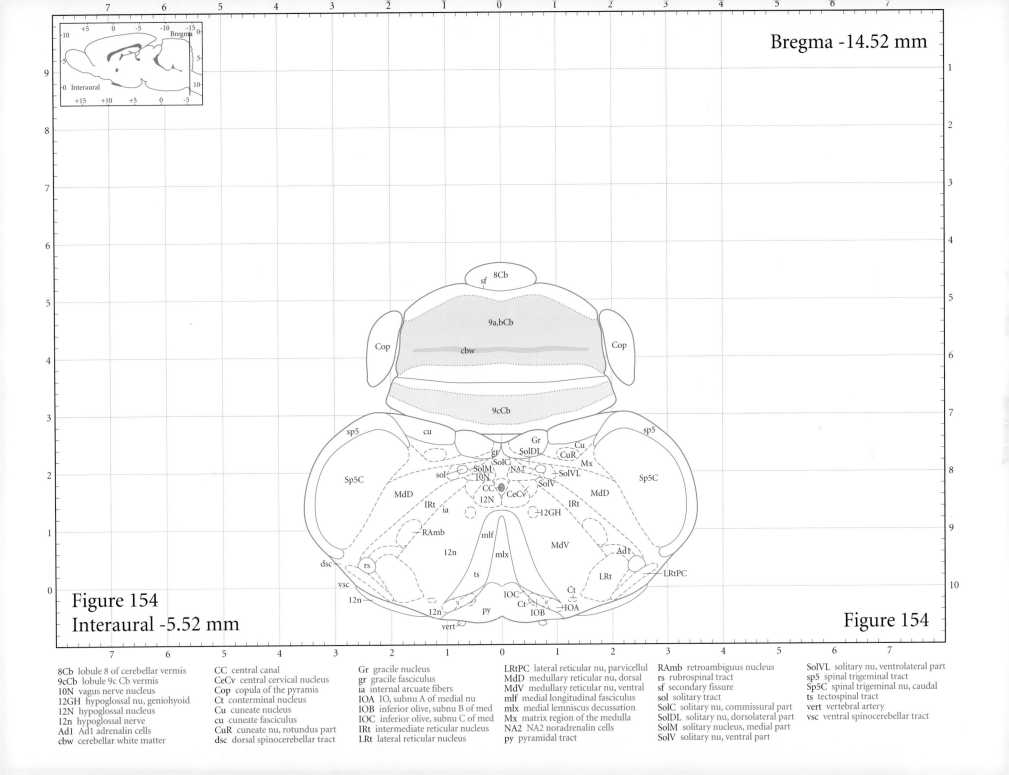

Plate 155

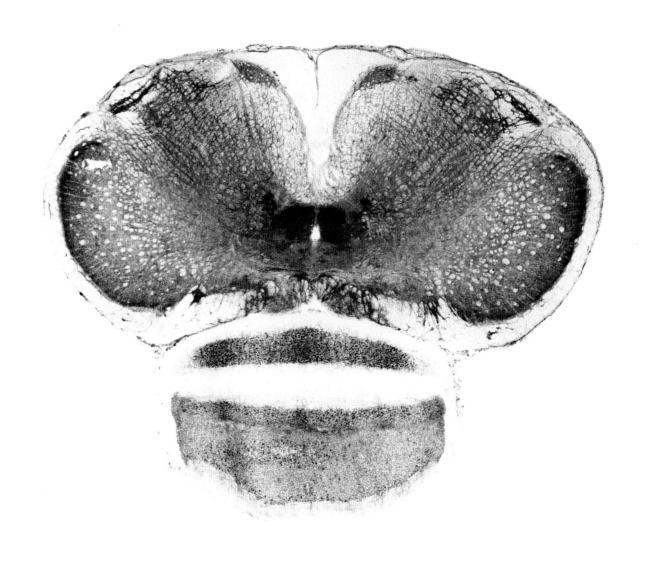

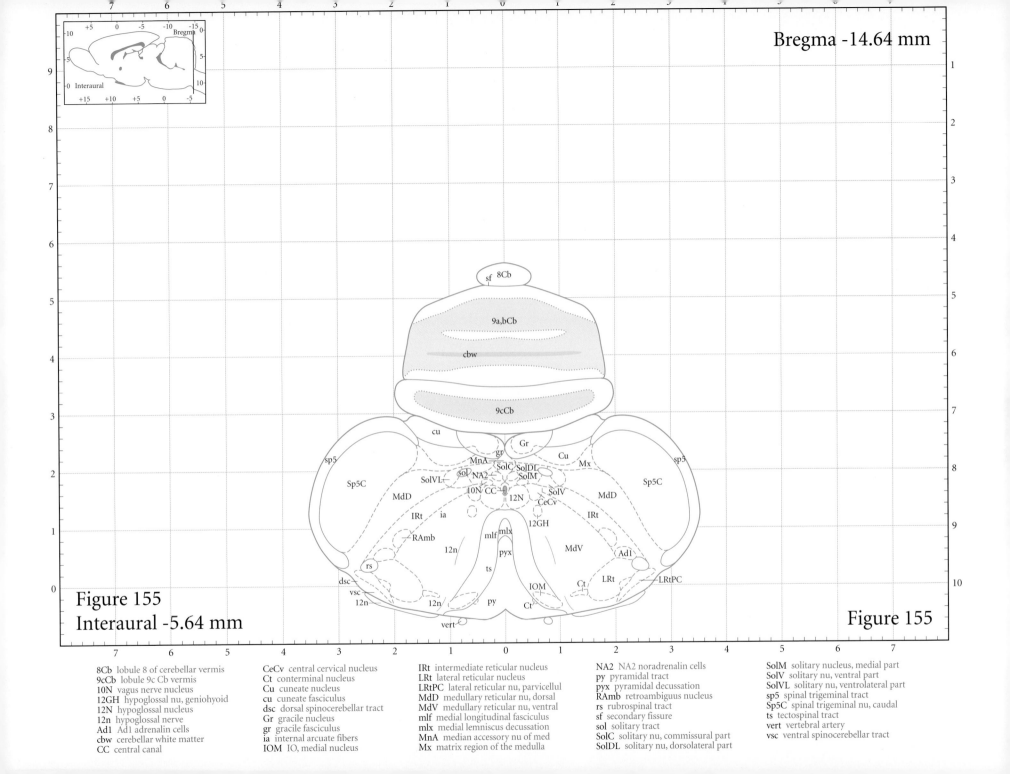

Plate 156

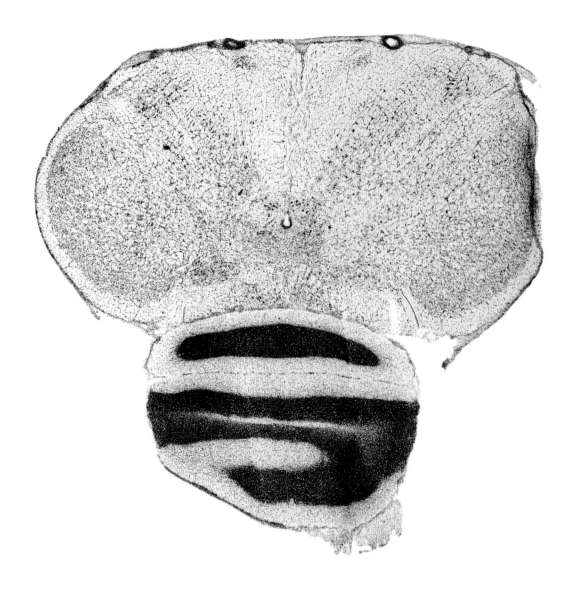

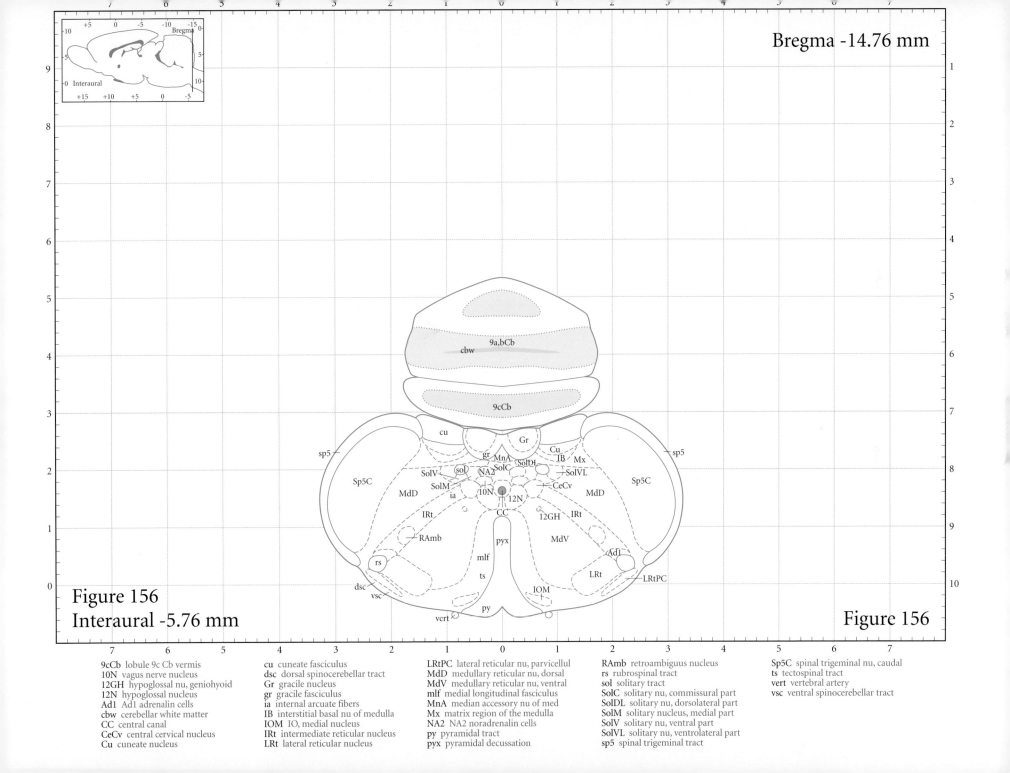

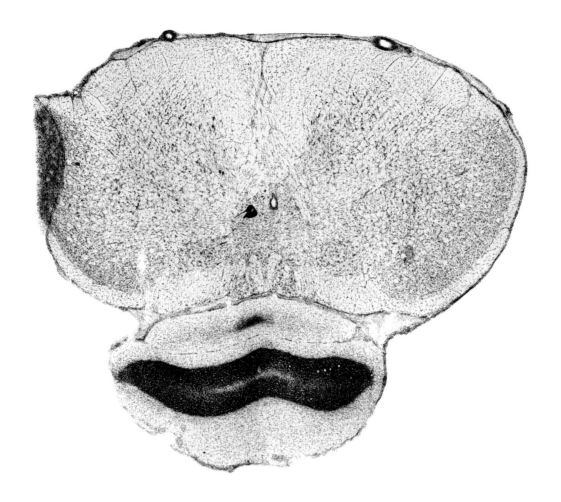

Plate 157

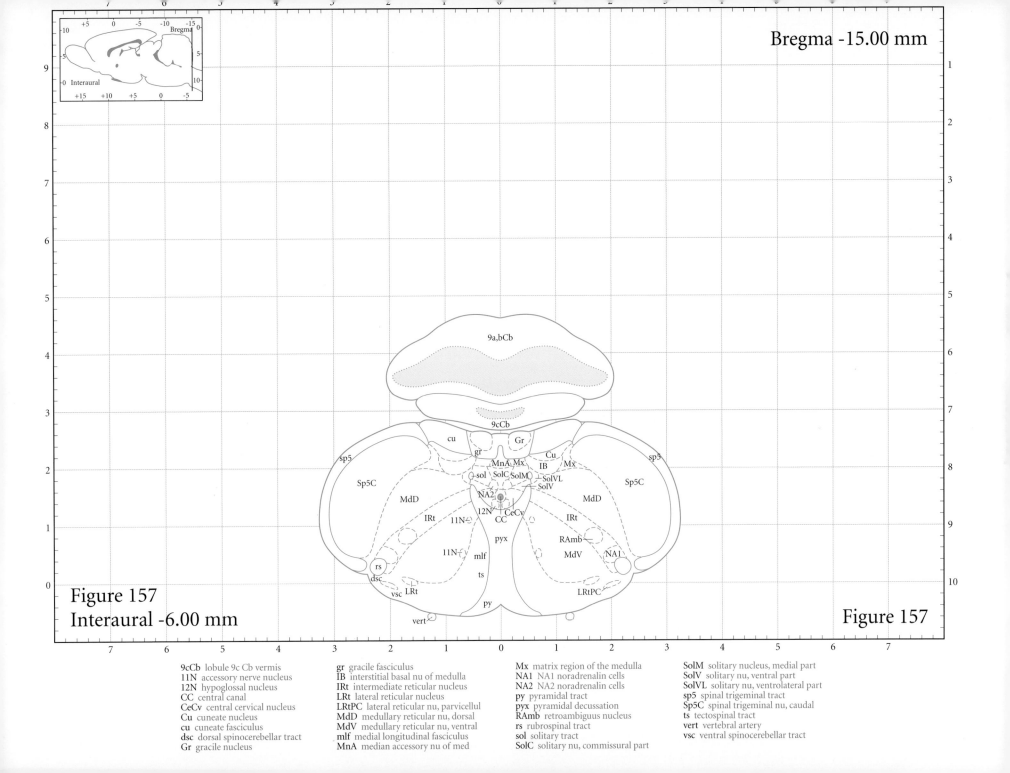

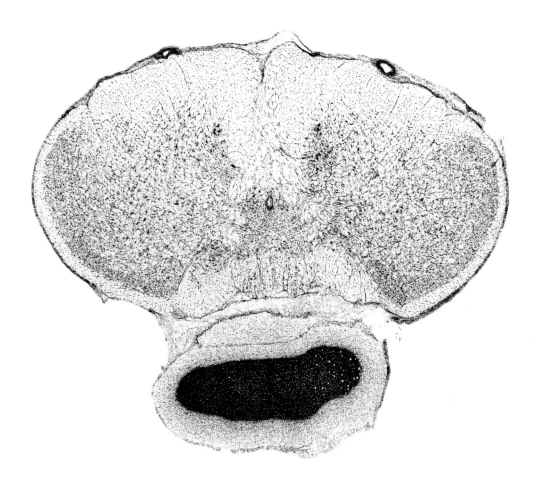

Plate 158

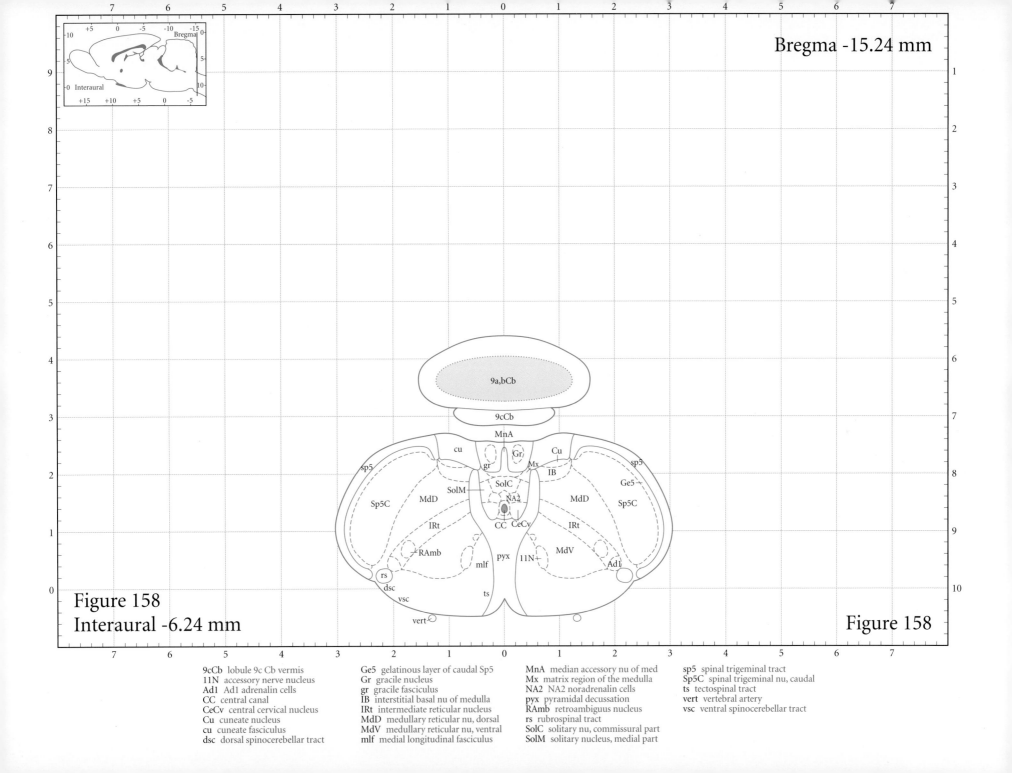

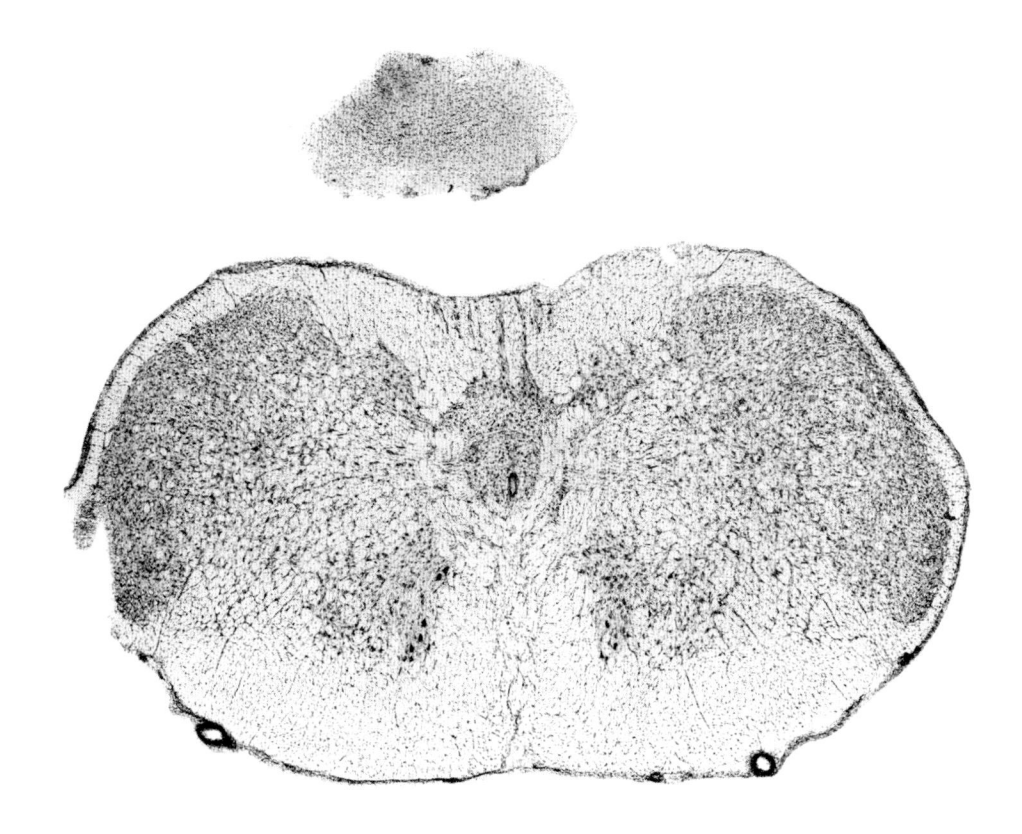

Plate 159

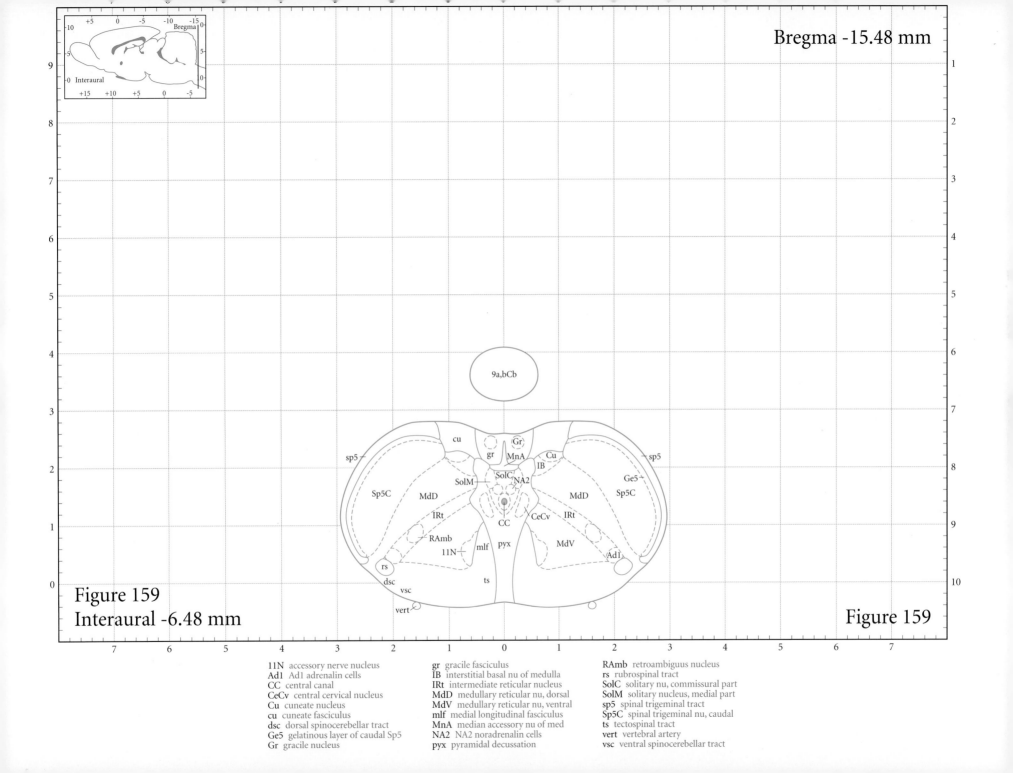

Plate 160

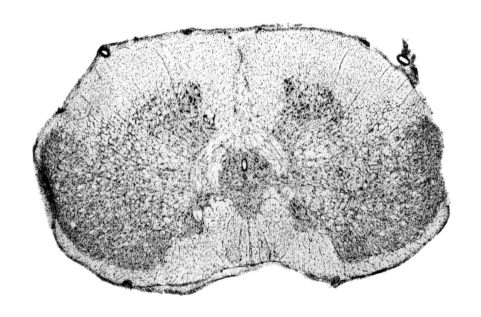

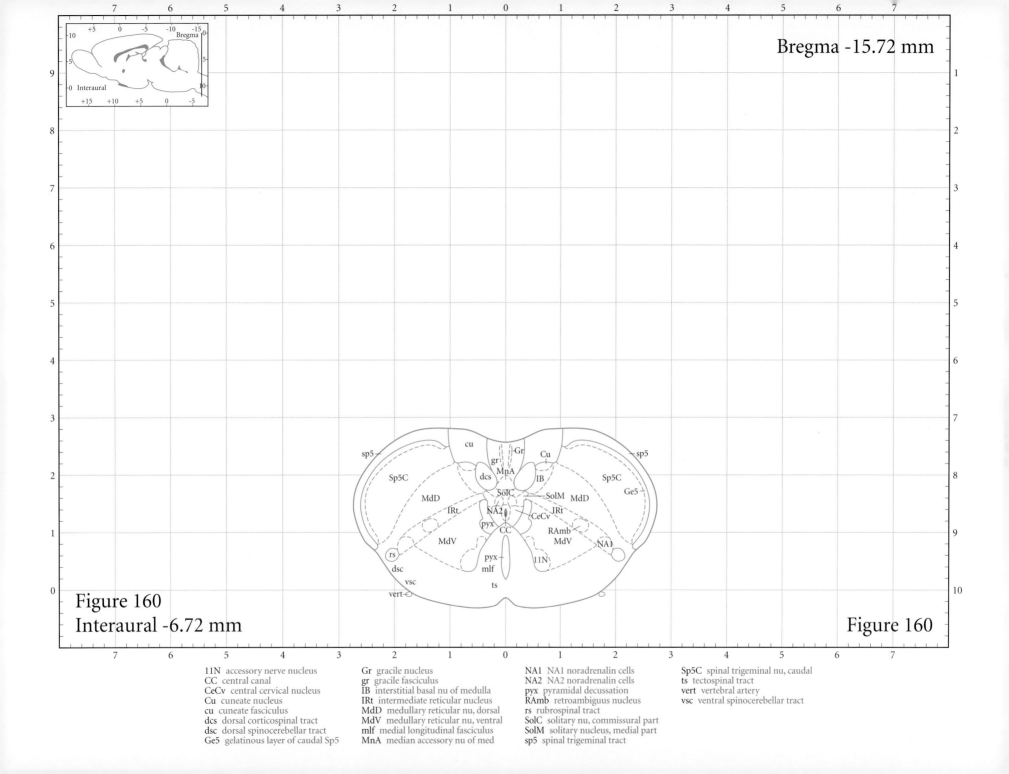

Plate 161

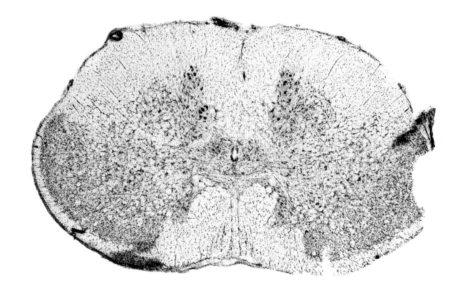

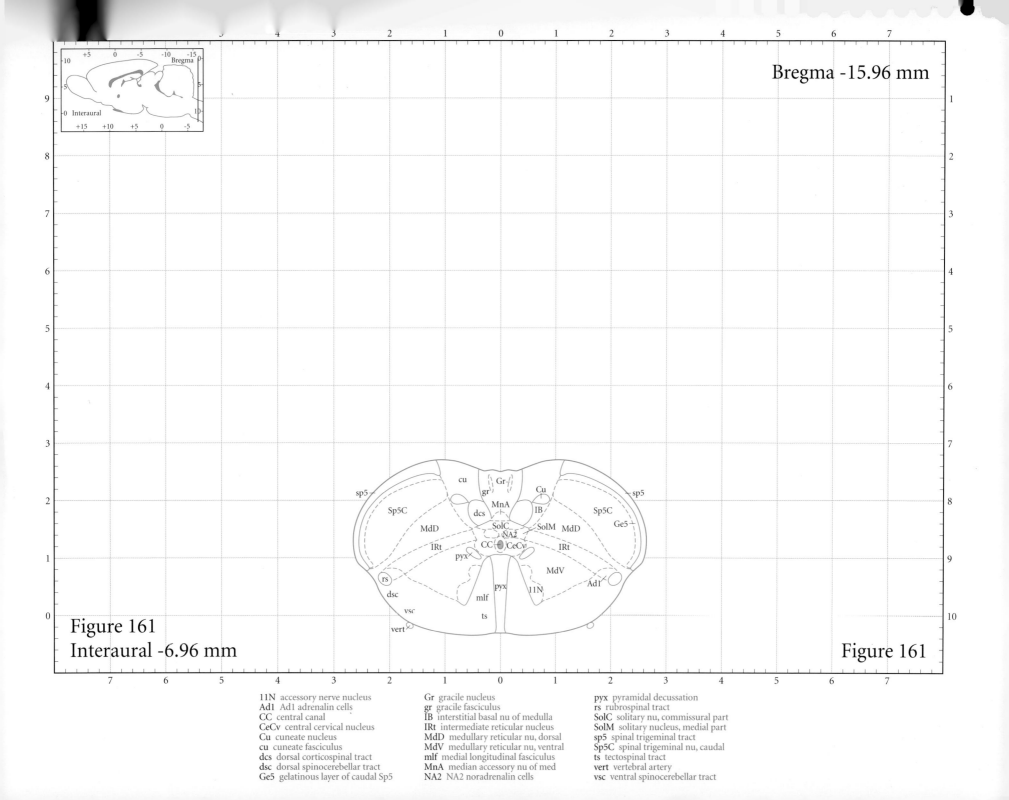